A STUDY OF WAR

A

STUDY

OF

WAR

QUINCY WRIGHT

Abridged by
LOUISE LEONARD WRIGHT

THE UNIVERSITY OF CHICAGO PRESS
CHICAGO AND LONDON

ISBN: 0-226-90999-9 (clothbound); 0-226-91000-8 (paperbound)

THE UNIVERSITY OF CHICAGO PRESS
CHICAGO 60637
The University of Chicago Press, Ltd. London
© 1942 and 1964
by The University of Chicago
Published 1942
Abridged Edition 1964
Third Impression 1970
Printed in the United States of America

TO

ALISON, PETER, MALCOLM, AND DIANA

IN THE HOPE

THAT THEY MAY LIVE

IN A WORLD

WITHOUT WAR

FOREWORD

A study of the causes of war was initiated at the University of Chicago in 1926. It involved consultations with numerous faculty members and the preparation of over fifty studies by research associates and faculty members. Several of these studies have been published in articles and books. The object of the study was to stimulate research in the field, but it was hoped that the results might eventually be summarized and, if possible, co-ordinated with one another and with the vast literature in the field. Quincy Wright, who had general supervision of the study, undertook this task and the results were given first in a series of lectures and then published in two volumes in January, 1942.

This volume, prepared at the request of the University of Chicago Press, is an abridgment of the two-volume edition. The original text has been used, with a few additions necessitated by significant developments since 1941. To come within the allotted number of pages, there has been some reorganization of the material. All the footnotes and all the appendixes, many of which contained original research results, had to be eliminated. This academic heresy can only be atoned by urging the reader to go to the original work, recently reprinted in one volume with a commentary on war since 1942, for references to the literature and confirming data.

In this abridged form, *A Study of War* becomes an extended essay on the phenomenon of war, its history, its causes, its control. It is to be hoped that this version will be useful to those who want to know why, at a time when individuals and statesmen agree that war as an instrument of national policy is "obsolete," military budgets are of unprecedented magnitude and every year since the advent of the nuclear bomb has witnessed armed hostilities—civil, guerrilla, or international.

LOUISE LEONARD WRIGHT

AUTHOR'S NOTE

In the Foreword to the original edition published in 1942, I wrote: "This investigation, begun in the hopeful atmosphere of Locarno and completed in the midst of general war, has convinced the writer that the problem of preventing war is one of increasing importance in our civilization and that the problem is essentially one of maintaining adaptive stability within the world-community, only possible if larger sections of the public persistently view that community as a whole. . . . So rapid has been the shrinking of the world as a result of inventions in the means of travel, transport, and communication, so rapid has been the acceleration in the rate of social change as a result of the conscious organization of technical and political invention itself, that the problems of functional synchronization and international adjustment have become increasingly difficult. Conflicts are more frequent, more difficult to resolve, more likely to spread."

These observations are even more appropriate as this abridged edition goes to press almost a quarter of a century later. The invention of nuclear weapons; the development of jet planes, missiles, satellites, and telecommunications; the breakup of empires and the doubling of the number of sovereign states; the nuclear arms race between the leaders of contending ideologies; and the rise of under-developed and unaligned states in world affairs have made the problem of war more exigent and more difficult to solve. There is a growing awareness on the part of governments and peoples that a solution must be found. Even the advocates of "realism" in international politics now admit that nuclear war is not a rational instrument of policy, that lesser wars are likely to escalate, and that a policy can only be realistic if it can be achieved by peaceful means.

This abridgment of the fifteen hundred pages of the first edition has been made by my wife, Louise Leonard Wright, with skill and patience, and with persuasive power to overcome my objections to necessary eliminations.

QUINCY WRIGHT

QUINCY WRIGHT'S CONTRIBUTION TO THE STUDY OF WAR

Karl W. Deutsch

Man's history has long been a story of struggle against suffering and death. This struggle began when hunger and illness were no longer accepted as irresistible and foreordained by fate—when men began to act against them. Today millions of men and women in medical work and medical research carry on this struggle against death. But today war is a greater threat to human life than famine or disease. And in the entire world only a few hundred or thousand men and women are engaged in serious professional research on what causes war and on how war could be abolished.

Nothing less than this—the understanding of war and the possible ways to its abolition—is on the agenda of our time. War usually has differed from simple murder in that it has been large scale, highly organized, long prepared in advance, and carried out with more costly and effective equipment, and in that it has killed more widely and with less foresight and discrimination. Moreover, war has been considered legitimate by many millions in most countries, who have seen it as good and right, or as a necessary means to a good end, or at least as a normal, expectable part of human life, recurrent in the past and unavoidable in the future. But, today, as we know, man's instruments of death have become incomparably quicker and more powerful than his instruments of life. Limited and local conflicts with conventional weapons—guerrilla and counter-guerrilla wars, police actions, and civil strife with marginal support from abroad for one side or both—all these may well continue to occur for many years before they are brought under control. But an all-out war today would kill more people in a few hours than could be replaced in many years and generations. In the age of nuclear weapons, if we do not abolish war, war is likely to abolish most of us. In our own time, research on the causes of war and ways to contain, control, and finally abolish all-out war has become an expression of mankind's will to live.

War, to be abolished, must be understood. To be understood, it must be studied. No one man has worked with more sustained care, compassion, and levelheadedness on the study of war, its causes and its possible prevention, than has Quincy Wright. He has done so for nearly half a century, not only as a defender of man's survival, but as a scientist. He has valued accuracy, facts, and truth more than any more appealing or preferred conclusions; and in his great book, *A Study of War*, he has gathered, together with his collaborators, a larger body of relevant facts, insights, and far-ranging questions about war than any other man has done.

A Study of War first appeared in 1942, during World War II, much as Hugo Grotius' book *On the Law of War and Peace* appeared in 1625 during the Thirty Years' War. As Grotius' book became a basis for the study of what later became known as "international law," so Quincy Wright's book marks the beginning of much that nowadays has become known as "peace research."[1] Although the number of people engaged in this research is still small, the public and private agencies concerned with it still few, and the material

---

[1] See, for example, *Journal of Peace Research* (Oslo: Peace Research Institute, 1964——); Quincy Wright, William M. Evan, and Morton Deutsch (eds.), *Preventing World War III: Some Proposals* (New York: Simon & Schuster, 1962). See also *Research Papers*, Nos. 1–4, published by the Institute for International Order (New York, n.d.); these include: Bernard T. Feld, Donald G. Brennan, David H. Frisch, Garry L. Quinn, and Robert S. Rochlin, *The Technical Problems of Arms Control;* Kenneth E. Boulding (director), *Economic Factors Bearing upon the Maintenance of Peace;* Arthur Larson, *The International Rule of Law;* Richard C. Snyder and James A. Robinson, *National and International Decision-making;* Ithiel de Sola Pool, *Communication and Values in Relation to War and Peace;* and *Current Thought on Peace and War,* a quarterly digest published by the Institute.

means devoted to it still scant, they are all considerably larger than they were only a few years ago. If modern civilization is to survive, these all will have to become much larger in the future. I believe that this in fact will happen.

As this happens, the importance of Quincy Wright's *A Study of War* is likely to increase. It summarizes carefully and clearly a large body of scholarship, research, and thought devoted during the half-century between 1890 and 1940 to the study of war and its causes. In particular, it gathers together the essentials of the discussions of these decades on the balance of power, on armaments and arms races, and on the role of the press and other mass media in maintaining or increasing warlike or peaceful moods. It brings together the intellectual work of those decades on the importance of international law for preventing or mitigating war and on the actual and potential contributions of international organizations.

In all these fields, *A Study of War* offers an unparalleled collection of relevant data and facts and a unique survey of the important literature. All this iformation is clearly organized and focused to one purpose: the understanding and control of war and its eventual prevention. This book offers the reader the best single foundation for the advanced study of international conflict that has appeared so far.

WRIGHT'S THEORY OF WAR

Quincy Wright has done more than pile up information about war. He has developed a basic theory of war. Summarized and in drastically oversimplified form, it might be called in effect a four-factor model of the origins of war.[2] Put most simply, his four factors are (1) technology, particularly as it applies to military matters; (2) law, particularly as it pertains to war and its initiation; (3) social organization, particularly in regard to such general-purpose political units as tribes, nations, empires, and international organizations; and (4) the distribution of opinions and attitudes concerning basic values. These four factors correspond to the technological, legal, sociopolitical, and biological-psychological-cultural levels of human life, respectively. At each level, conflict is likely, and violent conflict becomes probable whenever there is an overloading or breakdown

[2] See pp. 1284 ff., and the new chapter, "Commentary on War since 1942" (pp. 1501 ff.).

of the mechanisms or arrangements that have controlled the inter-play of actions and actors at any level and that previously have preserved some non-violent balance or equilibrium.

Violence and war, according to Quincy Wright, are probable and natural whenever adequate adjustments or controls on one or more of these levels are lacking. Peace, as he sees it, is "an equilibrium among many forces." It is unlikely to come about by itself. It must be organized in order to bring it about, to maintain it thereafter and to restore it after it has broken down.

Whenever there is a major change at any level—culture and values, political and social institutions, laws, or technology—the old adjust-ment and control mechanisms become strained and may break down. Any major psychological and cultural, or major social and political, or legal, or technological change in the world thus increases the risk of war, unless it is balanced by compensatory political, legal, cul-tural, and psychological adjustments. Peace thus requires ever new efforts, new arrangements, and often new institutions to preserve the peace or to restore it after its partial or world-wide breakdown.

The decades since 1942, the first appearance of *A Study of War,* have seen unparalleled changes sweep the world. These have been changes at all levels—in demography, in technology, in law, in cul-tures and values, and in social systems and in politics—and, conse-quently, the basic risk of war is now greater than ever. It follows that we must increase our efforts to create international organiza-tions and practices capable of reducing this mounting risk of war to very low proportions.

Wright's conception of these factors is such that the changes in each are conceived of as, in principle, measurable. Technological change can be measured by statistical data about the explosive power of bombs, about the speed and range of delivery systems, and about the total energy supply of the national economies behind each military establishment. Changes in attitudes and values held by the masses of the populations, and in the possibly different values held by the elites of political decision-makers, may be measured by means of public opinion data and by the content analysis of major newspapers or by policy statements. Changes in the number and size of states of various types and in the number, scope, and observ-ance of international laws, treaties, and organizations could all, in principle, likewise be noted. From such data, inferences could be drawn to estimate the speed and scope of processes increasing or decreasing the likelihood of uncontrolled large-scale conflicts and hence the size and power of the forces making for war or peace.

These forces are seen as working behind and beneath the health or illness and the wisdom or folly of individual statesmen, leaders, or commanders. The decisions of such individuals still count for much in Quincy Wright's view of the world, but they must govern —either against or with—the current of large events made up of the changes of large systems and the changing values and actions of hundreds of millions of people. In the present age of dangerous transition, the problems before statesmen and peoples are in some ways similar to the difficult adjustments that European peoples had to make in the great transitions of the fifth and the fifteenth centuries, each of which, as Wright reminds us, was made successfully.

As a pragmatically oriented thinker, Quincy Wright has sought more to be empirically comprehensive than to be mathematically elegant. At this stage of social science, his broad factors are not completely operational. They represent large categories and aspects of society and politics. Once we try to specify quantitative variables within each of these broad factors, their number soon becomes large and their analysis difficult. Much work is to be done here, but it will be aided and illuminated by Wright's grand conception. Details of this conception, applied to the historic past, as well as to the present and future, fill hundreds of pages of *A Study of War*. They still furnish suggestions for research for years to come. Indeed, seeing the world in this manner, Quincy Wright has necessarily become one of the chief pioneers of modern peace research. In due time, more explicit, detailed, and sophisticated models will doubtlessly follow upon his pathbreaking effort, but they will bear a debt to the work he has done.

QUINCY WRIGHT AND THE DEVELOPMENT OF PEACE RESEARCH

But this book offers more than a fundamental education. It was and is a pathfinder in matters of substance; and its substantive concerns have been carried forward since the time of its first appearance in 1942. Quincy Wright himself did this by editing a volume on *The World Community* in 1948 and by writing his important text on *The Study of International Relations* in 1955, which marked a significant advance in the use of quantitative data in a larger frame-

work of analysis.[3] His work on the study of conflict has been the
pioneer for such later work as the continuing research by many
scholars appearing in the journals *Conflict Resolution* and the *Journal of Peace Research* and for such books as Kenneth Boulding's
Conflict and Defense and Anatol Rapoport's *Fights, Games and Debates* and *Strategy and Conscience*.[4]

His chapter in *A Study of War* on the balance of power showed
the way in which a balance-of-power system may gradually turn
into an international or supranational community. His work on the
nature and formation of supranational communities pointed the way
both directly and indirectly to the whole stream of research and
discussion that has led to such major works as Ernst Haas's books
The Uniting of Europe and *Beyond the Nation State*[5] and to the
work on international systems and on community formation of David
Singer, Harold Guetzkow, and Amitai Etzioni;[6] and he has had a
substantial influence on my own thinking.[7]

Quincy Wright has also been a pioneer in the wide-ranging and
concerted use of different methods of research. He has broadened
and deepened the intellectual unity of the study of international

[3] Wright (ed.), *The World Community* (Chicago: University of Chicago Press,
1948); Wright, *The Study of International Relations* (New York: Appleton-Century-
Crofts, 1955).

[4] Boulding, *Conflict and Defense* (New York: Harper & Bros., 1962); Rapoport,
Fights, Games and Debates (Ann Arbor: University of Michigan Press, 1960); Rapoport, *Strategy and Conscience* (New York: Harper & Row, 1964).

[5] Haas, *The Uniting of Europe* (London: Stevens, 1958); Haas, *Beyond the Nation
State* (Stanford, Calif.: Stanford University Press, 1964).

[6] Cf. J. David Singer, "The Level-of-Analysis Problem in International Relations,"
World Politics, XIV (October, 1961), 77–92; and Singer's forthcoming historical study
of quantitative indicators of international conflict, devoted in part to the testing of
some of Wright's variables. See also Harold Guetzkow, *Multiple Loyalties* (Princeton,
N.J.: Princeton University Center for Research on World Political Institutions, 1954);
Guetzkow, "Isolation and Collaboration: A Partial Theory of Inter-Nation Rela-
tions," *Journal of Conflict Resolution*, I (July, 1957), 48–68; Etzioni, "Paradigm for
the Study of Political Unification," *World Politics*, XV (October, 1962), 44–74; Etzioni,
"Dialectics of Supranational Unification," *American Political Science Review*, LVI
(December, 1962), 927–35; Etzioni, "Epigenesis of Political Communities at the International Level," *American Journal of Sociology*, LXVIII (January, 1963), 407–21;
Etzioni, "European Unification: A Strategy of Change," *World Politics*, XVI (October,
1963), 32–51.

[7] Karl W. Deutsch, *Political Community at the International Level* (Garden City,
N.Y.: Doubleday & Co., 1954); K. W. Deutsch *et al.*, *Political Community and the
North Atlantic Area* (Princeton, N.J.: Princeton University Press, 1957): K. W.
Deutsch and J. David Singer, "Multipolar Power Systems and International Sta-
bility," *World Politics*, XVI (October, 1963), 390–406.

relations. *A Study of War* draws upon a prodigious volume of historical facts and judgments and a large arsenal of legal learning. At the same time, *A Study of War* was one of the first works to pay serious attention to the mathematical work of Lewis F. Richardson on the analysis of foreign policies and of arms races, wars, and other forms of "deadly quarrels."[8] Quincy Wright was one of the first scholars to make significant use of data from the *quantitative* analysis of communications, particularly of the press.[9] Since then this work has been carried forward by Harold Lasswell, Ithiel de Sola Pool, Daniel Lerner, and their associates.[10] Wright pioneered in using large amounts of research results from what are now called the "behavioral sciences," particularly from psychology. He led the way toward systematic efforts at measurement in international relations. Above all, he never ceased in his efforts to integrate *all* kinds of evidence—historical, legal, behavioral, statistical, and mathematical—with the best political judgment to arrive at a balanced and comprehensive understanding of reality. And he never ceased insisting quietly, by the very implications of his work, that realistic political insight and responsible political ethics are profoundly relevant to one another.

It is high time for the members of the new generation in American political science—and, indeed, in political science around the world—to become more deeply familiar with the work of Quincy Wright. They may have to put a little work into the effort, but they should find it worthwhile. If he were writing *A Study of War* today, he would probably abbreviate some of the historical and legal arguments and integrate his behavioral and quantitative data into his main text rather than segregate them in appendixes where they were thought to be less of a stumbling block to some of his first readers in 1942. We have become a little less fearful of numbers and data

8 See, for example, pp. 1268–69, 1327–28, Appendix XLII (pp. 1482–83), and thirteen other references. See also L. F. Richardson, *Statistics of Deadly Quarrels,* ed. Quincy Wright and C. C. Lienau (Pittsburgh: Boxwood Press, 1960; Chicago: Quadrangle Books, 1960); L. F. Richardson, *Arms and Insecurity,* ed. Nicolas Rashevsky and Ernest Trucco (Pittsburgh: Boxwood Press, 1960; Chicago: Quadrangle Books, 1960); Anatol Rapoport, "Lewis F. Richardson's Mathematical Theory of War," *Conflict Resolution,* I (September, 1957), 249–99.

9 See Appendix XLI (pp. 1472–81) and pp. 215, 1245, 1269.

10 See Lasswell, Lerner, and Pool, *The Comparative Study of Symbols* (Stanford, Calif.: Stanford University Press, 1952); Pool *et al., The Prestige Papers* (Stanford, Calif.: Stanford University Press, 1952); and others in the series. An enlarged and revised reissue of the Stanford series is being prepared by the M.I.T. Press.

since those years. But we can now appreciate all the better the boldness of Quincy Wright's design, and we can easily make the small effort to adjust ourselves to his form of presentation. As for me, I have found the effort abundantly rewarded.

A Study of War is now available in two versions. The present edition, abridged by Louise Leonard Wright, aims at presenting the pith and essence of his thought, stripped of appendixes and hence of much of the supporting evidence. It should be useful for junior undergraduates, for all those who want to become acquainted with the outline of a major edifice of ideas, and for those who wish to use the abridged version as a guide on their way to the understanding of the full-length work.

The second version is a reissue of *A Study of War* in full length, with all the data and appendixes and with the full indexes. This version, in my view, is indispensable for the serious graduate student and, of course, the scholars, teachers, and researchers in political science, international relations, and all sectors of the social sciences that deal with war and peace. Serious undergraduates might also profit from using the full-length version. In any case, the full-length version should be owned, not merely borrowed. You will want to go into it often, mark it up, to consult it for specific problems, and to reread it for thinking through once more your understanding of the field.

Until now, to the best of my knowledge, the Nobel Peace Prize has never been given to a social scientist. In contrast to the policy of other Nobel Prize committees, in other fields, the Norwegian Parliament has awarded mankind's highest honor for contributions to peace only to men of political action or to other persons engaged in popular persuasion. Recipients of the prize thus have usually been statesmen of national governments or international organizations or else writers, educators, or natural scientists trying to influence popular attitudes. In regard to the social sciences, the pursuit of more knowledge about peace thus far has gone unnoticed and unhonored at the highest level. On the day on which this changes—on the day when the crucial role of knowledge and of social science in the search for peace and will be more fully appreciated than it has been in the past—mankind may well remember the pioneering contributions of Quincy Wright.

November 1964

CONTENTS

C. Cultural Diversity of Nations

D. The People and War

PART ONE

THE
PHENOMENON
OF WAR

CHAPTER I

CONCEPTS OF WAR

To different people war may have very different meanings. To some it is a plague which ought to be eliminated; to some, a mistake which should be avoided; to others, a crime which ought to be punished; and to still others, it is an anachronism which no longer serves any purpose. On the other hand, there are some who take a more receptive attitude toward war and regard it as an adventure which may be interesting, an instrument which may be useful, a procedure which may be legitimate and appropriate, or a condition of existence for which one must be prepared. To people of the latter type war is not a problem. They take it for granted, whether with eagerness, complacency, or concern. Its details may prove unexpected or disagreeable, but they are not interpreted as presenting a problem of war-in-general. They can be satisfactorily handled by the professional historian, diplomat, international lawyer, or strategist. To the first group, however, war-in-general is a problem, and it is clear that this group has increased in the past century, and especially in the last twenty-five years, until it constitutes a majority of the human race, although in some countries and regions it may be in the minority.

1. WAR AS A PROBLEM

This growth of the opinion that war is a problem may be attributed to four types of change: (*a*) the shrinking of the world, (*b*) the acceleration of history, (*c*) the progress of military invention, and (*d*) the rise of democracy.

a) *The Shrinking World.*—Modern technology has made the world of today smaller in travel and transport time than was Europe or the United States in 1790 and smaller in communication time than was or is the House of Commons or Independence Hall. The result has been that people in every section of the world have become interdependent in their economy, culture, and politics. They have become more aware of, and more affected by, all wars, even distant ones. Formerly wars were unknown to the average man unless near at hand. Now any war interferes with almost everyone's normal way of life.

b) *The Acceleration of History.*—The progress of science and invention and the rapid intercommunication of ideas and techniques have conspired to accelerate the speed of social change. Whereas formerly a man might expect the technical and economic skills, the social and moral code, and the scheme of values which he received from his father to last through life, today each of these may change several times in a single life. Education must emphasize the processes of learning and living rather than traditional techniques and dogmas. But even with modern education the rapid and radical changes required are difficult for both individuals and institutions. Tempos differ between regions, classes, and groups, with the result of greater tensions and more conflicts and wars than in more leisurely centuries.

c) *The Progress of Military Invention.*—The introduction into the modern world since the eighteenth century of universal military service, efficient national propaganda, and centralized totalitarian government; the industrialization of military transport and equipment; and the inventions of the submarine, aircraft, missiles, and nuclear bombs, rendering national commerce, industry, and population generally vulnerable to attack, have given war a totalitarian character unprecedented in history. As a result of this change in the character of war and of the increased economic interdependence of peoples, war has tended to spread more rapidly, to destroy larger proportions of life and property, and to disorganize the economy of states more than ever before. Either the preparation for, the waging of, or the recovery from war has tended to dominate the political, economic, and social life of peoples.

d) The Rise of Democracy.—The growth of communication and literacy and the general rise in the standard of living have tended to create a national consciousness among the various peoples. This has meant that a favorable public opinion has become a necessary condition of successful foreign policy and that increased participation of people in government has been widely insisted upon. Foreign policies and wars have ceased to be mysteries but have become human acts which people can influence if not control. While the responsibility for war may be difficult to locate, war is commonly looked upon as human rather than as a visitation of either God or the devil. Democracy has stimulated the will of people to eliminate war, although it has not yet enlightened their intelligence as to the means.

Because the world is getting smaller, because changes occur more rapidly, because wars are more destructive, and because peoples are more impressed by the human responsibility for war, the recurrence of war has become a problem for a larger number of people, an increasing number of whom have come to believe that the elimination of war from international relations is not only desirable but also possible.

2. DEFINITIONS OF WAR

In the broadest sense war is a *violent contact* of *distinct* but *similar* entities. In this sense a collision of stars, a fight between a lion and a tiger, a battle between two primitive tribes, and hostilities between two modern nations would all be war. This broad definition has been elaborated for professional purposes by lawyers, diplomats, and soldiers and for scientific discussion by sociologists and psychologists.

International lawyers and diplomats have usually followed Grotius' conception of war as "the condition of those contending by force as such," though they have often excluded from the conception duels between individuals and insurrections, aggressions, or other conditions of violent contention between juridical unequals. Furthermore, they have insisted that "force" refers to military, naval, or air force, that is, to "armed force," thus excluding from the definition contentions involving only moral, legal, or economic force. Grotius criticized Cicero's definition of war as simply "a contending by force" because, he said, war was "not a contest but a condition." Modern dictionaries, however, have followed Cicero, and sociologists have accepted the same popular conception with the qualification that violent contention cannot be called war unless it involves actual conflict and constitutes a socially recognized form or custom within the society where it occurs. From the sociological point of view war is, therefore, a socially recognized form of intergroup conflict involving violence.

Legal and sociological definitions suggest that "states of war" are separated by exact points of time from "states of peace" which precede and follow them. International lawyers have attempted to elaborate precise criteria for determining the moment at which a war begins and ends, but they have not been entirely successful, and, furthermore, they have been obliged to acknowledge the occurrence of interventions, aggressions, reprisals, defensive expeditions, sanctions, armed neutralities, insurrections, rebellions, mob violence, piracy, and banditry as lying somewhere between war and peace as those terms are popularly understood. The recognition of such situations casts doubt upon the reality of a sharp distinction between war and peace and suggests the utility of searching for a variable of which war and peace are extreme conditions. Such a variable might be found in the external forms or the internal substance of international relations.

Philosophically minded military writers have sought the first, emphasizing the degree in which military methods are employed. Thus, Clausewitz defined war as "an act of violence intended to compel our opponents to fulfill our will," and elsewhere he emphasized the continuity of violence with other political methods. "War," he wrote, "is nothing but a continuation of political intercourse, with a mixture of other means."

Psychologists, ignoring the form, have found the substance of war in the degree of hostile attitude in the relation of states. Thus,

Hobbes compared the oscillations of war and peace to the weather: "As the nature of foul weather lieth not in a shower or two of rain, but in an inclination thereto of many days together; so the nature of war consisteth not in actual fighting, but in the known disposition thereto during all the time there is no assurance to the contrary." As the weather may manifest many degrees of fairness or foulness, so the relations of any pair of states may be cordial, friendly, correct, strained, ruptured, hostile, or any shade between.

We may thus conceive of the relations of every pair of states as continually varying and occasionally passing below a certain threshold, in which case they may be described by the term "war," whether or not other states recognize the situation as juridically a "state of war" and whether or not the precise form of conflict which sociologists designate "war" has developed. Subjectively there might be war, although objectively there might not be.

Whatever point of view is selected, war appears to be a species of a wider genus. War is only one of many abnormal legal situations. It is but one of numerous conflict procedures. It is only an extreme case of group attitudes. It is only a very large-scale resort to violence. A study of each of these broader categories when applied to the specific characteristics of war—abnormal states of law between equals, conflict between social groups, hostile attitudes of great intensity, and intentional violence through use of armed force—may throw light upon the phenomenon of war, although war itself does not exist except when hostility and violence contemporaneously pass beyond a certain threshold producing a new situation which law and opinion recognize as war.

Combining the four points of view, war is seen to be a state of law and a form of conflict involving a high degree of legal equality, of hostility, and of violence in the relations of organized human groups, or, more simply, the legal condition which equally permits two or more hostile groups to carry on a conflict by armed force.

It is to be observed that this definition implies sufficient social solidarity throughout the community of nations of which both belligerents and neutrals are members to permit general recognition of the behaviors and standards appropriate to the situation of war. Although war manifests the weakness of the community of nations, it also manifests the existence of that community.

3. MANIFESTATIONS OF WAR

A definition of war might be constructed not from an analysis of the literature but from an analysis of wars. The historical events which have been called wars have been characterized by (*a*) military activity, (*b*) high tension level, (*c*)abnormal law, and (*d*) intense political integration of each belligerent.

a) *Military Activity.*—The most obvious manifestation of war is the accelerated movement and activity of armies, navies, and air forces. Although modern states are at all times engaged in moving such forces around, in constructing warships, aircraft, missiles, and munitions, in organizing and training forces, and in making military appropriations, war is marked by a great acceleration in the speed of such activities. Such phenomena as mobilization, conscription, blockade, siege, organized fighting, invasion, and occupation may all occur without war; but they occur more frequently and on a larger scale during war. Each of the terms "battle," "campaign," "war," "arms race," and "normal military activity" designates a certain intensity of military activity. The type of events or conditions designated by each successive term manifests a lesser intensity of military activity but a wider space and a longer period of time in which such activity is occurring.

(1) *Battle.*—The most concentrated type of military activity is the battle. It may be taken as a generic term to cover a period of continuous direct contact of armed forces in which at least one side is engaged in a tactical offensive. There may be a battle of land forces, of naval forces, or of air forces. There may be a single battle combining all of these forces, as, for instance, in the siege of a port or a landing operation. In wars of past centuries, battles have usually been identifiable events, seldom lasting, except in the case of sieges, over a day, seldom covering over a score of square miles of territory, and seldom involving over a hundred thousand men. This is no longer true. The progress of invention with respect to instruments of communication, transportation, defense, and attack has made it possible for centralized military direction to be maintained over vastly greater numbers of men, operating through greater areas, for longer periods of time. Some of the episodes designated as battles in World War I lasted for several weeks, extended over tens of thousands of square miles, and involved millions of men. Because

of the immobility of trench warfare, they resembled sieges of the past rather than pitched battles. In World War II new techniques restored mobility, and battles covered even larger areas. Whereas earlier battles were named by towns (Saratoga, Waterloo, Gettysburg, Port Arthur), World War I battles were named by rivers or areas (the Marne, the Somme, Flanders) and World War II battles were named by countries or oceans (Norway, Belgium, France, Greece, Russia, the Atlantic). If a nuclear war should occur there might be only one battle, and it would probably be designated by a continent or hemisphere. The designation of a battle thus involves a judgment as to the continuity of contact, of attack, and of central direction of the opposing forces.

Within modern civilization, there appear to have been some 3,000 battles which involved casualties (killed, wounded, and prisoners) of at least 1,000 men in land battles or 500 in naval battles. Although most of these battles took place in wars, some of them did not, and there were many wars during the period without a single battle of this magnitude. If a lower casualty limit had been adopted, the number of battles would have been much greater. Of the 3,000 battles and sieges, the United States participated in 150, and, of that number, the United States Navy participated in only 15. Yet from 1775 to 1900 United States army units engaged in over 9,000 distinct battles and skirmishes. United States naval units engaged against hostile naval or land forces in 1,131 distinct episodes and, in addition, captured some 4,000 merchant vessels. It seems likely that the number of distinct hostile encounters between public armed forces has been more than a hundred times as great as the list of battles. There have probably been over a quarter of a million such hostile encounters in the civilized world since 1500, an average of over 500 a year.

(2) *Campaign.*—A less concentrated type of military activity than the battle is the campaign. This term is used to designate a group of military operations within a limited period of time connected by a strategic plan under the control of a single command. Several battles may be fought during a campaign, but a campaign may be conducted without any actual contact with the enemy. A campaign does, however, involve movements of actual armies, navies, or air forces, in which at least one side is engaged in a strategic offensive, such as an effort to occupy hostile territory, to acquire resources

from the enemy, to destroy hostile forces, to blockade hostile territory, to break civilian morale by military attacks, or to accomplish
other military objectives. A campaign is more likely than a battle to
combine both the army and the navy, but ordinarily it involves only
one force. In the past, campaigns have usually been identifiable
events, seldom lasting over six or eight months (the "campaigning
season" in European latitudes has often been terminated by winter
weather) and involving only two or three armies of from 50,000 to
100,000 men each. Naval campaigns sometimes covered very large
areas and continued over longer periods but usually involved fewer
men. The conditions which have increased the duration, area, and
number of participants of battles have done the same for campaigns.
In recent major wars it has been as difficult to distinguish and identify campaigns as it has been to distinguish and identify battles. In
minor hostilities—colonial wars, interventions, and insurrections—the
campaign is the normal unit of military activity. Thus many campaigns occur outside of recognized wars.

While in the twentieth century (1900–1964) there have been only
60 wars, there have been over 700 campaigns, of which more than
500 were outside of these wars. During this period there were over
1,000 battles of 1,000-casualty magnitude. It is probable that campaigns have been about as numerous as battles of this magnitude
during the entire modern period, although many included no battles at all and others a large number of battles.

(3) *War.*—From the military point of view, it is more difficult to
identify wars than either battles or campaigns. The unity of a war
derives more from legal or political than from military activities.
The beginning and end of the legally recognized state of war may
be evidenced by formal declarations, recognitions, and treaties, but
protracted hostilities on a scale large enough to be properly designated "war" may occur with no evidence of beginning and end
except the first and last "act of war."

Periods of war have been characterized by military movements of
abnormal size and frequency. The battles and campaigns of a war
are usually united through the continuity of the political direction
of each of the belligerents and the persistence of a grand strategical
objective of at least one of the participating states. These unifying
conditions, however, are not always present. From the military point
of view a war does not usually have such clear time and space

limitations as does a battle or a campaign. From the legal point of view its time limitations and its space limitations, at least with respect to land, are usually precise. The military activity of a war has seldom been continuous for over five years, but there has been a Hundred Years' War, a Thirty Years' War, and a Seven Years' War, and a number of other wars, such as the French Revolutionary and Napoleonic Wars, in which some military activity continued for more than five years. Usually, however, these periods were broken by long truces. Some of these wars continued through revolutionary changes in the political control of all or some of the belligerents, through the disappearance of old belligerents and entry of new ones, and through radical changes in the war aims or grand strategic objectives of most of the participants. Thus the time-space continuum, which in a legal sense is designated a war, has not necessarily been accompanied by a unity or uniformity of intense military activity. Although in international legal theory a state of war between two states begins and ends at definite moments of time, these moments have frequently been difficult to establish in practice.

At least 308 wars occurred from 1480 to 1964. These events ranged in size from minor episodes, involving only two small countries and lasting a few months, to such events as the Thirty Years' War, involving most of the European continent; the Seven Years' War, involving most of the European powers and including hostilities in America, India, and the high seas; and World War I, lasting, in the case of certain belligerents, for ten years, involving at times half of the countries of the world and including hostilities in Europe, Asia, Africa, and the high seas. World War II spread even wider. In the twentieth century, before 1964, there were 60 wars, and nearly every state which existed in this period participated in at least one.

(4) *Armament race.*—An even less precise type of military activity is the armament race. This is characterized less by military movements and hostile clashes, though such events may occur, than by acceleration in all countries involved of the rate of armament growth. Military budgets, military personnel, military equipment and stockpiles become steadily larger. A larger proportion of the productive energy of states is devoted to military affairs. Armament races have usually lasted for thirty or forty years. They have been characterized by an increasing frequency of small wars, imperial wars, and interventions, generally terminating in a balance-of-power war, during

which military building reaches a maximum. For ten or twenty years after such a war there has usually been a period of demobilization and decline of military building, sometimes stabilized by disarmament agreements. Armament races have resulted primarily from the political relations of states involved in a balance-of-power system, though the exigencies of arms-traders and of national economies may also have played a part.

In the seventeenth and eighteenth centuries distinct armament races sometimes occurred without precise simultaneity in western Europe, northern Europe, and southeastern Europe. During the nineteenth century Europe was a unity with respect to armament races, though North America, Central America, the La Plata area, the Andean area, and the Far East had distinct armament races. A European armament race began about 1787 and lasted until 1815. Another began about 1840 and lasted until 1871. In the twentieth century armament races have tended to be synchronous and simultaneous throughout the world, tending toward concentration of military power in two great alliances. There was a general armament race lasting from about 1886 to 1919. Another began about 1932 and continued through 1945. A third, designated the "cold war," began in 1946 and had not ended by 1964, with some evidences of *détente* after 1953. Probably a study of army and navy building coupled with a study of the balance of power would disclose twenty-five distinct armament races in modern history, though the boundaries either in time or in space could not be very clearly defined. These armament races were related to the tendency toward a fifty-year periodicity in the frequency of battles.

(5) *Normal military activity.*—This is a conception which can be ascertained only by studying the military history of a civilization over centuries to ascertain the size of military and naval budgets, the size of the standing army, the proportion of national effort directed toward military affairs, and the frequency of minor and major uses of military force usual among the states of that civilization. Though difficult to apply in the dynamic conditions of modern civilization, the conception of normal military activity theoretically constitutes a standard of comparison by which the more accelerated activity during armament races, wars, campaigns, and battles can be judged. If it is realized that the great powers of modern history have been formally at war nearly half of the time and have been engaged

in minor military campaigns or armament races a good share of the remaining time, it will be perceived that in modern civilization normal military activity is quite remote from an ideal conception of peace.

b) *High Tension Level.*—Another manifestation of war is the high tension level of public opinion within the belligerent states. Attention is concentrated upon symbols of the nation and of the enemy. Only favorable attitudes toward the former and unfavorable attitudes toward the latter are expressed. Graphs constructed from statistical analyses of numerous attitude statements taken from newspapers indicate that on the approach of war the opinions of the population of each country about the other become more hostile and more homogeneous. During war itself these opinions reach levels of extraordinary hostility.

Such graphs present the best picture of the changing direction, intensity, homogeneity, and continuity of the attitudes of one people toward another, but easily observable phenomena make possible a rough classification of the intensity of such attitudes. Five states of tension level may be expressed by the words "symbolic attack," "threats of violence," "discrimination," "disapproval," and "normal relations."

(1) *Symbolic attack.*—In time of war the press, public addresses, sermons, moving pictures, the radio, and other instruments of publicity frequently contain direct attacks upon the enemy, emphasizing his satanism and urging his destruction. Such sentiments may appear not only in unofficial but in official utterances. The latter were formerly rare except in time of war, but with the development of the radio, breaking down the distinction between domestic and foreign communication, they have become more common.

(2) *Threats of violence* against another state may be publicized in times of strained relations short of war, but if they proceed directly from high officials of the government they are likely to lead to a breach of relations or to war itself. Overt threats, mobilizations, displays of force, and ultimata have always been considered much more serious than formal diplomatic protests, though the latter may carry an implication of eventual resort to force. The United States resented the prediction of "grave consequences" in the Japanese ambassador's note on the immigration crisis in 1923, interpreting it as a threat of war. The abusive comments of Hitler toward President

Beneš of Czechoslovakia in his address of September 12, 1938, indicated that hostilities might be near at hand. "Incidents" concerning nationals, vessels, or officials of one country for which another country is considered responsible, but which might be of little political importance in normal times, are often interpreted as threats in times of high tension and may result in a breach of diplomatic relations.

(3) *Discrimination.*—Private boycotts and official discriminations in tariff rates, export, import, and navigation embargoes, and prohibitions against loans and concessions are an evidence of strained relations; but they frequently occur without war and are usually considered less serious than threats and displays of force. Such economic discriminations are always intensified between enemies in time of war.

(4) *Disapproval.*—Official expressions of disapproval of the policy or behavior of a foreign state manifest a serious strain in relations if they concern the internal policy of that state or its relations with third states. References to the policy of another government are not, however, deemed as serious as utterances disrespectful or contemptuous of the personality of high officials or of the state itself. The attitudes of governments toward such criticism have varied with respect to the degree of resentment which should be felt and with respect to the responsibility of states for hostile utterances made by private individuals or in private publications. Autocracies are likely to be much more sensitive on such matters than democracies.

(5) *Normal relations.*—In the normal relations of states, formal protests are usually confined to cases where the state, its government, or its nationals have been injured because of a breach of international obligations by another state. Objections to the policy of another state are not formally protested, although they may be made the subject of representations. Even in normal times the private press sometimes abuses other states, but unless excessive or unless the press is controlled by the government, such license does not indicate a strain in relations. The normal level of respect manifested by the government of one state for another varies greatly among different states and at different times.

c) *Abnormal Law.*—A third manifestation of war is the entry into force of new rules of law, domestic and international. Contracts with alien enemies are suspended. Resident alien enemies are interned or placed under supervision. Trading with the enemy is prohibited. Many treaties with the enemy are terminated or suspended. Military forces are free to invade the enemy territory and to attack its armed forces, limited only by the rules of war. Neutrals are obliged to prevent the use of their territory or vessels for military purposes by belligerents. Neutral vessels at sea are liable to visit and search and to capture if they assist the enemy.

In the case of war, recognized in the legal sense, all these abnormal rules come into force. There are other situations in which a modified form of abnormal law prevails. The legal situation consequent upon an outbreak of hostilities differs accordingly as the violence occurs in a state's home territory, in a colonial area of different culture, or in

Relative Status of Combatants	International Strife	Colonial Strife	Civil Strife
Equality	International war	Imperial war	Civil war
Moderate disparity	Aggression—defense	Colonial revolt —punitive expedition	Insurrection—military suppression
Great disparity	Disorder—intervention	Native unrest —pacification	Mob violence —police

the relations of two recognized states. It may also differ accordingly as the two parties are equal or are moderately or greatly disparate in status. The nine categories listed in the accompanying chart may, therefore, be distinguished with respect to the abnormal legal situation which results, although, according to the Geneva Convention, the humanitarian rules of war apply to all.

(1) *Civil war, imperial war, and international war,* if recognized as such, imply that both sides are to be treated as equals by other states designated as neutrals. Both are entitled to the rights and powers of belligerents as long as the war lasts. In civil war and often in imperial war the revolt is in violation of the municipal constitution and laws of the state, and if the legal government is successful, it may, of course, apply its own law to punish treason after hostilities are over. In international war one of the parties may be acting

in violation of its obligations under international law, and this fact may influence the settlement, even though the states have generally recognized the situation as "war" by proclaiming neutrality.

(2) *Insurrection, colonial revolt, and aggression* not recognized as legal war do not imply a duty of third states to treat the two parties as equal. In the case of insurrection or native uprising the recognized government has often been favored by third states. Although if the insurrection is of a magnitude to make the result uncertain, third states should not intervene in behalf of either faction, as illustrated in the treatment of the Spanish Loyalists on a parity with the insurgents under the nonintervention agreement of 1936. If a state engaged in international hostilities has been found by appropriate international procedures to be an aggressor, in the sense that it resorted to force in violation of its international obligations, third states may discriminate in favor of its innocent victim engaged in defense. Such discrimination was required by the League of Nations Covenant and the United Nations Charter and is permissible for parties to other antiwar treaties such as the Pact of Paris. An international agency found Japan an aggressor in its hostilities against China (1931, 1937) and the United States (1941); Italy against Ethiopia (1935); Russia against Finland (1939) and Hungary (1956); North Korea and China against South Korea (1950); and Germany against Poland (1939), the Netherlands (1940), and Norway (1940).

(3) *Mob violence and native unrest* within the state's domain and *intervention* in a dependent state do not usually involve international law or the rights of third states. Municipal law may recognize a state of siege or martial law in such situations. The case of a great power intervening to deal with disorders or international delinquencies in a much smaller independent state has often been treated in a similar manner. In law, however, the justifiability of the intervention is properly an international question to be decided by international procedures according to international law. As treaties now generally prohibit forceful intervention except for defense against armed attack, there is a presumption against the legitimacy of such action unless expressly permitted by a protectorate, mandate, trusteeship, or other treaty relation with the state in whose territory the action is taken or unless that state has been found guilty of an aggression which withdraws it from the benefits of antiwar treaties and permits military sanctions against it.

d) *Intense Political Integration.*—A further manifestation of war consists in legal, social, and political changes within the belligerent community toward intensive integration, eliminating internal conflicts and facilitating the conduct of the international conflict. Legislation regulates industry and directs it toward war production. Censorship comes into effect, and important instruments of communication are taken over by the government. Consumption may be rationed in many directions. Loyalties to church, party, or profession are subordinated to loyalty to the state. The normal degree of government control of the activities of individuals varies greatly among states; but, however intense or loose the normal control, it becomes more intense in time of war.

In time of war or threat of war the armament industry and the production of raw materials for armament manufacture are usually the first economic activity to be regulated or taken over by the government. This is soon followed by the taking-over of agencies of transportation and communication, education, and propaganda. A more general control of business and consumption may follow. Finally, uniformity of ideology or religion may be required. The intensive preparedness required by modern war tends to bring about many of these changes long before war begins. Totalitarian states exhibit this intensive political integration as a permanent characteristic.

This description of the military, psychological, legal, and sociological manifestations of war suggests that all may be regarded as variables which reach a certain threshold of intensity in actual war. War may therefore be regarded from the standpoint of each belligerent as an extreme intensification of military activity, psychological tension, legal power, and social integration—an intensification which is not likely to result unless the enemy is approximately equal in material power. From the standpoint of the group, including all belligerents, war may be considered a simultaneous conflict of armed forces, popular feelings, jural claims, and national cultures so balanced as to lead to an extreme intensification of each. This definition corresponds to the earlier one suggested. War is a legal condition which equally permits two or more hostile groups to carry on conflict by armed force.

To say that war implies a legal condition means that law or custom recognizes that when war exists particular types of behavior or attitudes are appropriate. War does not imply a sporadic or capricious or accidental situation but a recognized condition. The thinking in any culture recognizes many different conditions, each with its appropriate behavior pattern. War implies one of many such recognized conditions characterized by the equality of the belligerents in law and their freedom to resort to violence.

To say that this condition pertains to hostile groups implies that the attitudes involved are social rather than individual and at the same time hostile rather than friendly. This expression therefore implies a differentiation between the in-group and the out-group. The individual loves his own group and hates the enemy group. This definition excludes duels or other fights between individuals from the conception of war and also excludes friendly armed contentions, as in a tournament or a fencing match.

To say that the groups are carrying on a conflict means that the pattern of behavior is an instance of the type of group interrelationship which sociologists have termed "conflict." This pattern includes competitive games, forensic litigation, political elections, family brawls, feuds, sectarian strife, and other situations in which opposing but similar entities aware of and in contact with each other are dominated by sentiments of rivalry and expectation of victory through the use of mutually recognized procedures. The pattern therefore involves a combination of separation and unity: separation in the fact of antagonism and hostility between entities; union in the fact of recognition by all entities concerned of a common objective (victory) and the procedure by which it is to be obtained (armed force). War does not, therefore, exist where the participants are so self-centered that each fails to recognize the other as a participant but treats it merely as an environmental obstacle to policy, as men treat wild animals or geographical barriers. War is thus distinguished from armed activities such as the chase among primitive peoples or colonial development among modern nations. As a conflict, war implies that attitudes and actions within each participating group are influenced by intergroup or international standards.

To say that the conflict is by armed force excludes forms of contentious procedure which permit only persuasive argument, intellectual skill, or friendly physical encounter, as in judicial trials, parliamentary debates, and athletic games. The technique of arms implies the use of weapons to kill, wound, or capture individuals of the opposing side. War is thus a type of violence. The word "violence," however, includes also activities which are not war, such as assassination and robbery, riot and lynching, police action and execution, reprisals and interventions. War, on the other hand, may involve activities other than violence. In modern war the propaganda, economic, and diplomatic fronts may be more important than the military front; but, if the technique of armed violence is not used or threatened, the situation is not war.

War is thus at the same time an exceptional legal condition, a phenomenon of intergroup social psychology, a species of conflict, and a species of violence. While each of these aspects of war suggests an approach to its study, war must not be identified with any one of them.

It must not be assumed that all relationships between sovereign groups are war or that all conflicts or all resorts to violence are war. Such assumptions, frequently made, render the control of war hopeless. The anarchists, striving to eliminate all legal coercion; the isolationists, striving to eliminate all intergroup relations; the idealists, striving to eliminate all conflicts; and the extreme pacifists, trying to eliminate all violence, are engaged in a hopeless task. On the other hand, it is possible that appropriate modifications of international law and procedure, of national attitudes and ideals, of social and economic conditions, and of the methods by which governments keep themselves in power may prevent the recurrence of war.

CHAPTER II

THE HISTORY OF WAR

1. ORIGIN OF WAR

The human species was probably biologically united in its origin and will probably be socially united before its end, but from the wandering of the first group of men from the ancestral home, sometime in the late Miocene or early Pliocene, to the modern era of world communication, human evolution has moved in a number of separate channels. These separations were so complete in prehistoric times as to produce distinctive races. The less complete separations of historic times produced distinctive civilizations. Such separations as have existed in modern times account for the distinct nations. The history of war can, therefore, be divided into four very unequal stages, dominated, respectively, by animals, primitive men, civilized men, and men using modern technology. The evidences available for studying war are very different for each of these stages.

For the first or prehuman stage, evidence is confined to the structure of the few paleontological remains of man's prehuman ancestors and the behavior of contemporary animals. The latter are not in the direct line of human descent, and their behavior merely suggests what may have been the nature of war among man's actual ancestors.

The second stage, that of primitive man, began with the emergence of primates, able to communicate with one another by a definite language, a million to a half-million years ago. The later developments of this stage continued in limited areas of Africa, Asia, Oceania, and America down to recent times. Evidence of the condition with respect to war and peace in this stage is to be found in archeological remains of the ancestors of the civilized communities and in the observations of contemporary primitive peoples, although the latter type of evidence must be used with caution. Contemporary primitive men have in most cases borrowed much from neighbors of a higher culture. In some instances their present culture may be a degeneration from a higher civilization.

The third or historic stage began in the valleys of the Nile and Euphrates, six or perhaps even ten thousand years ago, in the valleys of the Indus and Yellow rivers four or five thousand years ago, and in Peru and Mexico perhaps three or four thousand years ago. Whether this stage, which, except in Peru, can best be dated from the use of writing, originated autochthonously at several points or by transmission of major elements from a single or small number of centers is debated. It continues in many places to the present time. Evidence of the nature of war in this period is to be found in contemporaneous and older writings, in inscriptions of a descriptive, chronological, and analytical character, and in archeological remains.

The fourth stage, that of world contact, may be said to have begun with the invention of printing in the fifteenth century, soon followed by the voyages from western Europe establishing continuous contacts between the centers of civilization in Europe, the Near East, America, and the Far East. Since that time nearly all areas of the world have been brought within the orbit of continuous world contact through printed communication. Such contacts have become notably more intense with the steam, electrical, and atomic inventions of the nineteenth and twentieth centuries. Evidence with respect to the history of war and peace in this period is to be found in descriptive writings, much more voluminous than those available for the earlier historic period, and, in addition, in a wealth of legal, economic, and statistical materials, contemporaneously organized for the purpose of political, economic, and sociological record and analysis.

It has been contended that war originated at a certain stage of civilization and that, in so far as war exists among primitive peoples, it has been learned by them from their civilized neighbors or has been retained by them, while in other respects they degenerated from civilization to savagery. This theory is supported by the extreme "diffusionist" or "historical" school of anthropology represented by W. H. R. Rivers and G. Elliot Smith and has been most elaborated by W. J. Perry. These writers contend that war was invented in predynastic Egypt, along with agriculture, social classes, and human sacrifice. This "archaic" civilization was diffused by widespread travels of the Egyptians during the pyramid-building age. The nomadic barbarians on the outskirts of this civilization learned war from it and developed war methods in attacks upon its centers. The majority of anthropologists decline to accept this theory.

Borrowing from neighbors and independent invention both occur, and evidence must be adduced to explain the presence of each particular trait in each particular group. The evidence of contemporary primitive cultures, of contemporary apes, and of the remains of prehistoric man suggests that forms of violence have always been widespread among men, though there has always been much variation in warlikeness among groups. No golden age of peace existed at any stage of human history, nor did any iron age of continual war. Neither the Rousseauan nor the Hobbesian concept of natural man is adequate. Man was and is a complex compound of inherited tendencies and social conditionings, crystallizing at different times and places into numerous cultures exhibiting varying forms and degrees of violence.

To decide, therefore, whether war was spontaneously practiced by human groups everywhere or was borrowed from one or a few societies, it is necessary to study the evidence from many groups. In order to do that, it is necessary to consider in what sense war is meant.

If by "war" is meant the use of firearms to promote the policy of a group, it must be admitted that the contemporary primitive peoples borrowed warfare from people of modern civilization. The dispersion of many modern war techniques—weapons, formations, tactical movements, and strategic ideas—can be demonstrated from historical evidence. Doubtless, in this technological sense, war was invented in Europe only about five centuries ago and subsequently diffused throughout the world.

If, however, by "war" is meant the reaction to certain situations by resort to violence, the assumption of borrowing seems more doubtful. Animals of the same species quarrel, and quarrel violently, and many of the things they quarrel about—food, territory, females—are the things men quarrel about. In this psychological sense war is a mode of behavior which belongs to most men and animals and probably to all children.

If by "war" one has in mind a period of time, initiated and ended according to law, during which, and during which alone, violence may be legitimately resorted to as an instrument of group policy, doubtless animals and many primitive men do not practice it, although there are primitive peoples with highly formalized belligerent practices. It appears to be in this sense that Perry uses the term "war," and anthropologists of other schools admit that war as a legitimate instrument for plunder or conquest was little known among primitive peoples. In this legal, political, and economic sense war probably originated among civilizations, accompanying the development among them of political organizations involving subordination, property, dense populations, and codified law. It then diffused to their less civilized neighbors.

Finally, if "war" means a social custom utilizing regulated violence in connection with intergroup conflicts, it appears to have originated wth permanent societies. Such societies are found among the social insects and were probably characteristic of man from the beginning. War in this sociological sense is found in nearly all existing human groups, however primitive.

There are, thus, senses in which war is an organic phenomenon, others in which it is a human phenomenon, others in which it is a phenomenon of civilization, and others in which it is an achievement of very recent times.

War in the psychological sense began with organic nature. The most primitive protozoa were endowed with drives adapted for obtaining food, for reproduction, and for self-preservation, and these drives, when stimulated by circumstances arising more or less frequently in the animal's environment, resulted in violent behavior of the organism as a whole. Although among the more specialized animals the circumstances causing violent behavior have varied greatly in type and frequency, and on the whole violence between animals of the same species has not been common, yet it is doubtful whether there are any animals which cannot be provoked into fighting by some stimuli. There are rudiments of war in the sociological sense especially among the social insects and among some of the higher mammals; but on the whole, war does not exist among animals other than man, except in the sense of violent behavior by the individual animal induced by the appropriate stimulus to an organic drive.

War in the sociological sense could not exist as a distinct phenomenon before the emergence of human societies, permanently constituted through communication by language and the accumulation of traditions. Under these conditions it was first possible for the individual to identify himself with a group represented by a symbol and to distinguish his group thus represented from other groups like it but bearing a different relationship to himself. Thus, morals began, and they generated a conscience in the individual and the possibility of belligerent behavior in response not to organic drives but to tribal mores, to the demands of the superego. While there are tribes that fight rarely, as there are animals that fight rarely, it seems probable that this is because of environmental circumstances which seldom stimulate the belligerent mores which exist. No tribes have been adequately described that will not fight as units under certain circumstances, and in most tribes the mores prescribe violent behavior in a variety of circumstances connected with tribal solidarity, religion, magic, marriage, and sport. War as a regular means of livelihood, however, to get food, slaves, or booty or to expand hunting grounds, seems to exist among the food-gatherers mainly through borrowing from peoples of a higher culture, although these types of war have been developed among a few of the social insects.

War in the sense of a legal situation equally permitting groups to expand wealth and power by violence began with civilization. Not until the arts of writing, agriculture, and animal husbandry had developed was it possible to organize a permanent human group or state larger than the primary or man-to-man-contact group, with a distinction of ruler and ruled, a clear conception of property, and a body of law, distinct from the mores, to regulate these relationships, to preserve internal order, and to formulate social interests. Only under these conditions could war become institutionalized as a ra-

tional means to political and economic ends. War as a legitimate
procedure for acquiring territory, cattle, slaves, and political prestige
has existed among civilized peoples and has been transmitted by
them to their more primitive neighbors. Only among civilized people
has war been an institution serving political and economic interests
of the community, defined by a body of law which states the circumstances justifying its use, the procedures whereby it is begun and
ended, and the methods by which it is conducted.

War in the modern technical sense began with the period of world
civilization. All belligerent entities—animals, primitive peoples, and
the historic civilizations of the past—have, of course, had war techniques—weapons, tactics, and strategic ideas—but in the modern
sense, "war" means the use of firearms, chemicals, and nuclear weapons for striking and of steam, gas, and electrical engines for military
movement by land, sea, or air. Use of sources of power other than
those of man and beast in hostile operations has transformed the
character of such operations and made them war in the modern
sense. It is true that human power had been converted in form and
direction in the past by mechanical devices such as the bow, arque-
bus, and siege engine, but the force of these instruments was limited
by the power of the human arm to bend the spring upon which the
device depended. Until recent periods man had no reliable methods
for releasing power stored by other than human or animal muscle
for the purpose of advancing toward or striking an enemy. This
change has made war more destructive, more likely to spread, and
consequently of more general interest. Resort to war anywhere has
tended to become a matter of concern to all governments, and consequently the use of this technique must be justified in terms of the
world-order, whether to sanction the status quo or to effect revisions
deemed to be desirable. Animals have fought from inherited drives,
primitive men from group custom, people of historic civilization for
group interests, but people of contemporary world-civilization fight
for what they deem to be a better world-order.

Thus the origin of war depends upon the definition used. It began
with animals in the psychological sense. It began with primitive
people, untouched by civilized neighbors in the sociological sense.
It began with civilization in the legal sense. Only since the advent
of continuous world-cultural contacts in the fifteenth century has
war existed in the modern technological sense.

In all its stages war can, of course, be approached from the sociological, legal, and technical points of view as well as from the psychological. In each of these four stages violent behavior has served superindividual functions, has exhibited formal regularities of recurrence and conduct, and has proceeded by describable techniques as well as from understandable psychological drives. Even animal warfare has functions, a theory, and a technique, but they are not the functions, theory, and technique which characterize modern human warfare. While the history of modern psychological drives goes back to the animals, that of modern sociological institutions goes back only to primitive man, that of modern law only to early civilizations, and that of modern technology only to the inventions of the late Middle Ages. Animal sociology rests on different foundations from human sociology, primitive law rests on different foundations from civilized law, and modern technology rests on distinctive foundations. War has changed its character with each of these great transitions. It may be that an equally momentous change has been ushered in by the invention of nuclear fission, intercontinental missiles, and artificial satellites.

2. ANIMAL WARFARE

The study of animal warfare may contribute toward understanding the organic bases and social tendencies of war and the influence of particular military techniques and of war in general upon the survival of societies and races. Human beings are but a small element of the organic population of the earth. The great symplasm, whose history began in pre-Cambrian times, is composed of protoplasmic cells, each adapted to a definite environment but similar to one another in origin, chemical composition, organization, and behavior; in exhibiting reactions of movement, repetition, response, and irritability; in engaging in activities of nutrition, reproduction, rivalry, and protection; and perhaps in experiencing feelings of hunger, affection, dominance, and fear.

a) *Drives.*—The psychological causes of war lie ultimately in the characteristics of protoplasm, and study of the simpler animal forms gives better evidence of the basic pattern of these characteristics than the study of such a complex form as man. Such a study suggests a classification of fundamental drives, in terms of the end object, as food, sex, dominance, self-preservation, home territory, activity, independence, and society.

Among individual animals violence is motivated most commonly by the drives for food when between animals of different species and by the drives of sex, territory, dominance, and activity when between animals of the same species. All animals have means of self-defense, but flight is more common than hostile action against the aggressor. Animals with highly organized societies, like the bees and ants, fight primarily from the societal drive. The society as a whole is driven to aggressive hostilities by the needs for food or territory and in some instances by the urge for migration or parasitic dominance. Such societies often have specialized members or castes to defend them when attacked. The need of defense has played an important role in developing animal aggregations and societies.

Among the animals biologically nearest to man, the drive for dominance is usually at the root of fighting, though frequently the drives of activity and sex play a part in such incidents. Because of the relatively weak social organization of apes, the dominance and activity drives in combination may lead to alliances against the dominant leader, especially when the capacity of the latter is declining with age. These occasionally result in balance-of-power wars like those among sovereign states.

Detailed studies of group behavior of monkeys and apes in captivity and of young children indicate that the situations precipitating fights are similar. The aggression precipitating a fight usually involves several drives, and fighting once begun tends to spread throughout the group. Aggressive behavior usually arises from rivalry for *possession* of some external object, from *intrusion* of a stranger in the group, or from *frustration* of activity.

Possessiveness may be manifested in respect to food, territory, objects of curiosity such as toys, or another member of the species. Jealously from possessiveness of the latter type leads to fighting more often than does rivalry for food. The desire for possession seems often to be increased among both apes and children by awareness that another of the group desires the same thing. Hostility against an intruder may arise from apprehension that a stranger may interfere with the satisfaction of other drives, particularly that he may become a rival for valued possessions. It may therefore be considered a hypothetical form of possessiveness or perhaps possessiveness toward the existing group situation as a whole. Rage, aggressiveness, and fighting may arise from frustration of the normal activity associated with any drive and may be directed against any person or object believed, often erroneously, to be guilty of the interference. This type of aggression is somewhat less characteristic of fighting among apes than among children. Among the latter the frustration may even be attributed to the child's own incapacity, leading sometimes to self-punishment.

b) Functions.—The sociological causes of war are to be found by analyzing the function of war in the life of the larger whole. Among colonial insects, fighting habits of certain members of the society function to preserve the society, as among men; but among animals in general, fighting habits, though varying in intensity among individuals, are characteristic of the entire species. Animal fighting, therefore, must ordinarily be interpreted functionally in relation not to a society or a culture but to a race or species. A tendency toward deadly intraspecific fighting would be a serious disadvantage for the race and would usually be eliminated by natural selection. For this reason intraspecific fighting among animals is seldom lethal. Differing from human war, which is always intraspecific and is often most serious between peoples of the same race, animal fighting declines in deadliness with closeness of genetic relationship of the combatants. The really deadly animal violence is between widely separated species, as between the lion and the antelope, and resembles human operations in the hunt or the slaughterhouse rather than in war. The competition for a limited food supply among animals of the same species results not in lethal conflict and combat but in dispersion and starvation of the least fit. It usually takes the form of aggressiveness by the male, sometimes aided by the female in defending the home territory, the nesting and feeding area, from others of the species. Its human analogy is economic competition between individuals or firms rather than warfare.

Animal individuals and species in a neighborhood exhibit widespread dependencies upon one another. Unmitigated predaceousness and parasitism may have a suicidal effect. Survival of the species depends upon wise maintenance of the balance of nature, and natural selection has shown a persistent tendency to limit parasitism and predaceousness even between unrelated species. The species with the largest number of individuals and the widest range usually has neither of these characteristics. Both ant and human societies have gone through hunting, pastoral, and agricultural stages, and the latter has proved to have the greatest survival value.

c) *Techniques.*—The modes of attack and defense—the specializations in mobility, striking power, armor, co-operation, and mass attack—are so diverse and extreme among animals that it is easier to see their relations to the incidence of fighting, and their effects on the preservation of species, than in the less extreme variations found in human history.

Among individual animals specialization in mobility, as among birds, deer, and monkeys, makes for a war of maneuver and is particularly favorable to intraspecific war, which, however, results in dispersion rather than in death. Tenacity, making for a war of attrition, as among boa constrictors and bears, is particularly unfavorable to intraspecific war as it results in the death of the victim and so is disadvantageous to the species. Specialization in striking power, as among lions and cobras, making for a war of pounce, is also unfavorable to intraspecific war, though it acts as an incitement to aggression against weaker species. Specialization in protective armor, as among tortoises, armadillos, and clams, makes war unlikely, unless the armor is accompanied by considerable striking power and moderate mobility, as in the elephant, rhinocerous, and swordfish. In that case a war of shock may occur even within the species, though more rarely than in the less heavily protected and more mobile animals. Genetic lines specializing in heavy protective armor tend to increase in size and to decrease in mobility and adaptability, sometimes to a suicidal extent, as in the dinosaurs.

The advantage of an animal in battle depends upon the particular combination of all these types of military equipment. It appears that genetic lines specializing in mobility and tenacity have prospered most, although the first has maximized and the second has minimized the frequency of intraspecific hostilities. Clumsiness, resulting from specialization in protective armor, and predaceousness, resulting from specialization in striking power, have not characterized the most numerous species, especially among the higher animals.

Certain animals, like ants, termites, and buffaloes, have developed collective military techniques, but these are more often for defense than for aggression. Animal societies which specialize in striking power and mobility, like the driver and slave-taking ants, tend to be predaceous and parasitic, characteristics not favorable to rapid multiplication. Specialization in protective walls, as among the termites, while avoiding intraspecific war, stunts the possibilities of adaptation to changing conditions. On the other hand, specialization in protective group loyalty, as among the ants, tends to maximize intraspecific war. The great body of ant colonies, however, with fifty million years of social experience behind them, generally keep to their own nests and feeding areas and engage in hostilities only when attacked by the parasitic or predaceous minority of the ants.

A study of these techniques suggests that long survival of a species has resulted from a balance between the efficiency which comes from integration of the entire structure and behavior of the animal about a specialized technique and the flexibility which comes from avoiding such complete specialization and integration that adaptation to new conditions becomes difficult or impossible. Violent changes of climate, food supply, or habitat have resulted in the elimination of the narrowly specialized species and genera, particularly those specializing in size, armor, and predaceousness.

d) Theory.—The theory of animal war is the theory of organic evolution—the non-survival of the unfit. The balance of organic nature is maintained principally through the process of one species preying upon another, especially upon the young, and of one species crowding another out of an area which forms for it a suitable habitat. These modes of elimination are, of course, counterweights to reproduction, which, when sexual, permits a tremendous multiplication of combinations from gene mutations arising in an individual. Climatic, geologic, and geographic change may at times suddenly alter the balance and exterminate populations or even species and genera; but in a constant physical environment, being preyed upon and being crowded out of a food supply are the modes of eliminating the superabundant population provided by the extraordinary fertility of most species—fertility such that almost any species would, if all survived and reproduced, occupy the world or the solar system in a short time. Only the social insects which confine reproduction to a single female in the society, the workers being made sterile, have adopted a process of limitation through birth control.

The normal modes of elimination may, at times, be greatly exaggerated or decreased through invasions of an area, especially by man; but in nature, their adjustment to reproduction is often so precise that from year to year the population of each species in a given area may vary very little. However, these populations usually undergo gradual quantitative changes, sometimes of a cyclical character. Such quantitative changes of populations are accompanied by evolutionary changes of type, the speed of which depends upon the balance of such factors as random variation, mutation, migration, crossbreeding of different races of the species, and the intensity of selection measured by the proportion between those succumbing and those surviving from year to year.

The rate of evolution of a biological community, or biocoenosis, will be augmented by intense selection among its constituent species, and such selection will be intensified by radical change in the physical environment or encroachment of neighboring biological communities. The rate of evolution of a species, however, is not determined by the intensity of selection among individuals of the species, as suggested by some interpretations of Darwinism, but by selection among comparatively isolated races which have drifted apart as a result of local inbreeding. As an evolutionary factor, selection must operate upon communities, races, subspecies, or species rather than upon individuals. But, whether between individuals or groups, the struggle for existence is not a conscious conflict resembling war but an unconscious competition for food supply.

Among colonial insects and perhaps other species, mutual aid, co-operation, and specialization of function appear to be of significance for regulating the survival and evolution of the group. All animals live in groups, using that term in the broadest sense to include communities manifested by symbiotic relations among different species in the same area, aggregations or close masses of animals of the same species, families united by sex and parental relations, as well as societies of every degree of integration and duration. Relatively few animals have developed social co-operation and specialization of function within groups smaller than the community and larger than the family.

While the propriety of identifying subhuman with human societies is controversial, the influence of co-operative relations among animals upon both their reproduction and their elimination is important in organic evolution but less so than selection through competition for a living.

With respect to the survival of individual animals, the role of war is indeterminate. Among the carnivores the most skilful in the use of violence will survive. Among the herbivores the most speedy and alert will survive. With respect to species, the gregarious herbivores have had an advantage over the predaceous carnivores. Skill in lethal violence has not been a characteristic of the most numerous species. Aggressiveness, especially of males, to defend the family and home territory against intrusion by others of the species has been common among both birds and mammals. This type of warlike behavior has been of value to the species in dispersing its members over a wide area and preventing their extinction. Animals lacking this characteristic, like the American bison and passenger pigeon, have tended toward excessive aggregation and are at a disadvantage when confronted by new enemies. With respect to biological communities, interspecific hostilities, preying and being preyed upon, are a major factor in preserving equilibrium among the numerous species composing the community. If most species were not the natural food of others, the great variety of animal life, valuable for the stability of such a community, could not continue. A few species would soon crowd all the others out. Thus, while herbivorous species are at an advantage in the interspecific competition for a living, from the standpoint of a biological community the existence of predaceous species is important.

War has played an important role in the preservation of the societies of many species of colonial insects. Among some such societies it may have been an agency for promoting internal solidarity, and it has undoubtedly served for external defense and for acquiring food.

The study of animal war has much to contribute to an understanding of the psychology of human war, and in this respect the role of dominance, activity, and sexuality among the primates, man's nearest relatives, is most instructive. The greatest difference lies in man's superiority in communications through his possession of language and, as a result, his vast superiority in social organization. In the latter respect the ants most resemble man, and the analogy of their wars for predation and defense with those of nations has often been insisted upon. There are, however, great differences. The members of a human society can communicate at a distance and so the society may expand over ever increasing areas. Whereas ant societies are composed of the children of one queen, human societies are genetically heterogeneous, assuring them a greater variability and duration of life. The members of a human society, moreover, lack the degree of hereditary and structural specialization, differentiation, and stratification characteristic of ants. Human society thus compensates for its difficulties in maintaining internal social order by the possibilities of progress and of eventual universal co-ordination of the species. Although the problem of civil war will always be more serious in human than in ant societies, the problem of external intraspecific war is soluble among men but not among ants.

It is to be anticipated that man, having organized his societies toward intellect and progress, will not converge toward the ant's "societies," emphasizing instinct and stability, though despotic totalitarianism would lead in that direction. The mechanism of formic social solidarity throws light, however, upon the irrational foundations of human societies. The history of both types of society indicates that there is survival value in miminizing predation, parasitism, and other forms of violent behavior. In this respect convergent evolution of the human and insect types of society may be expected.

3. PRIMITIVE WARFARE

Psychologists and sociologists seldom deal with the subject of war without at least a preliminary chapter on primitive war, and sometimes they seem to feel that the subject of war has been adequately treated without getting beyond the primitive stage. Davie writes at the end of his study of primitive war:

> In our study of the evolution of war in early societies, we have surveyed the greater portion of the whole history of the institution, for civilization is as yet in its infancy as compared with the vast expanse of primitive times. In the light of the perspective which we have acquired, what may be predicted about the future? The underlying causes and motives of war were present at the beginning and for the most part still exist.

Strategical writers and jurists, on the other hand, do not deal with primitive war at all or introduce merely decorative, inaccurate, and unconvincing illustrations from the field. The official code of the United States Army (Art. 381) refers to the "internecine war of savages" as the unspeakable condition to which unjust or inconsiderate retaliation, by removing the belligerents farther and farther from the mitigating rules of regular war, will by rapid steps bring civilized belligerents. Strategical writers insist on the need of "more brutal" methods in dealing with savages who do not observe "the individual decencies of civilized regular soldiers." Even when dealing with the specialized topic of "small wars," that is, operations of civilized against uncivilized people, such writers do not give proper consideration to primitive warfare but only to the technique and rules which have been or should be used by civilized peoples in such operations.

This difference among writers suggests that, if a study of primitive war has anything to contribute to knowledge of contemporary war, it is to its psychological foundations and sociological functioning rather than to its law and technique. Yet primitive peoples have usually observed rules in the initiation and conduct of war and have utilized a variety of technical and strategical methods.

Primitive war, like animal war, has been evolving through a vastly greater period of time and among a much greater variety of social organizations than has civilized war. Thus, if the data were on hand, it would present superior opportunities for comparison, for correlation of the incidence of war with varying social and material conditions, and for estimating the variability or persistence of the elements of warfare.

a) The Conception of Primitive Warfare.—The study of primitive warfare at once confronts two formidable difficulties: Who are primitive people? and What part of their behavior is warfare?

It is difficult to distinguish primitive man from civilized man. There are very few of the present "primitive," "preliterate," "simpler," "natural," or "savage" people who have not received some elements of their culture from civilized people. It appears that war practices, weapons, and techniques are among the first things to be borrowed by primitive people, although the rapidity of such borrowing varies greatly among different primitive tribes. It cannot be assumed, therefore, that the war practices of any contemporary primitive people have any close resemblance to the war practices of man in the hundreds of thousands of years before there was any civilization.

As a convenient even if rather arbitrary rule, primitive people may be defined as human beings who live in self-determining communities which do not use writing. The absence of writing and of recorded history usually involves other features of culture. The community is usually confined to a group which can be reached for purposes of administration and leadership by general assemblies addressed by word of mouth or by runners carrying messages in memory. The absence of writing also limits early education to that which the family and neighbors can pass on to the child from memory. Law is limited to customs carried in memory and passed by tradition from generation to generation. Scientific generalization is limited to that which can be developed from evidence within one man's memory of his own experience, of the experience of his acquaintances orally and uncontrollably repeated to him, and of the even less reliable tradition of the group passed orally from generation to generation. The methods, tools, utensils, and machines for carrying on the practical affairs of life are limited to those which have been invented by a trial-and-error process without aid of general ideas. Some primitive people use domesticated animals, but except for the dog, such use has been borrowed from civilized people in historic times. The Plains Indians, for example, did not use the horse until the Spaniards had introduced it to America. Most preliterate people are limited to instruments operable by manpower.

Primitive people are as a rule politically integrated in relatively small clans, villages, or tribes which speak a common language. Blood relationship plays a major part in their organization. They form their pattern of life by relatively fixed tribal customs and attempt to control their environment through magic ritual, through propitiation of supernatural beings, through hostility against neighbors, and through practical techniques mainly utilizing the power of the human individual. Because of the inefficiency of these controls, they are in the main bound to adapt their way of life to the surrounding physical, animal, and especially vegetable environment. Thus the group customs manifest great variety according to the differences in this environment, but within each group the behavior patterns are more uniform and less complex than among civilized people.

Less than ten thousand years ago all people were primitive in this sense. However, the total number of the human race, though scattered over all the continents, may then have been less than that of a moderate-sized city of today. Civilization affected only a minority of the world's population for thousands of years. At the beginning of the Christian Era probably half of the world's population was still "primitive"; at the time of the discovery of America a quarter was probably in that condition. Civilization, however, has spread rapidly in the recent era, and today less than one in a thousand of the world's three billion people is still primitive. In a sense, even these are not primitive within our definition because they are nominally subject to states where writing is used. But, in so far as they still enjoy practical autonomy, they may be classified as primitive. Under the influence of missionaries, administrators, and traders, however, they are rapidly becoming eliminated, assimilated, civilized, or deprived of all autonomy.

Added to the difficulty of identifying primitive peoples in general and in particular is the difficulty of identifying their wars. Primitive peoples only rarely conduct formal hostilities with the object of achieving a tangible economic or political result. Their hostilities are seldom conducted by a highly organized professional military class using distinctive instruments and techniques regulated by an intergroup law applicable only during periods of "war" and designed to render war an efficient instrument of policy. These elements which go to make up the concept of war today are products of civilization, and only their rudiments can be found among primitive peoples.

Though broader than the concept of civilized war, the concept of primitive war is narrower than that of animal warfare. Animal war includes violence against animals of other species and violence against members of the group whether prohibited or commanded by the group.

These three types of violent behavior—the hunt, crime, and punishment—although considered "war" among animals, are among all primitive groups so distinct from violence sanctioned by the group as a whole against other human beings external to the group that they can be excluded. The line between privately initiated external violence sanctioned by the group, such as feuds, and head-hunting, and action for which the group as a whole is responsible is less easy to draw. In most cases these two types of activity, which may be denominated, respectively, "reprisals" and "war," can be distinguished. They are, however, closely related, and it seems advisable to include all external, group-sanctioned violence against other human beings in the conception of primitive war.

b) *General Characteristics of Primitive War.*—Primitive peoples may be classified racially, geographically, culturally, and sociologically. There are relationships between these classifications. Peoples in the same area are likely to be racially and culturally similar, and peoples of similar race and culture are often similarly organized. But there are numerous exceptions, and correlations cannot be assumed. All these classifications may provide evidence with respect to the evolution or diffusion of particular cultural traits such as war, but great caution is necessary in generalization.

Although the functions, drives, techniques, and formalities of war vary greatly from tribe to tribe, it will be convenient, first, to classify primitive war according to the general degree of its development as an institution. There are primitive people who fight not at all or rarely and in an unorganized and unpremeditated manner; war is not a definite institution of the mores. There are others who fight frequently in well-recognized circumstances and with well-established rules and techniques; war is definitely within the mores. There are, of course, line cases. Most peoples can, however, be rather definitely divided into the warlike and the unwarlike. The familiar distinction between the industrial and military types of political organization, emphasized by Herbert Spencer, T. H. Buckle, and others in comparing civilized as well as uncivilized states, conforms in some degree to this distinction. Among primitive peoples the distinction is more emphatic, since all civilized people have war to some degree in their mores. For purposes of correlation this dual classification has been refined by distinguishing the most unwarlike peoples who fight only in defense; the moderately warlike who fight for sport, ritual, revenge, personal prestige, or other social purposes; the more warlike who fight for economic purposes (raids on herds, extension of grazing lands, booty, slaves); and the most warlike of all who, in addition, fight for political purposes (extension of empire, political prestige, maintenance of authority of rulers). Classification of peoples according to their peaceableness or warlikeness may be correlated with other ways of classifying primitive peoples.

Geographically, people may be divided according to the continent in which they live. Among primitive people war as an instrument toward rational ends has been least developed in Australia and most developed in Africa. European civilization seems to have sprung from very warlike primitive peoples. America and Asia exhibit both very warlike and very unwarlike people.

More significant geographical classifications can be made according to the climatological and topographical environment of peoples. Primitive peoples in extremely cold and extremely hot climates tend to be unwarlike, although the very warlike Bering Sea Eskimo lives in as cold a climate as the very unwarlike Greenland Eskimo, and the warlike Bantus and unwarlike Pygmies both dwell in the tropics of Africa. In general, however, a temperate or warm, somewhat variable, and stimulating climate favors warlikeness. However, it also favors civilization. These favored regions have developed civilization or have been occupied by civilization, leaving the primitive people only the less satisfactory environments. Among contemporary primitive people the largest proportion of the warlike live in hot regions of medium climatic energy.

Primitive people inhabiting deserts or the seashore are more likely to be warlike than those in forests and mountains, and those in the grasslands are the most warlike of all. Warlikeness appears to be related to the stimulating character of the climate and to the lack of barriers to mobility rather than to the economic difficulty of the environment. The primitive nomad of desert and steppe has a hard environment to conquer, but he may have a stimulating climate. His terrain, adapted to distant raids and without natural defenses, leads him to institutionalize war for aggression and defense. The seashore dweller, because of easy opportunities to travel, is encouraged to piracy, as the nomad is encouraged to raid. The Eskimo of the north, with an equally difficult economic problem but with too severe a climate and with the protection of isolation and impediments to travel, is often but not always peaceful. The hunters of forest and mountain, protected by natural barriers, tend to be peaceful. But where the climate is stimulating as with the eastern American Indian, they may be warlike. The forest dwellers of the Andaman Islands, Africa, Malaya, and Indonesia, with a less stimulating climate, are more peaceful.

Race is not closely related to war practices among primitive peoples although Pygmies and Australoids seem to be the least warlike; Negroes, Hamites, and whites the most warlike; with the red, yellow, and brown races in an intermediate position. Some subraces belonging to the more warlike races, such as Papuans, Dravidians, Arctics, and Eskimos, are quite unwarlike.

Culturally, primitive peoples have often been divided into those who make their living by collecting shellfish, fruits, and nuts; by hunting animals; by herding domestic animals; or by agriculture. Those who ascribe an evolutionary significance to these stages regard the herdsmen, agriculturalists, and higher hunters and fishers as parallel developments from the lower hunting and agricultural cultures. The picture is not, therefore, one of continual progress but of a tree with different types of culture developing above a certain point. It seems clear that the collectors, lower hunters, and lower agriculturalists are the least warlike; the higher hunters and higher agriculturalists are more warlike; while the highest agriculturalists and the pastorals are the most warlike of all.

Sociologically, primitive peoples may be classified into those who are integrated in primary (clan), secondary (village), tertiary (tribe), and quaternary (tribal federations or states) groups. In general, the first are the least and the last the most warlike.

Primitive peoples may also be classified sociologically according as they utilize division of labor between the sex and age groups only, between involuntary classes (castes, serfs, slaves, nobles), and in addition between voluntary, professional, or occupational groups (farmers, herdsmen, various types of artisans, soldiers, priests, and rulers). The voluntary type of specialization is little developed among primitive peoples, although it appears that voluntary specialization may develop in groups that have never known compulsory classes. Professional soldiers, except as an age and sex group, exist only among semicivilized and civilized people. In general, the more the division of labor, the more warlike, the groups with compulsory classes being the most warlike of all primitive people.

Finally, primitive peoples may be classified sociologically according to the abundance of extra-group contacts with societies of widely different cultures. Some peoples are isolated by natural barriers or the frugality of the food supply; others are in continuous communication with civilized or semicivilized people; others are on highroads of migration and in frequent close contacts with such people. In general, the groups with the most varied and frequent contacts are the most warlike. Hoijer concludes a detailed study of the causes of primitive war with this statement:

The presence of many groups within a certain area offers—providing natural barriers do not interfere—opportunities for numerous cultural contacts. In striving to remain a tribal entity and to preserve itself physically, the group must perfect a strong social organization and a powerful war machinery. Needless to say, these strivings are unconscious. If they fail, they lose their group identity, if, indeed, they are not annihilated altogether. Those who succeed, establish strong tribal organizations whose lives can only be maintained by hostility—warfare becomes the necessary means of preserving group identity, in primitive society.

It would appear that the seriousness and degree of institutionalization of war among primitive peoples are related more closely to the complexity of culture, political organization, and extra-group contacts than to race or physical environment, although a warm but stimulating climate and an environment favorable to mobility over wide areas also seem to be favorable to warlikeness.

c) Causes and Consequences of Warlikeness among Primitive Peoples.—The conclusions arrived at above with respect to the static circumstances of warlikeness and unwarlikeness among primitive peoples suggest the following generalizations with respect to the dynamics of the situation.

Unwarlikeness has been the result of prolonged opportunity of neighboring groups to achieve equilibrium in relation to one another and to the physical environment. This opportunity has only been offered if the physical environment has been stable and if peoples of different culture have not interfered. The latter has resulted from natural barriers, lack of means of travel, or inhospitableness of climate.

Reciprocally, warlikeness has resulted from frequent disturbances of the equilibrium of a group with respect to its physical environment or its neighbors. The first has usually resulted from climatic changes, migrations, or the invention or borrowing of new types of economic technique. The second has usually resulted from migrations, invasions, or other influences bringing a group into continuous contact with a very different culture.

Among primitive peoples borrowing or invention of means of mobility or more efficient weapons promoting migration, invasion, or expansion of contacts increases warlikeness. Such borrowing or invention proceeds very slowly among primitive groups unless forced by contact with much more civilized peoples. Thus, the more primitive the people, the less warlike it tends to be.

Primitive warfare was an important factor in developing civilization. It cultivated the virtues of courage, loyalty, and obedience; it created solid groups and a method for enlarging the area of these groups, all of which were indispensable to the creation of the civilizations which followed. It must, however, be emphasized that primitive warfare was very different from warfare in historic civilizations and differed even more from contemporary warfare among the advanced peoples. Recognition of the progressive tendencies of primitive war (assuming that the movement from savagery to civilization is progress) does not imply that warfare at later stages is progressive. As primitive society developed toward civilization, war began to take on a different character. Civilization was both an effect and a cause of warlikeness. The custom of war provided a basis for wider aggregation and more secure defense. The rise of wider aggregation and division of labor, division of ruler and ruled, created conditions favorable for the development of war in the interests of the ruling class. The value of war in developing social virtues and social organization was more and more offset by its evils in eliminating human sympathy, in preventing co-operation beyond the warmaking group, and in increasing destructiveness; but on the whole, among primitive people its advantages for progress outweighed its disadvantages. Marrett writes:

It is a commonplace of anthropology that at a certain stage of evolution —the half-way stage, so to speak—war is a prime civilizing agency; in fact, that, as Bagehot puts it, "Civilization begins, because the beginning of civilization is a military advantage." The reason is not far to seek. "The compact tribes win," says Bagehot. Or, as Spencer more elaborately explains, "from the very beginning, the conquest of one people over another has been, in the main, the conquest of the social man over the anti-social man."

With primitive groups as with animal species, the survival value of war utilization has varied with the particular situation of the group or species; but with animals, on the whole, adaptations based upon peace and co-operation have proved more favorable to multiplication of the type than adaptations based upon predation and parasitism. Among primitive people, on the other hand, the warlike groups have multiplied both in individual and in type. The peaceful groups could not organize to a size sufficient for extensive division of labor without the military virtues and the sense of group solidarity created by the fear of an external enemy, and they could not protect wealth, herds, or agricultural lands from warlike neighbors unless they institutionalized war themselves. Out of the warlike peoples arose civilization, while the peaceful collectors and hunters were driven to the ends of the earth, where they are gradually being exterminated or absorbed, with only the dubious satisfaction of observing the nations which had wielded war so effectively to destroy them and to become great, now victimized by their own instrument.

4. HISTORIC WARFARE

The study of animal and primitive warfare throws light on the function of conflict in a living system and on the nature of the drives toward lethal conflict among the entities which compose this system. Only indirectly or by analogy do these studies aid in understanding the influence of changing military techniques and of changing theories or rationalizations upon the incidence and character of contemporary war. A study of historic war, by which is meant warfare within or between the literate civilizations from Egypt and Mesopotamia down to the age of discovery in the fifteenth century—a span of over six thousand years—may give a more direct understanding of these influences.

Efforts to tabulate the military events of history and to study their trends and fluctuations over long periods of time have not been particularly rewarding for a number of reasons. The historic record is, except for the most recent times, extremely fragmentary. A fair record is available of the wars of western Europe and of the Classical Mediterranean civilization since the fifth century B.C., but for an adequate statistical base the wars of the Egyptian, Mesopotamian, Syriac, Indian, Chinese, Mayan, and other civilizations should be available for comparison. Much of this material exists, but it has not as yet been made easily usable for any but specialized historians. Furthermore, even when records of the battles and wars exist, data as to the number of participants and casualties are unreliable.

Furthermore, no class of military incidents has the same significance in all periods of history. The battle has been the most persistent type of military incident. At some periods, however, battles have been isolated events from which flowed important political consequences. At other times a battle has been but an incident in a campaign consisting of complicated strategic operations over a season or in a siege or maritime blockade. In such circumstances political consequences cannot be attributed to the single battle but only to the whole campaign. Campaigns themselves have sometimes been but incidents in a war waged on many fronts with a number of distinct armies over a series of years. Neither the battle, the campaign, nor the war is entirely satisfactory as a unit for statistical tabulation.

It is difficult to rate the relative importance of any of these incidents. Sir Edward Creasy's *Fifteen Decisive Battles of History* suggests the inadequacy of any objective scheme of rating, such as duration, number of men engaged, and number of casualties, emphasized by Bodart, Woods, Sorokin, and Richardson. Creasy's list contains Joan of Arc's victory at Orléans and the French Revolutionary battle of Valmy, in each of which the casualties were only a few hundred, as well as Marathon, Syracuse, the destruction of the Armada, and Saratoga, where they were only a few thousand. In the rest of his battles—Arbela, Metaurus, Arminius' victory over Varus, Châlons, Tours, Hastings, Blenheim, Pultova, and Waterloo—the casualties reached tens of thousands, but he included none of the bloodiest battles with casualties asserted to have reached hundreds of thousands in ancient, medieval, and modern history. Creasy's rating was based on a subjective judgment of the political and social consequences of the battle, a sort of criteria unsuitable for a general statistical study of military incidents. Richardson, on the other hand, has made a compilation of "deadly quarrels" based solely on number of casualties. He has achieved some interesting statistical results, but his compilation goes back only to 1820 and so does not touch the period here considered.

Military events have varied greatly with respect to the purposes and nature of the combatants. Should one place in the same category civil wars between factions in the same state, international wars between states of the same civilization, and imperial wars between groups from different civilizations? Should the objectives (wars, interventions, reprisals), the legal characteristics (just or unjust, formal or informal), and the technical characteristics (battles, sieges, naval battles, blockades) of military events be distinguished? Finally, what time and space limitations should be adopted in a statistical tabulation of military incidents? The data are lacking for a single tabulation of all battles or wars between civilized peoples since civilization began, and it is doubtful whether a tabulation which lumped together Egyptian, Mesopotamian, Chinese, Indian, and Mayan wars would be illuminating. If there are regular trends or fluctuations of war and peace, it seems probable that they are relative to groups of people in more or less continuous contact with one another, that is, to a civilization.

a) Relation of War to Changes of Civilization.—What has been the relation between changes in civilization and changes in war? No precise correlation is possible, for a civilization has not been a simple thing. Each civilization has been, through most of its life, composed of many states which have been in ceaseless process of alliance, federation, union, disunion, or disintegration, and each state has usually been composed of factions and groups which also continually coalesce and separate. A model of history would be a time dimension rising above a two-dimensional map. This three-dimensional space would be divided into civilizations with fluctuating time and space boundaries within which are vaguely bounded states composed of vaguely bounded factions.

A complete history of civilized war would involve identification of all these units and description of the characteristics of war in each. In this section, however, historic units smaller than a civilization will not be considered. Although civil wars and interstate wars have occurred in all civilizations, intercivilization wars have been commonly regarded as most important and have figured most in history. For example, the bulk of the battles before 1500 in Harbottle's *Dictionary of Battles* occurred in wars of this category, as did eight of Creasy's fifteen battles, including each of the seven before Hastings.

Military activity has varied with respect to the magnitude of war, to the size of armies, to the character of operations, to the relations of belligerents, and to the objectives and justifications of hostilities. These variations may be related, using Toynbee's classifications, to the antiquity and type of civilization and to the stage of a particular civilization in which the military activity occurred.

(1) *Successive civilizations.*—The magnitude of war has oscillated in long waves during the historic period. In the West and Middle East, armies were bigger and major wars more frequent in antiquity and modern times than in the Middle Ages, although it is generally accepted that there has been much exaggeration in the descriptions of the size of armies engaged in battles among ancient Oriental nations. Furthermore, every civilization tends to rise to a maximum of military activity in its time of trouble and then to sink down until

there is a new high of military activity at its decline and fall. With allowance for these oscillations, however, it appears that, comparing successive civilizations, armies have tended to become larger absolutely and in proportion to the population, war has tended to become absolutely and relatively more costly in life and wealth, and military activity has tended to be more concentrated in time with longer peace intervals between wars. War has tended to become more extended in space, with fewer places of safety and more inconvenience to the civilian. War has ideologically and legally become more distinct from peace and has tended to be regarded as more abnormal and more in need of rational justification. The changes in war have on the whole tended to favor defensive rather than offensive operations. Consequently, war has tended to be less rapidly decisive and, relative to other institutions, less influential upon world politics.

Up to World War II, these trends exhibited a movement toward a situation of enormous armies and great war costs; of alternating periods of concentrated military operations over the entire world and relatively peaceful intervals between; of extensive rules for justifying, condemning, and restricting military operations; and of strategy tending toward protracted stalemates, the use of mutual attrition, and relatively indecisive results.

(2) *Types of civilizations.*—Because of the great variations in the military characteristics of each civilization during its long life, it is not easy to relate military characteristics to the geographic, economic, or other characteristics of the civilization as a whole. As all civilizations have tended to become less aggressive as they have become older, civilizations must be compared at the same stage of their development.

If warlikeness is regarded as a compound (i) of habituation in cruelty arising from bloody religious rites, sports, and spectacles, (ii) of aggressiveness manifested by frequency of active invasions in imperial or interstate wars, (iii) of military morale indicated by discipline of armies and reserves, and (iv) of political despotism manifested by completeness of territorial and functional centralization of authority with absence of constitutional and customary limitations, it appears that the most warlike civilizations were the Classic, Tartar, Babylonic, Syriac, Iranic, Japanese, Andean, and Mexican. The moderately warlike were the Hittite, Arabic, Germanic, Western, Scandinavian, Russian, and Yucatec. These lists appear to exhibit no high correlation with chronology, geography, or economy. The more warlike civilizations include primary, secondary, and tertiary civilizations; large and small populations; steppe, arable, maritime, and plateau environments; grazing, dry farming, and commercial economies. There was a tendency, however, for civilizations in a plateau or mountainous environment, civilizations dependent on grazing, and civilizations with a heterogeneous population and with close intercivilizational contacts to be warlike.

By the same criteria, the most peaceful civilizations were the Egyptian, Minoan, Orthodox (Byzantine), Mesopotamian (Sumerian), Nestorian, Irish, Indian, Hindu, Sinic, Chinese, and Mayan. These also are distributed among the chronological, population, geographic, and economic types, although it appears that homogeneous and isolated civilizations and those on rivers and dependent on irrigation agriculture tended to be peaceful. The isolated Japanese, perhaps, owed their warlikeness to their heterogeneous racial composition.

There was perhaps a tendency for secondary civilizations, such as the Babylonic, Syriac, Classic, Hindu, and Mexican to be more warlike than the primary civilizations from which they sprang—the Egyptian, Mesopotamian, Minoan, Indian, and Mayan—and for tertiary civilizations, such as the Western, Arabian, Iranian, and Russian to be the most warlike of all. This may have been due to the fact that each successive civilization began as a barbarian or external proletarian group seeking to gain the advantages of the established civilization by warlike activity. If it was eventually successful, having been born in military success, it usually continued warlike. Furthermore, the later civilizations with better means of communication were less isolated, while their greater inventiveness made them less homogeneous. Both of these factors tended toward warlikeness.

It is possible that maritime civilizations like the Scandinavian and nomadic civilizations like the Tartar and Arabic tended to be more warlike than agricultural civilizations such as the Egyptian, Mesopotamian, and Mayan because of greater mobility and more immediate responsiveness to climatic changes. This relationship, however, is far from clear.

It has been said that civilizations in cold climates have tended to aggress against those in hot climates, as illustrated by the raids of

Mongols, Aryans, Achaeans, Gauls, and Germans upon the civilizations to the south of them. It appears, however, that such aggressions have really proceeded from the steppes in many directions as a result of drought, of administrative collapse in the invaded area, or of trade stoppages.

There is perhaps some support for the suggestion that martial character has come from mountains and passivity from plains when one considers the relatively warlike disposition of the Hittites, Iranians, Andeans, and Mexicans. Such exceptions as the peaceful Nestorians may be due to the early stage at which that civilization was destroyed by conquest.

A civilization, however, has seldom been dominated by any one factor. Warlikeness has resulted from a conjunction of circumstances in the internal composition and external relations and conditions of a particular civilization. Its peculiar combination of social, religious, political, and military institutions and external contacts has been an adaptation to a variety of circumstances in its environment and basic economy and not a consequence of any limited set of conditions.

(3) *Stages of a civilization.*—The intensity of war measured by the frequency of battles and the number of participants has usually increased through the heroic age (the swarming period) when war has characteristically been utilized either for aggression against or for defense from other civilizations. Its intensity has continued to increase through the time of troubles, but during this period war has often been interstate and civil as well as imperialistic and defensive. The period of stability which follows has usually been more peaceful, though the size of armies has not greatly decreased. As this period has progressed, aggression from the outside has often occurred. Barbarians, or another civilization in the swarming stage, have assaulted the civilization with increasing capacity as their armies have learned more of the art of war from contacts with the older civilization. This was illustrated by the assaults of the Hyksos against Egypt (1760 B.C.), of the Phrygians against the Hittites and the Babylonians (1780 B.C.), of the Scythians against Persia (500 B.C.), of the Achaeans against the Minoans (1200 B.C.), of the Gauls and Germans against Rome in the fourth century B.C., and again with more success in the fourth and fifth centuries A.D., and of the Scandinavians against western and southern Europe in the ninth and tenth centuries. The intensity of war has thus increased in the final stage.

External attacks, often successful, have led to internal revolts until the civilization has disintegrated under joint pressure from outside and from within. These variations in the intensity of war have been typically accompanied by changes in the character of armies and of military operations.

b) Characteristics and Consequences.—The frequency and character of historic wars have varied primarily with the stages of a civilization, secondarily among the different civilizations and, to a limited degree, among the political groups within the civilization. Civilized war could be traced to the same organic drives which activated animal and primitive war, but these drives were combined in more complex and flexible behavior patterns which could be stimulated by leadership to more popular support for war with political, religious, or economic objectives. It differed from animal and primitive war in that its techniques changed more rapidly and tended to become more destructive with the advance of knowledge and the competition between offensive and defensive inventions. It differed from primitive war in that its rules were less prescriptive of custom and more rationally related to the conscious objectives of the state and conscious values of the civilization. It functioned to promote change rather than stability and, in the long run, to disintegrate rather than to integrate civilization.

Civilized war resembled animal war in that its influence was dynamic rather than static and primitive war in that its influence was social rather than biological. The dynamic influence of civilized war, however, was marked on a much shorter time scale than that of animal war. While animal war required hundreds of thousands of years to register important evolutionary changes, civilized war produced marked changes in the course of centuries—changes which were registered in the stages of the civilization. These changing stages caused by war reciprocally influenced the character of war. All civilizations indulged in similar types of war when young, when middle aged, and when old. The primary function of war was apparently to assure these successions in the life of a civilization.

There were, however, some differences in the warlike characteristics of different civilizations. War, therefore, played a dynamic part in the competition of civilizations. It spread the culture of some at the expense of others. Here, again, there was a resemblance to the dynamic function of animal war, that of giving a wider geographical distribution to organic species. The more belligerent civilizations were most successful in distributing their culture widely. The conquests of Alexander, of Caesar, of Islam, and of the sixteenth-century discoverers distributed Greek, Roman, Arabic, and Western civilization over wide areas.

Although there were some differences, all the political groups within a civilization usually had war practices and methods of similar type at the same stage in the history of the civilization. This occurred through the process of military competition and imitation, with the result of a military balance of power. The balance, however, was broken because civilized states, differing from primitive peoples, continually invented new offensive techniques, and one eventually conquered all the rest, creating a universal state; but the culture of the defeated sometimes prevailed in large areas, as did, for example, much of Greek culture in the Roman Empire.

The importance of this dynamic influence both between civilizations and within a civilization differentiated civilized from primitive war. Civilized war, however, resembled primitive war in performing a conservative function as well, that of preserving the political and cultural solidarity of existing political groups against internal sedition and external aggression. Observed over such short periods as decades, this function of war seemed most important. Statesmen often believed that resort to war was necessary to prevent or resist invasion and to preserve the status quo. With a longer view such beliefs frequently proved erroneous. The internal and external advocates of change usually gained more from war than the advocates of stability. The result of war was seldom completely predictable, so stability was immediately jeopardized when it began, because no one could predict the result. The more prolonged and destructive was the war, the more changed was the world after it was over. On the other hand, revolutionists and aggressors had little to lose and frequently won glory and achieved the political changes they sought by war. With such expectations they worked at war with more single-mindedness and eventually usually won. Even if their first effort resulted in a deadlock or in victory for the forces of the status quo, the destructiveness of the war itself often created a more fertile field for the future activities of the advocates of change. Thus wise defenders of the status quo usually preferred propaganda, economic controls, or legal argument as methods for settling controversies, avoiding violence and war to the utmost. In diplomacy and in popular thinking the function of war in a given stage of a civilization was the achievement of ideals, the reform of the status quo.

If civilized war was dynamic favoring change, did it in the entire life of a civilization favor integration or disintegration? Did it favor the insurgent who wished to divide political structures or the imperial conqueror who wished to combine them? Although vast empires were built by war, they were preserved only by organizing peace. Because of its destructiveness, wide external conquest typically brought internal discontent and sedition in parts at least of the new empire, and did not provide a suitable milieu for constitutionalism in spite of imperial efforts to create a regime of law. Conquest provided the material for civil war. Thus war in the long run favored political disintegration rather than political integration.

States and civilizations were built up by war but eventually disintegrated through war. Realization of this may in part account for the frequent decline in warlikeness of states and civilizations as they got older. But because all old civilizations of the past had younger civilizations beside them or within them, they were not able to escape war altogether, and their inflexibility made it difficult to cope with attacks effectively.

The paradox set forth in connection with primitive war may be recalled. Moderate war is socializing, whereas too much war and war that is too destructive is disintegrating. With primitive man war made for stability and gradual progress. With civilized man the threshold was passed. A rising civilization developed war too much and sealed its own doom. The time of trouble, as Toynbee points out, marked the beginning of the end of a civilization. Those civilizations in which war was a relatively unimportant political instrument, such as those of ancient Egypt, of Sumeria, of ancient China, lasted the longest. The military civilizations of Babylonia, classical antiquity, western Europe, Arabia, and Turkey were relatively short-lived.

Although war had the function of insuring change in civilization, its ultimate effect was to produce oscillations in the rise and fall of states and civilizations. What persistent evolution there was in human history was not due to war but to thought. The Alexanders, Caesars, and Napoleons produced oscillations; the Aristotles, Archimedeses, Augustines, and Galileos produced progress. Yet, paradoxically, the very persistence of general progress produced progress in the art of war, facilitating conquests and oscillations of increasing amplitude and decreasing length.

CHAPTER III

MODERN WARFARE

1. FLUCTUATIONS AND TRENDS

Modern civilization emerged in the European Renaissance with the effective use of explosives, of clocks, and of printing; the discovery of America and of new routes to the East; the rise of vernacular literature, the rediscovery of ancient literatures and the renascence of art; the fall of eastern Europe, the reformation of Western Christianity and the rise of strong national dynasties in England, France, and Spain; and the acceptance by the European leaders of the ideas of scholarship, of science, of territorial sovereignty, and of business accounting. The interaction of civilizations, the pioneering spirit, and the shattering of ancient beliefs and authorities released human energies and established the conditions for a universal pluralistic civilization, based on values of humanism, freedom, science, and toleration.

The ultimate drives to war did not change but with new techniques augmenting the impact of war on the life of the people, the gap between the objectives of the leaders and the attitudes of the people widened. Since popular support became more necessary with the new technology, the management of opinion became increasingly important. Development of the sentiment of nationalism, identifying the citizen with the state, made it possible to arouse popular enthusiasm for war by suggesting that it was required for the nation's security, power, and prosperity. National sovereignty, defined by the new international law, became the prevailing value, the dominant sentiment, the political objective, and the leading cause of war in the modern period.

The intensity of modern war may be measured by the frequency of battles, of campaigns, or of wars. These military incidents may be weighted according to the absolute number of combatants engaged, to the number engaged relative to the supporting population, to the absolute number of battle casualties of various types (killed, wounded, prisoners), or to the number of casualties relative either to the number of combatants or to the number of the supporting population.

The intensity of war may also be measured by the absolute or relative losses of life attributable to it in the military and civil population during a given time or to the absolute or relative losses of wealth attributable to it. In estimates of this type there is always a question as to the extent to which indirect losses should be counted. The population suffers not only from casualties in battle but also from disease in the army, from war-induced famine and epidemics in the civilian population, and from decline in the birth rate and increase in the death rate from causes more or less related to the war. Prosperity suffers not only from direct destruction of property and workers by military action but from dislocations of production and trade which may cause serious depressions long after the war is over. Our interest here is not in the consequences of war but in the relative intensity of military activity at different times and places. For this purpose intensity must be measured by units of military activity occupying as limited a time and space as possible. The frequency of battles seems to conform more closely to this requirement than any other of the readily available indexes. The number of casualties and other immediate consequences of military activity will sometimes be taken into consideration.

a) Warlikeness of States.—Countries differ greatly in the frequency with which they have been at war. From 1480 to 1940 there were about twenty-six hundred important battles involving European states. Of these twenty-six hundred battles, France participated in 47 per cent, Austria-Hungary in 34 per cent, Germany (Prussia) in 25 per cent, Great Britain and Russia each in about 22 per cent, Turkey in 15 per cent, Spain in 12 per cent, the Netherlands in 8 per cent, Sweden in 4 per cent, and Denmark in 2 per cent. These percentages are for the whole period of four hundred and sixty years. When tabulated by fifty-year periods, it appears that the percentage of participation by France, Austria, Great Britain, and Turkey has been relatively constant, that by Prussia and Russia has tended to increase, and that by Spain, the Netherlands, Sweden, and Denmark has decreased in the last three centuries. The results are similar if participation in wars rather than in battles is taken as the criterion, though the differentiation between great and small states is less extreme. Clearly, the great powers have been the most frequent fighters.

The same conclusion is suggested by an analysis of the proportion of war years in the history of states. F. A. Woods concludes such an analysis with the statement: "It is the stronger nations since 1700 that have devoted the most time to war. Moreover, the lesser nations were once the great powers. Spain, Turkey, Holland, and Sweden were active in warfare at the same period that they were politically great." His figures indicate that the powers which could be classed as great during the whole modern period had averaged twice as many wars as the smaller states, though the wars of the latter often were of longer duration. Dutch wars, for example, averaged 5.4 years each and French only 1.8. The French, however, had fought 147 wars compared to only 29 by the Dutch. Consequently, the French had been engaged in fighting a much larger proportion of the period.

The fighting propensity of the great powers is illustrated by an analysis of the participants in the military campaigns (wars, interventions, suppression of insurrections) from 1900 to 1930. The seven great powers averaged 46 campaigns each during these thirty years and each campaign averaged fourteen months. Eight secondary powers of Europe and Asia averaged 19 campaigns each of an average duration of eight months. The Balkan, Latin-American, and minor African and Asian states averaged 10 campaigns each of an average duration of six months. The remaining states, nine non-colonial small powers of northern Europe, averaged only one campaign each of five months' average duration. Several of these—Denmark, Sweden, Norway, Switzerland, Estonia—did not fight any campaigns at all in this period.

b) Size of Armed Forces.—The size of armies has increased during the modern period both absolutely and in proportion to the population. In the sixteenth century the mercenary armies seldom reached over twenty or thirty thousand. In the seventeenth century armies began to be nationalized and often reached fifty or sixty thousand. The European population during this century attained about the level it had reached in the Roman Empire and about as large a proportion of the population was under arms, some three in a thousand. From then on there has been a steady rise in the size of European standing armies, absolutely and relatively. In the eighteenth century Marlborough, Prince Eugene, and Frederick the Great had armies of eighty or ninety thousand men. Napoleon had as many as two hundred thousand men in certain battles, and at times he may have had a million men or 5 per cent of the French population mobilized.

There was some diminution in the size of armies in the period of tranquillity after 1815, but after 1870 there was, among the great powers, a steady growth in the size and cost of armies and navies. Before the nineteenth century was over the eight great powers averaged five hundred thousand men each in the army and navy, and before the outbreak of World War I another one hundred thousand had been added. The military establishments did not increase, however, much faster than national populations and national budgets. On the average about five in a thousand of the population were in the military services and about a third of the national budget was spent for their maintenance. Budgets, however, increased more rapidly than did populations so the per capita military cost for the great powers advanced on the average from $1.70 per year in 1870 to $4.27 in 1914. After World War I military costs diminished under the influence of poverty and disarmament agreements, but the state of tension which prevailed after 1931 and again after 1947 carried them to an average of $25.00 per capita in 1937 and $200.00 in 1963 for the great powers other than China.

In 1937 the world as a whole had about eight million men in its standing armies, or four to one thousand of the population—considerably above the ratio maintained by the Roman Empire of Augustus. Immediately mobilizable reserves would add two million more and trained reserves some thirty million more. Of the soldiers in standing armies more than half were in Europe, which, however, had less than one-quarter of the world's population. Europe in 1937 kept nearly three times as large a proportion of its population under arms as it did in the days of Augustus. France, with less than half the population of the Roman Empire, maintained almost twice as big an army.

In the period between World War II and 1964, the armies of western European powers had not reached the levels of 1937, but the armies maintained by China, Russia, and the United States, and the large armies of many lesser states in eastern Europe and Asia meant that some twenty million men or seven per thousand of the world's population were under arms, not to mention the even larger number of civilians engaged in war industries.

Furthermore, modern states enlarge their armies far more in times of war than was formerly the case, and in addition the bulk of the adult civilian population is mobilized for some war work. The relative size of the war army has, therefore, increased even more rapidly than that of the peace army. While in the seventeenth century countries rarely mobilized 1 per cent of their population for war, the original belligerents of World War I mobilized 14 per cent of their populations and the major belligerents even more. It is clear that during the modern period there has been a trend toward an increase in the absolute and relative size of armies whether one considers the peace army, the number mobilized for war, the number of combatants engaged in battle, or the number of the military and civil populations devoting themselves to war work.

c) Length of Wars.—Another general trend has been toward a decrease in the length of wars and in the proportion of war years to peace years.

While major wars from the sixteenth to the nineteenth centuries often lasted over ten years, there were many short minor wars. The average duration of a war during these centuries was about five years, compared with three years in the nineteenth century and four years for the first sixty years of the twentieth century.

In the sixteenth and seventeenth centuries the major European states were formally at war about 65 per cent of the time. In the three succeeding centuries the comparable figures were 38 per cent, 28 per cent, and 18 per cent, respectively. This refers only to recognized wars. If the colonial expeditions and interventions in America, Asia, and Africa were counted, most of the great powers would have been "at war" a large proportion of the time even in the twentieth century. The United States, which has, perhaps somewhat unjustifiably, prided itself on its peacefulness, has had only twenty years during its entire history when its army or navy has not been in active operation during some days, somewhere.

d) Battle Frequency.—A third trend has been toward an increase in the length of battles, in the number of battles in a war year, and also in the total number of battles during a century. Reference has been made to the increasing number of battles over a day's duration in recent centuries. The number of battles in a war has also tended to increase. In the sixteenth century less than two important battles occurred on the average in a European war; in the seventeenth century, about four; in the eighteenth and nineteenth centuries, about 20; and in the twentieth century, over 30. The number of war years per century has declined, but the number of battles per war year has increased more rapidly. As a result, the total number of battles fought in a century has tended to increase. Harbottle lists 106 battles in the sixteenth century. Bodart lists 231 battles in the seventeenth century, 703 in the eighteenth, and 730 in the nineteenth. There had already been 882 battles in the twentieth century by 1940. The intensity of war measured by frequency and duration of battles has certainly increased.

This conclusion is confirmed by Sorokin, who has compared by centuries the number of wars weighted to take account of duration of war, size of fighting force, number of casualties, number of countries involved, and proportion of combatants to total population. His indexes for the principal European wars from 1100 to 1938 are:

12th	13th	14th	15th	16th	17th	18th	19th	20th
18	24	60	100	180	500	370	120	3,080

The intensity of war seems to have been exceptionally high in the seventeenth and exceptionally low in the nineteenth century. Apart from these two centuries the index rises continuously, and extraordinarily in the twentieth century.

e) Expansion of Wars.—A fourth trend has been toward an increase in the number of belligerents in a war, in the rapidity with which a war spreads, and in the area covered by a war.

In a list of 308 wars from 1450 to 1964 the average number of participants in wars of the late fifteenth and sixteenth centuries was 2.4 and in the following centuries 2.6, 3.7, 3.2, and 5. Thus, apart from the nineteenth century, in which there was a larger number of imperial and civil wars, the trend was toward an increase in the number of participants. Wars between great powers involving the balance of power have had the largest number of participants. Hitherto wars between non-European states or imperial wars of European powers overseas have frequently involved only two powers, and this type of war was especially prevalent in the nineteenth century. Experience in the twentieth century suggests that in the future even those types of war are likely to spread.

Wars have often begun as civil wars or bilateral wars. Consequently, the fact that they have, in increasing degree, ended with a larger number of participants indicates the increasing tendency of wars to spread. Except during the nineteenth century, the position of nonbelligerency has become increasingly difficult to maintain.

This seems to be a consequence of the increasing interdependence of states with respect to commerce, opinion, and politics and of the development of techniques which have made possible an extensive interference in these fields by the belligerents. More and more belligerents have tried to destroy or regulate enemy and neutral commerce by naval and other means; to control enemy and neutral opinion by propaganda; and to influence the foreign policy of non-belligerent governments by appeals to alliances, the balance of power, or collective security. This had developed to such a point that in World War II nearly all states found their commercial interests seriously injured, their publics excited and divided by propaganda, their mortality increased by war-spread diseases, and their security menaced by immediate aggression or by changes in the balance of power which might result from the war. The legal obligations of neutrals had also increased.

Small states have usually preferred to endure these evils rather than to enter a war which would probably make their situation worse and to whose result they could contribute little. They have, however, succeeded in doing so only if they were distant from the scene of battle or if the major belligerents preferred to have them neutral. The great powers, on the other hand, have generally, if the war lasted long, considered it to their advantage to enter the war. They have believed that they could defend their frontiers and contribute to the results of the war and that, if participants, they would have an opportunity to exert influence at the peace conference.

Since the Thirty Years' War there have been fourteen periods in which war existed with a great power on each side for over two years. There were only three of these major war periods—those containing the War of the Polish Succession (1733–38), the American Revolution (1775–83), and the Crimean War (1854–56)—in which a single one of the great powers remained at peace throughout the period. A closer analysis indicates that the difficulty of a great power's maintaining neutrality has become progressively greater during these three centuries.

In the twentieth century the world war of 1914 spread within a year to all the European great powers and Japan, and to the United States in two and one-half years. Thirty-three states, half those of the world, including several from Asia, Latin America, and Africa, eventually became belligerents. Others became quasi-belligerents and also participated in the peace conference. The war was renewed in 1931 with spasmodic and relatively isolated hostilities in China, Ethiopia, Spain, Czechoslovakia, Lithuania, and Albania, but with the German attack on Poland in September, 1939, World War II gave signs of becoming general. France and Great Britain and the dominions, except Ireland, entered immediately. Before the end of 1940 Russia attacked and made peace with Finland; Germany invaded Denmark, Norway, the Netherlands, Belgium, Luxembourg, and Rumania; Italy entered the war, attacking Egypt and Greece; and Japan, already fighting China, joined the axis and attacked the United States at Pearl Harbor in December, 1941. After this, all but six of the remaining states of the world became belligerents.

The tendency of wars to spread can also be illustrated by charting the distribution of battles in these wars. The battles of the Thirty Years' War were all concentrated in central Europe; those of the War of the Spanish Succession, in the Low Countries, central and western Europe, and America; those of the Seven Years' War, in various parts of Europe, India, and America; those of the Napoleonic Wars, in all sections of Europe, the Near East, and America; those of the two world wars, in all sections of Europe, the Near East, the Far East, Africa, and the waters of the Atlantic and the Pacific.

The tendency of the war system of modern civilization to be less localized may also be statistically indicated by noting that Bodart's list of battles, deemed important for this system, includes no battles outside of Europe prior to 1750. Since 1600 the percentages of extra-European battles in succeeding centuries were 0, 2, 13, and 25.

f) Cost of War.—A fifth trend has been toward the increased human and economic cost of war, both absolutely and relative to the population. The human cost of war is a difficult problem to get data upon. The proportion of persons engaged in a battle who are killed has probably tended to decline. During the Middle Ages 30–50 per cent of those engaged in a battle were often killed or wounded. In the sixteenth century 40 per cent of the defeated side might be killed or wounded and about 10 per cent of the victors. The latter cut down the members of the defeated army as they ran away. Thus at the beginning of the modern period the average casualties in battle were probably about 25 per cent of those engaged. In the three succeeding centuries the proportion has been estimated as 20, 15, and 10 per cent, respectively, and in the twentieth century about 6 per cent. Prior to 1900 about a quarter of the battle casualties died, and in World War I about a third; thus the proportion of those engaged in a battle who die as a direct consequence of the battle seems to have declined from about 6 per cent to about 2 per cent in the last three centuries.

The proportion of the population engaged in the armies, however, has tended to become larger, and the number of battles has tended to increase. As a result, the proportion of the population dying as a direct consequence of battle has tended to increase. The losses from disease in armies have declined. Dumas gives figures of the Napoleonic period suggesting that 80 or 90 per cent of the total army losses were from disease. Bloch states that in the nineteenth century this proportion averaged 65 per cent. In World War I, while disease accounted for 30 per cent of the losses in the Russian army and 26 per cent in the American army, in the German army only 10 per cent of deaths were from this cause. It has been estimated that, of 1,000 deaths in the French population in the seventeenth century, about 11 died in active military service. The corresponding figure for the eighteenth century is 27; for the nineteenth, 30; and for the twentieth, 63. For England the corresponding figures for these four centuries are 15, 14, 6, and 48. The exceptionally heavy losses of the seventeenth century because of the civil wars and the exceptionally light losses of the nineteenth century because of the dominance of British sea power obscure the trend in this last case. Adequate figures were not obtainable for other Continental powers, but indications are that they would disclose an upward trend, as does France, although, because of the Thirty Years' War, German losses in the seventeenth century were exceptionally heavy and, because of their relative frequent participation in war, the military service losses of France and England were probably above the general average for European countries in both centuries.

The civilian losses from the direct ravages of war were much less than the service losses until World War II. Furthermore, until that war, civilian losses had tended to decrease since the seventeenth century. Air raids were serious during World War I, but they did not kill as large a proportion of the civilian population as were killed in the Thirty Years' War, when sieges sometimes resulted in the slaughter of all the inhabitants of the city. The civilian losses from air bombardment greatly increased, however, in the Ethiopian, Spanish, Chinese, Polish, and Finnish hostilities of the 1930's and reached a high point in the battle of Britain in 1940. An even greater proportion was killed in the destruction of Dresden by air raids in 1943 and of Hiroshima and Nagasaki by atomic bombs in August, 1945—some three hundred thousand people in these three cities. Taking all factors into consideration the proportion of deaths attributable to military service and to hostilities has probably increased among European countries from about 2 per cent in the seventeenth to about 3 per cent in the twentieth century.

War has always resulted in serious losses to the civilian population from decline in the birth rate, and this has probably increased because of the increased proportion of the population engaged in war. Death rates reached absolute maximums and birth rates reached absolute minimums in most of the belligerent countries during World War I. In the past, wars have assisted in spreading epidemics among the civilian population. Whether the superior preventive medicine of modern times has decreased these losses is difficult to say. The influenza epidemic of 1918 made serious ravages in all countries, belligerent and neutral. It has been estimated that during World War I losses of population from indirect causes were as great as the direct losses in Europe. Each was about ten million. Outside of Europe the indirect losses were much greater because of the ravages of influenza in Asia and America. The total deaths from military action and war-distributed disease attributable to World War I have been estimated as over forty million and those attributable to World War II as over sixty million. The losses from all causes in World War II were greater. It is probable that the total of deaths indirectly due to war has been three times as great as direct war deaths in twentieth-century Europe and that the proportion of such losses outside Europe and in Europe in earlier centuries has been greater. Probably at least 10 per cent of deaths in modern civilization can be attributed directly or indirectly to war.

While it is difficult to be certain of the increase of war's destructiveness in regard to the quantity of population, there is little doubt that war has been progressively more detrimental to the quality of population. Vernon Kellogg, continuing the arguments of Herbert Spencer, David Starr Jordan, and John Bates Clark, has demonstrated the race-deteriorating influence of modern war by studies of the statistics concerning the recruiting of soldiers, by study of the measurable physical effects of the Napoleonic Wars upon the French population, and by study of the influence of war upon the spread of race-deteriorating diseases.

Closely related to the racial cost of war but even less susceptible to objective measurement are the social and cultural costs of war in the deterioration of standards. Wars of large magnitude have been followed by anti-intellectual movements in art, literature, and philosophy; by waves of crime, sexual license, suicide, venereal disease, delinquent youth; by class, racial, and religious intolerance, persecution, refugees, social and political revolution; by abandonment of orderly processes for settling disputes and changing law; and by a decline in respect for international law and treaties. The standards of some people and groups have, however, been stimulated by war in the opposite direction. The measurement and evaluation of such postwar movements are highly subjective, but probably standards have tended seriously to deteriorate.

There is also little disagreement respecting the increasing economic cost of war in direct burdens on the government and indirect losses from maldirection of productive forces. War has become so thoroughly capitalized that it is necessary to mobilize the entire resources of the country. Debts of astronomical magnitude are incurred, so great that they cannot be paid. The resulting default and the readjustments necessary because of the malapplications of capital during the war bring depressions long after the war.

It is true that some military experts think the progress in utilizing expensive machines for war may limit war's destructiveness because of the anxiety the high command will be under to safeguard these devices which are so expensive that they cannot easily be replaced. This assumed moderating influence of military capitalization, however, has not yet been demonstrated in practice; instead, increasing capitalization has increased the destructiveness of war. There is a consensus that a war fought with nuclear-headed missiles would introduce a new magnitude of cost and destruction in war.

From the standpoint of the loss of human life, the deterioration of racial stock, and the loss of economic wealth, the trend of war has been toward greater cost, both absolutely and relative to population. It is to be observed that these trends are most obvious if data are confined to strictly international wars. Civil wars such as the French Huguenot wars of the sixteenth century, the British War of the Roses of the fifteenth century, and the Civil War of the seventeenth century, the Thirty Years' War from the standpoint of Germany, the Peninsula War, from the standpoint of Spain, the American Civil War, and the Chinese Taiping Rebellion were costly both in lives and in economic losses far in excess of contemporary international wars. This fact is not surprising when it is considered that a single country bears all the loss, that both sides usually employ the same techniques, that large levies of untrained troops usually figure on both sides, and that defenses have not been prepared, with the result that campaigns cover large areas of territory, making the civilian losses exceptionally heavy. This observation, of course, applies only to civil insurrections reaching the stage of recognized war. Many rebellions, revolutions, and insurrections are suppressed before going to such extremes.

War has during the last four centuries tended to involve a larger proportion of the belligerent states' population and resources and, while less frequent, to be more intense, more extended, and more costly. It has tended to become less functional, less intentional, less directable, and less legal. Before World War I the despotic states attempted a more efficient utilization of war as an instrument of policy and led the nations to a more complete organization of the states' resources, economy, opinion, and government for war even in time of peace. States became militaristic and war became totalitarian to an unparalleled extent.

These trends of war have been related to the ideological, economic, social, and political trends of modern civilization. The acceleration of technological and social change in the modern world, the more rapid geographical diffusion of ideas and methods, the increasing economic and political interdependence of separated areas, the growth of population and standards of living, the rise of public opinion and popular initiative in politics, have together tended to concentrate military activity in time and to extend it in space; to make it less easy to begin, to localize, and to end; to make it materially more destructive and morally less controllable; to make it appear psychologically more catastrophic and less rational; to make it more difficult for any state to isolate itself from militarization in time of peace and from hostilities in time of war once the controls of international law and organization have been successfully defied.

2 . TECHNIQUES

Throughout the long history of war there has been a cumulative development of military technique. Invention of defensive instruments has usually followed close on the heels of the invention of offensive weapons. This balance of technology has tended to support a balance of power, but the balance has not tended toward increasing stability. Consequently, the political effect of military invention has not been continuous. There have been times when inventions have given the offensive an advantage, and conquerors have been able to overcome the defenses of their neighbors and build huge empires. At other times the course of invention and the art of war have favored the defensive. Local areas have been able to resist oppression, to revolt, and to defend themselves from conquest. Empires have crumbled, local liberties have been augmented, and international anarchy has sometimes resulted.

During the last five centuries military invention has proceeded more rapidly than ever before. Important differentials in the making and utilization of such inventions have developed. In general, the inventions have favored the offensive, and there has been a tendency for the size of political units to expand. This tendency was, however, arrested during much of the nineteenth century by inventions favoring the defensive, and many self-determination movements were successful. In spite of the tremendous and probably permanent advantage of the offensive in the atomic age, the self-determination of new states has proceeded with increasing rapidity. The suicidal effect of an offensive with nuclear weapons against a state similarly equipped has created a stalemate similar to that created by the advantage of the defensive in the past.

a) *Development of Modern Military Technique.*—"Until within the last few years," wrote Rear Admiral Bradley A. Fiske in 1920, "the most important single change in the circumstances and methods of warfare in recorded history was made by the invention of the gun; but now we see that even greater changes will certainly be caused by the invention of the airplane." Modern civilization began in the fifteenth century with the utilization of the first of these inventions and has witnessed the steady improvement of this utilization through development of accuracy and speed of fire of the gun itself; penetrability and explosiveness of the projectile; steadiness, speed, and security of the vehicle which conveys it over land or sea toward the enemy; and adaptation of military organizations to such utilization.

While the airplane continued this development by providing an even swifter vehicle for carrying the gun, it also introduced the third dimension into warfare. This made possible the use of gravitation to propel explosives, more extensive and accurate scouting, and military action behind the front, over vast areas, and across all barriers of terrain. Both of these inventions, after their use was thoroughly understood, greatly augmented the power of the offensive, though, in the case of the gun, the defense immediately began to catch up, and the general trend of war between equally equipped belligerents was toward a deadlock. A similar tendency was observable in the case of the atomic bomb and the missile, although the deadlock developed not from the invention of defenses but from the unacceptable destructiveness of nuclear warfare.

These inventions are but the most striking of the numerous applications to war of the technical advances characteristic of modern civilization. In historic civilizations men and animals provided the power for military movement and propulsion. In the modern period, wind and sail, coal and the steam engine, petroleum and the internal combustion engine, jet propulsion and missiles, have successively revolutionized naval, military, and aerial movement, as gunpowder, smokeless powder, high explosives, and nuclear bombs have successively revolutionized striking power. The history of modern military technique falls into four periods, each initiated by certain physical or social inventions and leading to certain military and political consequences:

(1) *Adaptation of firearms* (*1450–1648*).—During the period of discoveries and wars of religion, medieval armor was being abandoned; pikemen, halberdiers, and heavy cavalry were going. The organization of the Turkish Janizary infantry, well disciplined, equipped with cutlass and longbow, and supported by light cavalry and artillery, was being copied throughout Europe. Heavy artillery had begun to reduce feudal castles in the early fifteenth century and the *Wagenburg* revolutionized field tactics. Hand firearms first used by Spaniards, Hussites, and Swiss in the fifteenth century were adopted by all in the general wars of the early sixteenth century. The experience of the Thirty Years' War ended this period of experimental adaptation of firearms by the mercenary armies, and modern armies began to emerge.

Naval architecture was greatly improved during this period. The clumsy galleons of the Spanish Armada, differing little from those of Columbus a century earlier and resembling the oar-driven galleys of the Middle Ages, were superseded in the mid-seventeenth century by longer, swifter, and more heavily armed "broadside battleships" which differed little from those of Nelson, nearly two centuries later.

Equipped with the new technique of firearms, Europeans had occupied strategic points in America, Africa, and Asia, readily overcoming the natives whom they found there. The tendency of this new technique was toward political integration inside and expansion outside of Europe.

(2) *Professionalization of armies* (*1648–1789*).—The seventeenth and eighteenth centuries witnessed the development of the professional army loyal to the king and ready to suppress internal rebellion or to fight foreign wars if paid promptly and if the officers were adequately rewarded by honors and perquisites of victory. Louis XIV and Cromwell contributed greatly to the development of this type of army, which, however, in the eighteenth century tended to be more concerned with safety and booty than with victory. Consequently, military invention emphasized defense and fortification. The art of war prescribed elaborate rules of strategy and siegecraft. Rules also

dealt with the treatment of prisoners, with capitulations, with military honors, and with the rights of civilians. The Prussian army with its vigorous discipline, aggressiveness, and new strategic ideas under Frederick the Great to some extent broke through this defensive technique and brought this type of army to the highest point.

The destructiveness of war was limited by the general exemption from the activities of land war of the bourgeois and peasants, who constituted the bulk of the population. The bourgeois were anti-military in attitude and of little influence in the politics of most states. The monarch preferred to leave his own bourgeois and peasants to production, provided they paid taxes, and to recruit his armed forces from the unproductive riffraff, officered by the nobility, whose loyalty could be relied upon. With the existing techniques the army could not easily attack the enemy's middle classes, unless his army was first destroyed and his fortifications taken. In that case such attack was unnecessary because these classes would usually accept whatever peace might be imposed. Lacking in patriotism or nationalism, they were little concerned if the territory on which they lived had a new sovereign, provided they could retain their property.

Naval vessels reached the limit in size possible for wooden ships in the seventeenth century and underwent very little change until the steel ship was developed two hundred and fifty years later. The problem of adequate raw materials for war instruments was sharply presented in England during the latter part of this period as a shortage of oak for the hull beams and of huge pines for the masts developed. The United States profited in the Revolution by blocking the British from their Canadian source of mast timber. The British never met this problem by a consistent policy of planting until after the Napoleonic Wars, when oaks were planted too late to become ripe until wood was superseded by steel in shipbuilding.

This negligence, however, indicated no lack of naval interest in Great Britain during the period. The increasing importance of commerce, the vulnerability of the British Isles to blockade, and the invulnerability of Britain to land attack induced Britain to adopt a policy of naval superiority and to rely upon control of the seas as the main instrument of warfare. By such control, at a time when land transport by wagons over bad roads was very meager, military supplies and the raw materials for making them could be withheld from the enemy's forces, sieges could be assisted by maritime blockades, and the bourgeois, in so far as they had influence, might be induced to exert it in favor of peace in order to escape loss of property and profits. The British took the lead in insisting upon the right to visit and search all merchant vessels and to capture and condemn all enemy vessels and property and such neutral vessels and property as were found to be assisting the enemy. They were, however, ready to surround these activities with the judicial safeguards of prize-court procedure which not only preserved the king's share in prizes but also might prevent privateering from degenerating into piracy and from inflicting such hardships upon neutrals as to make them enemies.

(3) *Capitalization of war* (*1789–1914*).—The French revolutionary and Napoleonic period developed the idea of the "nation in arms" through revolutionary enthusiasm and the conscription of mass armies. The idea of totalitarian war was developed in the writings of Clausewitz, rationalizing Napoleonic methods. After these wars the issue between professional long-service aristocratically officered armies and conscript short-service democratic armies was debated on the continent of Europe with a general relapse to the former type during the long peace of Metternich's era. The rise of nationalism, democracy, and industrialism and the mechanization of war in the mid-century re-established the trend toward the nation in arms and totalitarian war.

The use of steam power for land and water military transportation developed in the first half of the nineteenth century and was given its first serious test in the American Civil War. Moltke appreciated the military value of these inventions, and his genius in using railroads for rapid mass mobilization won Bismarck three wars with extraordinary rapidity against Denmark, Austria, and France. The ironclad and heavy naval ordnance were also tested in the American Civil War. The era of military mechanization and of firearms of superior range and accuracy progressed rapidly, adding greatly to military and naval budgets and to the importance of national wealth and industry in war. The new methods were given a further test in the Spanish-American, Boer, and Russo-Japanese wars.

The great nineteenth-century naval inventions—steam power, the screw propeller, the armored vessel, the iron-hulled vessel, heavy ordnance—were at first favorable to British maritime dominance because British superiority was more marked in iron and coal resources and a developed heavy industry than in forests and wooden ship-builders. But this advantage did not continue. The new battleships were more vulnerable than the wooden ships because ordnance gained in the race with armor, and repair at sea was impossible. Furthermore, the mine, torpedo, submarine, and airplane added new hazards to the surface fleet, especially in the vicinity of the enemy's home bases. Warships, therefore, became more dependent upon well-equipped and secure bases for fueling and repair, and approach to even a greatly inferior enemy became hazardous. With the industrialization of other powers and their development of naval strength, Britain found it increasingly difficult to maintain a three- or even a two-power superiority in the ships themselves, and its distant bases became less secure.

Britain abandoned the effort to dominate the Caribbean after the Venezuelan controversy with the United States in 1896, acquiesced in American seizure of the Spanish islands, and agreed to an American fortified Panama Canal. It also welcomed American acquisition of the Philippines and in 1902 made an alliance with Japan, indicating doubt of its capacity to maintain its Far Eastern position by its own forces. The entente with France indicated awareness that British Mediterranean interests could no longer be defended single handed.

Britain thus recognized that the development of naval techniques had tended toward a regionalization of sea power, and as a result it reduced its commitments for unilateral sea control from the seven seas to those seas controllable from bases on the British and Portuguese isles and from Gibraltar, Suez, and Singapore. The far-flung British empire, the highways of the Mediterranean, the Caribbean, the China Sea, and the Pacific, could no longer be defended by the British navy alone. They must be defended by the dominions themselves and by alliances and friendships, especially with the United States, France, and perhaps with Japan. It was clear that the British capacity to maintain reasonable order, respect for law and commercial obligations, and to localize wars by maintaining the balance of power in Europe had been greatly reduced. Naval inventions and the spread of industrialization had ended the *Pax Britannica*.

This situation was realized by the Continental powers. They developed their armies and navies with increasing speed after observation of the Russo-Japanese War and after the failure of the Hague conferences to achieve disarmament. They paid particular attention to the potentialities of the improved rifle, machine gun, and artillery as well as to the art of intrenchment. The possibilities of the mine, torpedo, and submarine were developed, especially by France, pointing the way for German utilization of these weapons in World War I. Beginnings were made, especially by France and Germany, in the adaptation of the airplane and dirigible to military purposes. The results anticipated by the Polish banker, Ivan Bloch, in his book published in 1898 occurred. War became deadlocked in the machine-gun-lined trenches and in the mined and submarine-infested seas of World War I. This deadlock was not broken until attrition had ruined all the initial belligerents, and new recruits and resources on the Allied side from the United States made the cause of the Central Powers hopeless.

(4) *Totalitarianization of war* (*1914*——).—The advent of aerial war in the twentieth century ended the relative invulnerability of the British Isles to invasion. The weakening of surface control of the sea by the use of mines, submarines, and airplanes further impaired the position of Great Britain, and that country during the 1920's accepted the thesis that the integrity of the empire depended upon collective security. The possibilities of the airplane and tank, neither of them fully exploited during World War I, supported hope in some quarters and fear in others that the power of the offensive would be increased, that mobility in war would be again possible, and that the deadlock would be broken.

These possibilities encouraged aggression by Japan, Italy, and Germany after 1930. Dissatisfaction with the political results of World War I, resentment at the self-centered economic policies of the democracies, serious deterioration of the middle classes, and the spread of revolutionary ideologies engendered by the costs of war and widespread unemployment flowing from the great depression of 1929 provided motives for aggressions; but if collective security had been better organized and the airplane and tank had not been invented, the prospects would hardly have been sufficiently encouraging to induce action. As it was, the initial success of Japan in Manchuria, of Germany in the Rhineland, and of Italy in Ethiopia encouraged these countries to consort and to continue aggression in weak areas, utilizing aviation with rapid success, while all phases of the national life were organized for total war.

As the development of the gun by the European great powers in the sixteenth and seventeenth centuries extended their imperial control to the overseas countries, followed by the latter's imitation of their techniques and eventual revolt, so the development of the airplane by the totalitarian states in the twentieth century first extended their empires and then compelled the democracies to adopt their techniques. Thus the great powers, whether with a democratic or an autocratic tradition, whether relying on the army or navy, whether European or American or Asiatic, have in a disorganized world felt obliged to follow the lead of that one of their number most advanced in the art of war.

The trend toward general militarization initiated by the gun was, however, checked in the eighteenth and nineteenth centuries through the rise of the naval, commercial, industrial, and financial power of a relatively liberal and antimilitary Britain; through the increasing indecisiveness and destructiveness of war; through the professionalization of the armed forces; and through the antimilitaristic philosophies of the rising bourgeois. It is possible that the universal fear of nuclear war, the sense of human solidarity through universal mass communication and the organization of peace and international cooperation in the United Nations may have a similar influence in the latter part of the twentieth century.

b) Characteristics of Modern Military Technique. (1) *Mechanization.*—The outstanding characteristic in which modern war has differed from all earlier forms of war has been in the degree of mechanization. The use of long-range striking power (rifles, machine guns, artillery, missiles), of power-propelled means of mobility (railroads, motor trucks, battleships, tanks, airships), and of heavy protective covering (armor plate on fortresses, tanks, and warships) has meant that the problem of war manufacture has risen to primary importance. In historic civilizations the soldier provided his own equipment, and it generally lasted as long as the soldier. Now a dozen men must be engaged in production and transportation services behind the lines to keep one soldier supplied.

(2) *Increased size of forces.*—A second important change has been in the size of the armed forces, both absolutely and in proportion to the population. It might seem that if each soldier needs such a large amount of civilian help there would be fewer soldiers, but this has not proved to be the case. Power transport and electrical

communication have made it possible to mobilize and control from the center a much larger proportion of the population than formerly. The men can be transported rapidly by railroad, motor lorry, or airplane and canned food can be brought to them. Where formerly 1 per cent of the population was a large number to mobilize, now over 10 per cent can be mobilized, of which a quarter may be at the front at one time. But 10 per cent mobilized requires most of the remaining adult population to provide them with the essentials for continuing operations. Thus instead of 1 per cent engaging in war and the rest pursuing their peacetime occupations of trade or agriculture, the entire working population must now devote itself to direct or indirect war service.

(3) *Militarization of population.*—A third change, consequent upon the second, has been the military organization of the entire nation. The armed forces have ceased to be a self-contained service apart from the general population. The soldiers, sailors, and airmen are recruited from those men whose services can be most readily supplied by women, children, and the aged. The experts in transportation and industrial services must be largely exempted in order that they may continue their "civilian" services which, under modern conditions, are no less essential to war. Such a gearing-in of the agricultural, industrial, and professional population to the armed forces requires a military organization of the entire population. Since the perfection of such an organization after the outbreak of war has been impossible, the conditions of war have more and more merged into those of peace. The military organization of the entire population in peace has become necessary as a preparation for war, most notably in the "cold war" period after World War II.

Such a militarization of the population must be distinguished from the militia system, illustrated in Switzerland. In this system the duty of military service, though considered a universal burden of citizenship, has involved only a limited training which has not withdrawn persons from normal civilian occupations for long periods. Furthermore, under the militia system the civilian activities have at all times been considered normal and the military abnormal. Both systems may be called "the nation in arms," but whereas the first has involved a militarization of the entire population, the second has involved a civilianization of the military services. The difference has depended upon the degree in which military has dominated over ci-

vilian government in peacetime, the degree in which military training has dominated over civilian activity in the life of the individual,
and the degree in which preparedness for war has dominated over
general welfare in national policy.

Both of these defense systems may be distinguished from the professional army system characteristic of the United States and Great
Britain and of most European states in the eighteenth century. In
this system the army has been voluntarily recruited for long service
and has existed with its own organization, discipline, law, and professional standards quite apart from the civilian population. Because
of the emphasis upon professional qualifications, its size has not been
greatly augmented in time of war, although as emergencies develop
voluntary recruitment has often given way to conscription and the
press gang.

While all three systems have played their part in the history of
most modern states, the development of modern military technique
has tended toward militarization of all states.

(4) *Nationalization of war effort.*—A fourth change, characteristic
of modern military technique, has been the extension of government
into the control of the economy and public opinion. The military
state has tended to become the totalitarian state. Other forces of
modern life have, it is true, had a similar tendency. Democracy, under the influence of nationalism, has induced the individual to identify all phases of his life with that of the state, while state socialism,
under the influence of depression, has induced the state to intervene
in all phases of the life of the individual, but the needs of modern
war have led and accelerated the process. Modern war has required
propaganda and civil defense programs to sustain morale among the
civilian population, which can no longer expect to be exempt from
attack. Modern war has also required an adjustment of the nation's
economy to its needs. A free-market system, depending on profits,
has proved less adequate than military discipline for reducing private consumption and directing resources and productive energy to
war requirements. Since transition from a free economy to a controlled economy would be difficult in the presence of war, preparation for war tends toward such a change in time of peace. Furthermore, autarchy is necessary as a defense against blockade. The controls necessary to confine the nation's economic life to those regions
whose resources and markets will be available in time of war must

be applied before the war. The modern technique of war has, therefore, influenced the rise of autarchic totalitarian states and has tended to impair free economy and free speech in all states.

(5) *Total war.*—A fifth change, characteristic of modern war technique, has been the greatly increased destructiveness of all modern weapons and the breakdown of the distinction between the armed forces and the civilians in military operations. The moral identification of the individual with the state has given the national will priority over humanitarian considerations. The civilian's morale and industry support the national will. Thus the civilian population, manufacturing, and transport centers have become military targets. Bombing aircraft and starvation blockades have made it possible to reach these targets over the heads of the army and fortifications; consequently, the principle of military necessity has tended to be interpreted in a way to override the traditional rules of war for the protection of civilian life and property.

The seventeenth-century writers on international law, while admitting that the entire population of the enemy was in strict law subject to attack, distinguished between the combatants and the noncombatants, asserting that approved usage should in general exempt the latter. With the progress of modern military technique in the nineteenth century, the "armed forces" came to include numerous noncombatants such as transport workers and trench-diggers, but the civilians outside of the armed forces were in general exempted from direct attack, though their property at sea was liable and in occupied areas they and their property were subject to requisition. Military practice and the rules of war were also, to some extent, influenced by the general distinction between the political and the economic life of the state, a distinction which developed particularly as a result of the physiocratic and classical schools of economics and the increasing influence of neutrals. Private property on land was generally considered exempt from capture, and there was a strong movement, especially in the United States to extend this exemption to private property at sea—a movement which was accepted in 1856 with respect to such property on neutral ships. Although the total exemption of the economic life of states from the rigors of war was not accepted, because of the opposition of Great Britain and other naval powers, the idea that war should be directed solely against the military and political life of the state had considerable influence during the nineteenth century, especially in countries like Germany, vulnerable to blockade.

While these distinctions according extensive exemptions to the noncombatants, the civilian population, and the national economy may still be supported by reference to the sources of international law, especially the Geneva conventions of 1949, the practice of war has tended to become totalitarian. This has been especially true of the tactics of guerrilla warfare widely developed since World War II.

Starvation, bombardment, confiscation of property, and terrorization involving the destruction of entire cities were applied in World War II against the entire enemy population and territory. The danger of reprisals and the desire to utilize the population of occupied areas were the only inhibitions. The entire life of the enemy state came to be an object of attack. The doctrine of conquest was even extended by some states to the elimination of a population and its property rights in order to open the space it occupied for settlement. The prospects of nuclear war with intercontinental missiles, if deterrence fails, present the possibility of even greater destruction of all belligerents, if not the entire human race, from fall-out.

(6) *Intensification of operations.*—A sixth characteristic of modern war technique has been a great increase in the intensity of military operations in time and of their extension in space.

Operations of war have always sought to concentrate a greater military force than that of the enemy at a given point, the control of which is regarded as important. Such points might be fortified places, government or commercial centers, transport and communication gateways, or a battleground selected by the enemy or one to which his forces might be lured. The belligerent with inferior forces would try to delay action while it brought up reserves and improved its trenches, but if one acquired marked superiority at any moment it would usually begin a battle or siege. This episode would end in retreat or surrender by one side after a day or, in the case of siege, after several months and would be followed by months or years of maneuver during which another point of importance would emerge, forces would be concentrated, and another battle or siege would occur. The campaigns would thus be broken into distinct and separate episodes, but, because of the slowness of communication and the difficulties of winter fighting, campaigns in separated areas or in different years would be, in considerable measure, isolated from one another. War typically consisted of a number of distinct battles and campaigns separated by long periods and wide areas of relative peace.

The inventions in mechanization and mobility, the organization of the entire population, the increase in the number of important targets for attack, have made it possible to concentrate enormously greater forces at a given point, to supply reserves and to continue attack and resistance at that point for a much longer period, to increase the number of points being attacked simultaneously, to enlarge the theater of the campaign by mutual efforts at outflanking, and to co-ordinate operations on all fronts at all seasons for the entire course of the war. The result was that World War I tended to become a single and continuous campaign, and the campaign tended to become one long battle or a series of battles so overlapping and united as to be hardly distinguishable. The pattern of war, instead of a grouping of dots on a map, became a large black spot of ink on the map which spread rapidly until the entire map was blackened. While this pattern was not at first duplicated in the hostilities which began in 1931, the blitzkrieg tactics and the occupation, blockade, and bombing strategies of World War II eventuated in an intense and continuous battle over the entire front.

These six characteristics of modern military technique collectively tend toward totalitarian military organization of the belligerents and totalitarian military operations during war. Though a trend in this direction began in the sixteenth century, it has been more and more emphasized in the twentieth century, with a marked acceleration since the 1930's.

These changes have been most marked in the characteristics of weapons, less marked in that of organization and operations, and of uncertain significance in the fields of policy and strategy. The art of using superior preparedness, a reputation for ruthlessness, and threats of war for bloodless victory, utilized during the "cold war," is as old as history and was expounded by Machiavelli. Such policies, however, militate against stability through mutual deterrence, which the same powers profess to favor.

Some writers on modern strategy believe that if war occurs they can still draw lessons from the campaigns of Hannibal, Caesar, Frederick, and Napoleon, but others believe that there has been a change in basic strategic principles. The general object of war, they say, is no longer to disarm the enemy by destroying or capturing fortifications and armed forces but to evade them and to strike directly at the government, economic nerve centers, and morale of the enemy. Advocates of "finite deterrence," threatening the enemies' cities, and of "counter-force strategy," threatening only his retaliatory capability, argue the same points in the nuclear age but with the difference that the object is said to be deterrence of war rather than victory in war.

Finite deterrence, it is argued, will actually deter because the population of each state is a hostage in the hands of the other if both have adequate second strike capability, whereas counterforce strategy is dangerous because it suggests an intent to make a first strike, destroying the enemy's retaliatory capability and thus inducing him to make a pre-emptive first strike. On the other hand, counterforce strategy is said to be more humane, to preserve the credibility of nuclear threat as an instrument of policy, and to make victory possible.

3. FUNCTIONS

What has been the function of war in modern civilization? Wars have been initiated in the modern period by national governments or national parties and not by world institutions or world parties. They have been intended to serve sovereign states or lesser groups rather than the world-community. Should we not, therefore, ask, "What has been the function of war in French history? in British history? in German history?" These questions are undoubtedly relevant to a history of war, and they have been dealt with in numerous national histories. The function of an activity may, however, be broader than its intention. This history of war treats modern civilization as a whole. It must, therefore, consider the effect of war in maintaining existing values or in achieving new values in that civilization. To do so, it will be necessary, however, to give some consideration to the function of war in the history of particular nations because, if war has favored some nations or types of nations at the expense of others, it will thereby have affected the values dominant in the civilization as a whole.

Modern civilization has not become either uniform or unified, although it has at times manifested a tendency toward both uniformity and unity. It has changed continuously, and these changes

have proceeded at different rates in different areas. The characteristics of war have also changed greatly during the past few centuries,
especially during the past century. The relationship between war
and political, economic, social, and cultural change has, therefore,
been extremely variable in modern history. This variability continues to the present time among different nations and regions. War
among the great powers has functioned differently in the sixteenth,
the eighteenth, and the twentieth centuries. War has functioned
differently in Europe, in the Far East, and in the Americas, in Great
Britain, in Germany, in Poland, in Italy, and in Japan.

Because of this variability, the most obvious generalization about
the function of war in modern civilization is that it is difficult to
ascertain. There is, however, a more widespread opinion than in any
other period in history that war has not functioned well in the
twentieth century. From being a generally accepted instrument of
statesmanship, deplored by only a few, war has, during the modern
period, come increasingly to be recognized as a problem.

a) Uses of War in the Modern Period.—War has, however, been
the method actually used both for achieving the major political
changes of the modern world and for maintaining stability in that
world.

The monarchs in the fifteenth, sixteenth, and seventeenth centuries used war to compel small feudal principalities to accept a
common rule, and after establishing their authority in the following
centuries, they created nations by the power which military control
gave them over civil administration, national economy, and public
opinion. They were not, however, always successful in creating and
maintaining a national sentiment in the entire population subject to
their rule. In fact, they sometimes maintained that rule by promoting local division of sentiment. As a result dissident minorities
and nationalities sometimes developed and occasionally achieved
statehood, very seldom without war, often with the military assistance of outside states, never without the passive assistance of some.
Such separatist movements have required centers, out of the reach
of the government attacked, within which propaganda might be
organized and arms assembled. The American colonists established
such a center under Franklin in Paris. The Greeks had such a center
in England and the Irish in the United States. The Cubans had a
junta in New York, the Czechoslovaks in Pittsburgh, and the Syrians

in Cairo. These movements were generally for separation but sometimes for union, as in the Italian *risorgimento,* the German empire,
and the Arab movement. Beginning in France and England in the
Middle Ages, nationalism spread to central and eastern Europe,
America, Asia, and Africa in the following centuries, resulting in
over a hundred and twenty sovereign states in the 1960's.

The expansion of the culture and institutions of modern civilization from its centers in Europe was made possible by imperialistic
war. This proved a relatively easy process as long as European powers continued to enjoy a great superiority in war techniques over
extra-European peoples and as long as there was enough extra-
European territory for all. Quarreling between Spain and Portugal,
it is true, marked the first discoveries; and imperial expansion has
never since been unaccompanied by diplomatic and sometimes military quarreling. But the colonially ambitious had opportunities to
seize land still unappropriated by a European power down to the
seizure of Abyssinia by Italy following World War I. The opportunities, however, then narrowed, and at the same time the populations of the territories still unappropriated by European powers
began to adopt European military techniques. The Turks already
had these techniques even before the Europeans. But at first the
American Indians, East Indians, Chinese, and Japanese, though
usually able to present forces much more numerous than the expeditions sent against them from the European countries, lacked firearms and tactical organization and were often divided among themselves. Cortez, with four hundred men, sixteen horses, three cannon
and muskets, and a tactful stimulation of the Tlaxcalans against their
Aztec oppressors, conquered eight million Mexicans. Pizarro was
similarly successful in Peru and Clive in India. Later the British and
Americans opened China and Japan by only moderate uses of force.
It is true missionaries and traders had their share in the work of
expanding world-civilization, but always with the support, immediate or in the background, of armies and navies. Beginning with
the American Declaration of Independence, imperial disintegration
has outstripped empire building.

War has, furthermore, contributed to the historic transitions of
human interest and ideas during the modern period. It has usually
accomplished this through facilitating a synthesis of conflicting opinions rather than the victory of one. The Thirty Years' War resulted

in victory for neither Protestant nor Catholic but for the sovereign state and the family of nations. The Napoleonic Wars resulted in victory for neither hereditary absolutism nor revolutionary democracy but for constitutional nationalism. World War I resulted in victory for neither agrarian nationalism nor industrial imperialism but for new political structures. World War II resulted in victory for neither Naziism nor Western nationalism but in a "cold war" between American free democracy and Soviet communism and in the creation of the United Nations and many new nations in Asia and Africa.

War has also had a role in maintaining the established status of nations and the established international order. It has served the state by protecting its frontiers, symbolizing its unity, and recalling to the population the necessity of political loyalty as the price of immunity from invasion. This function of war has been more important in some states than in others, but there is none in which war or war preparations have not to some degree at some time been used as an instrument of national stability and order.

War has been one means for maintaining the balance of power upon which the political and legal organization of the world-community has in large measure rested. The balance of power is a system designed to maintain a continuous conviction in every state that if it attempted aggression it would encounter an invincible combination of the others. The manifest willingness of menaced states to go to war has assisted in maintaining that conviction. When any state has developed its armaments too little or too much, and when alliances have not been made promptly to rectify the balance, war has usually ensued. In a world expectant of violence, the maintenance of a constant relationship in the military potential of all has been the price both of state independence and of peace. Because of the variety of factors affecting military potential, of the difficulty of measuring their changes accurately, and of compensating variations promptly, the balance has hitherto been maintained with insufficient delicacy to preserve the peace. But with all its crudity the system has prevented any one of the modern states from getting sufficiently powerful to swallow or dominate all the others, as Rome swallowed the states of its time. Whether power balancing will serve a useful function in the atomic age is doubtful.

War or the danger of war has, therefore, contributed toward building the modern nation-states, toward spreading and developing modern civilization, toward preserving peace and stability within the states, and toward maintaining the international system of independent states. These results, however, are not entirely consistent with one another. Furthermore, war has at times contributed toward destroying nations, civilization, stability, and the international system.

The dynamic function of building new nation-states and of spreading and changing cultural ideas has often been incompatible with the static function of preserving the nation-states and the world-system which has existed at a given time. The divergence between the advocates of change and the advocates of stability has been continuous, and the fact that each has, on occasion, found war a useful instrument accounts, in some measure, for the continuance of war.

b) Stability and Change.—In the short view political stability is the absence of sudden change. Stability is compatible with gradual changes, even though in cumulation they may be very important and even though in a distant and unpredictable future they may result in violent reaction. Evolution may be a prelude to eventual revolution, and revolution may be a step in an evolution if a sufficiently long view is taken; but in a short view the two can be distinguished—revolution manifests instability and evolution manifests stability.

In this sense the stabilizing influence of war during the modern period appears to have been in inverse relation to its intensity. Wars of great intensity have destroyed existing political values, institutions, and standards, opening the way for radical changes. The destruction and hardships resulting from such wars have provided a suitable ground for revolutionary movements.

The intensity of war increased in the fifteenth, sixteenth, and seventeenth centuries, then declined in the eighteenth and nineteenth centuries, with a marked increase in the twentieth century. A. J. Toynbee, from general historical information, comments on the remarkable ferocity of the wars of religion in the sixteenth and seventeenth centuries and of the even greater violence of the recent wars of nationalism. While no precise correlation between these variations in the intensity of war and variations in the degree of political instability can be made, it would appear that the political order of Europe changed most radically and rapidly in the seventeenth and twentieth centuries when war reached greatest intensity. The seventeenth century witnessed the supersession of feudalism and the Holy Roman Empire by the secular sovereign states as the dominant political institutions of Europe. The twentieth century appears to be witnessing the supersession of the secular sovereign states by a new order.

Important changes took place, it is true, during the less warlike eighteenth and nineteenth centuries. Science and technology developed; many areas were industrialized; international trade, international communication, and population increased. But these changes, though sometimes referred to as revolutionary and involving occasional violence, proceeded by such gradual steps that they could better be described as evolutionary. The significant political changes of these centuries—the expansion of modern civilization overseas and the rise of democracy and nationalism—often proceeded with revolutionary speed, but such revolutions were usually closely associated with wars. The Seven Years' War marked the apex of colonial expansionism. The French Revolutionary and Napoleonic Wars signalized the advent of democracy in Europe. The Bismarckian wars marked the rise of nationalism.

Thus, while it cannot be said that there was less change in world-civilization during periods of comparative tranquillity than during periods of war, it appears that political institutions were more stable during such periods and that changes of a revolutionary character usually occurred in periods when war was intense, notably in the period of World Wars I and II.

c) Integration and Disintegration.—Among the earlier civilizations war, in the long run, favored political disintegration rather than political integration. States and civilizations built up by war were eventually destroyed through war. This seems to have been due to the tendency in the past for experience with war to augment the power of the defensive over the offensive, thus making the revolt of local groups possible; to increase the destructiveness of hostilities because of stalemate and attrition, thus promoting political instability; and to militarize all states, thus promoting inflexibility and incapacity to adapt political and social organization to new conditions. Efforts to integrate the civilization through general political organiaztion never kept pace with the disintegrating effect of war.

The present world-civilization is still in its early stages, and because of its universality and the superiority of its mastery of science, it is markedly different from earlier civilizations. The tendency of this civilization to the end of the nineteenth century was toward political integration. The thousands of feudal principalities of Europe made independent by the disintegration of the medieval church and empire and the numerous native states of America, Asia, Africa, and the Pacific were integrated, in large measure through the agency of war, into sixty-odd nation-states and empires.

There has, however, been a countertendency. Beginning with the American Revolution, there has been a tendency for the modern empires to disintegrate with increasing acceleration as nationalism has spread from modern Europe. The Spanish, Portuguese, French, Ottoman, Hapsburg, Russian, Chinese, Japanese, British, Dutch, Belgian, and American empires have given birth to new nation-states. By the end of the nineteenth century the process of disintegration became more rapid than that of integration.

In the interwar period, it is true that Japan, Italy, Germany, and Russia attempted to integrate a number of formerly independent states into empires by conquest but after World War II, the process of disintegration of empires accelerated and the number of independent states was doubled. At the same time political integration was proceeding not through empire building but through the peaceful establishment of continental and universal international organizations.

That civilization entered a "time of troubles" in the twentieth century was evidenced by two world wars, the propaganda of opposing ideologies, and a protracted period of "cold war." Up to that time, there was a tendency for military operations to become more concentrated with longer gaps of peace between. War and peace alternated in oscillations of increasing amplitude. In the 1890's anxieties developed about war, and efforts to eliminate it became active. Ivan Bloch, a Polish banker wrote:

. . . As the popularity of war decreases on all sides, it is impossible not to foresee that a time will approach when European governments can no longer rely on the regular payment of taxes for the covering of military expenditure. . . . These changes tend to make the economic convulsions caused by war far greater than those which have been experienced in the past. . . . But even if peace were assured for an indefinite time, the very preparations made, the maintenance of armed forces, and constant rearmaments, would require every year still greater and greater sacrifices. Yet every day new needs arise and old needs are made clearer to the popular mind. These needs remain unsatisfied, though the burden of taxation continually grows. And the recognition of these evils by the people constitutes a serious danger for the state. . . . The exact disposition of the masses in relation to armaments is shown by the increase in the number of opponents to militarism and preachers of the Socialist propaganda. . . . Thus side by side with the growth of military burdens rise waves of popular discontent threatening a social revolution. Such are the consequences of the so-called armed peace of Europe—slow destruction in consequence of expenditure on preparation for war, or swift destruction in the event of war—both events convulsions in the social order.

Bloch's suggestion, that actual or potential war had become costly beyond any value either to national states or to world-civilization, induced Tsar Nicholas II to call the first Hague Peace Conference. The idea was supported by economists such as Norman Angell and Francis Hirst before World War I and was widely endorsed after the war in the League of Nations. In spite of this growing opinion, the unwillingness of states to modify their economic, political, and legal sovereignty sufficiently to assure an adequate functioning of the League led to new tensions between the states favorable to the status quo, on the one hand, and on the other, the states insistent that territorial revision was necessary to their economic requirements in a world of increasing economic barriers. The result was the organization of despotism and totalitarianism within the revisionist states and their initiation of World War II.

Before World War II was over, the "peoples of the United Nations" who, after Pearl Harbor, had pledged themselves to defeat the axis, expressed their "determination to save succeeding generations from the scourge of war" and established a permanent organization "to maintain international peace and security" more effectively than had the League of Nations. But the same factors which had hampered the League induced the United States and the U.S.S.R. to emphasize ideological and political differences, to form opposing alliances, and to engage in a "cold war" manifested by provocative propaganda, an arms race, political conflicts, and several small wars.

Judged by the standards of modern civilization, war has tended to increase in costs and to decline in value. One would expect an increasing reluctance of statesmen to resort to it. The general increase in the length of time between wars as well as the utterances of statesmen suggested to many at the beginning of the twentieth century that this expectation had been realized. Yet it takes only one powerful state to start a war, and there were enough new inventions and rash statesmen to produce two general wars and many small ones before the century was half over. To many the prospect seems grim.

d) *Dictatorship and Democracy.*—In a world where states were convinced that they could maintain and advance themselves only through the use of war or threats of war, democracies found themselves at a disadvantage when dealing with despotisms. Consequently, they usually professed a desire to increase the role of law and discussion in international affairs and to reduce the role of war and threats to a minimum. The reluctance of democracies to curb the sovereignty of the nation, their distrust of other nations and of distant authorities, their frequent incapacity to perceive the international repercussions of measures undertaken for domestic purposes, often thwarted the realization of these professions in practice. The despotisms, however, realizing their relative advantage in the game of power, and aware that war itself would be less dangerous to their basic principles and ideals than to those of the democracies, exerted their efforts to diminish the role of law and discussion and to increase that of violence and menace. To them war continued to be useful both for internal and for external policy; in fact, as the number of democracies increased, the value of war to the despotisms increased. The greater the number of sheep, the better hunting for the wolves.

Periods of general war and tension, therefore, tended to increase the number of despotisms and to increase the influence in world-civilization of the standards appropriate to such regimes. The long periods of peace, especially those in the eighteenth and nineteenth centuries, on the other hand, witnessed a remarkable development of the idea and practice of democracy. Nations defended by maritime or other natural barriers often felt free to develop liberty and democracy, but nations with extremely vulnerable frontiers, especially when accompanied by memories of past invasions, tended in the opposite direction. The population of such nations habitually preferred discipline to liberty. Military preparation came to occupy more attention than popular welfare, with a consequent trend toward dictatorship and totalitarianism. The varying influence of military vulnerability, of national traditions, of the infiltration of liberal ideas, of industrial development and economic progress, produced different degrees and forms of democracy and authoritarianism among the great powers. The smaller states, defended by the jealousy of their great neighbors rather than by their own defenses, found it easier to abandon militarization and to accept democracy. The wars of the twentieth century discouraged the progress of democracy and increased the number and influence of autocratic states, but the development of the United Nations tended in the opposite direction in spite of the "cold war."

e) Traditionalism and Progress.—War has frequently been characterized as always destructive and never constructive. Doubtless war in itself has never constructed new political, economic, social, or cultural institutions or practices, and it has often destroyed old organizations and customs. By doing the latter, however, it has sometimes cleared a field in which the new could develop if the creative intelligence of man was present. War has been like a fire which, if not too severe, may facilitate the growth of new vegetation by removing accumulations of dead grasses, brush, and logs. If too intense, however, fire may destroy the roots, the seeds, and even the fertility of the soil. Perhaps better, war may be compared to the wrecking crew which facilitates the growth of the city by destroying obsolete buildings so that new ones may be built in their place. Such analogies, as also the analogy to the catabolic and anabolic processes in an organism, both of which are essential to its life, may lead to unwarranted conclusions.

War in modern civilization was not the only method of eliminating the obstacles to progress. Education and legislation were both used to destroy the old as well as to build the new, and they were less likely than war to get out of hand and become dangerous and destructive. They maintained a continuous relationship between the consuming and producing aspects of sound political progress.

War was contributing to unanticipated historical results, some of which were subsequently regarded as good and others as bad. It was also an instrument employed to bring about expected historical results, sometimes at more and sometimes at less cost. These facts suggest that war, as it existed in modern history, could not unequivocally be considered wholly destructive or wholly constructive.

It has been pointed out that war contributed to the building of the modern nation-states, to their organization in a European system, to the development of ideas, sometimes inconsistent, peculiar to that system, and to the planting of the seeds of those ideas all over the world. These contributions, however, were made largely in the fifteenth, sixteenth, and seventeenth centuries.

In the eighteenth and nineteenth centuries war was less intense in Europe, but the European states by means of war or threats of war extended their dynamic civilization at the expense of the traditional cultures of America, Africa, and the Pacific and injected the virus of their civilization into the ancient civilizations of China, Japan, and India. Traditionalism was in this period destroyed by war and progressivism profited by war, though there was a considerable give-and-take among the Occidental and Oriental states in which the latter contributed much to the world-civilization initiated by the former. War contributed to the rapid augmentation of world contacts and thus to the spread of European ideas of humanism, liberalism, science, and tolerance as well as to the spread of European ideas of strategy, imperialism, and nationalism.

In the twentieth century the military, diplomatic, economic, and propaganda aspects of war have undergone a transformation. Most states from fear of war have tended to increase armaments, to subject their people to more discipline, to organize their national economy and opinion in the interest of efficient war without realizing that their efforts decreased their security and increased their fear. The result was a shaking of confidence in the standards of world-civilization as they had been understood in the mid-nineteenth century. These standards were dealt severe blows by World War I. They survived, however, and appeared to achieve a wider extension and acceptance than ever before in the years immediately following. This phase, however, was short lived. Even more shattering blows were dealt by the great depression and the totalitarian aggressions of the 1930's, World War II of the 1940's, and the "cold war" of the 1950's. Departures from what have been considered civilized standards were notable not only in the totalitarian states but everywhere.

The preceding survey suggests that in the most recent stage of world-civilization war has made for instability, for disintegration, for despotism, and for unadaptability, rendering the course of civilization less predictable and continued progress toward achievement of its values less probable. In the 1960's there appeared some evidence that the very gravity of the situation was convincing man that, unless he destroyed war, it would destroy him. The anthropologist Malinowski wrote even before the atomic age:

Only with the formation of independent political units where military power is maintained as a means of tribal policy does war contribute, through the historical fact of conquest, to the building up of cultures and the establishment of states. In my opinion we have just left this stage of human history behind and modern warfare has become nothing but an unmitigated disease of civilization.

CHAPTER IV

CHANGES IN WAR THROUGH HISTORY

Among animals war between species was an element in the balance of nature and contributed to static equilibrium among species and societies in the biological community. Among men, war has tended since its beginning in primitive human history to be increasingly destructive of life and disruptive of social organization. But the increase was slow and gradual during the first epoch when war was a continuous and normal custom of primitive social life. With the development of pastoral and agricultural economies, changes in military technique became more rapid and tended to maintain society in dynamic equilibrium.

During the historic period war contributed to great fluctuations of history, marked by the rise and fall of civilizations, but successive fluctuations tended to increase in amplitude and to decrease in length. In the modern period of world-civilization fluctuations of war and peace have tended to become stabilized at about fifty years, although the severity of each war period has tended to increase.

While war has become increasingly destructive and disruptive, other periodic visitations which formerly upset human society, such as pestilence and famine, have tended to be controlled. Thus war has stood out more and more as a recurrent catastrophe in civilized human existence.

War was originally a function of the internal structure of each fighting unit, and as there were very many of these units, the probability of any unit of a class being at war in a given time might have been calculated from statistical averages. Change has been in the direction of reducing the number of fighting units so that there is less statistical basis for such calculations. Change has also been in the direction of integrating this smaller number of units into a single unit so that war has tended to become a function not of the fighting unit but of the entire human community of which all fighting units are parts. Thus the problem of war has shifted from that of classifying fighting units to that of analyzing the organization of human society as a whole. Accepting Mead's conclusion, "the more the process of nature can be described in terms of laws, the greater is man's freedom," the trend has been to reduce the freedom of the individual fighting unit to escape war through intelligence and to increase the freedom of the human race as a whole to escape war, provided the laws governing its present organization can be discovered. Prediction, from being based upon the analysis and measurement of numerous independent agencies insusceptible of central control, has come to be based upon the analysis of a few personalities exercising central control. The moral and subjective factors have tended to become more important than the material and objective factors. War can less and less be treated from a deterministic point of view. More and more it must be treated from a constructive point of view. The individual can less profitably be interested in studying the historical causes of war in order to decide a policy for himself or his group. He can more profitably be interested in the engineering of peace for the human race as a whole.

1. WAR AND CHANGING CONDITIONS

The history of war suggests certain general relationships of war to economic, political, military, and cultural change.

a) *Economic and Social Change.*—When independent groups,

utilizing markedly different military techniques, have come into close economic and social contact, continuous war has been usual until the group with the less efficient technique has been exterminated or conquered or has adopted a more efficient technique. Among animals the balance between carnivorous and herbivorous species in the same area has been maintained by the more rapid breeding of the latter, compensating for the predations of the former. A similar balance has often prevailed for long periods between aggressive nomads and peaceful agriculturalists. Where such a balance exists, hostilities have been continuous and of unvarying intensity. Among primitive peoples the development of new external contacts has broken such a balance and increased the amount of warfare sometimes resulting in important political and social changes. The development of intercivilization contacts has similarly stimulated imperial war by civilized states. Such contacts arose, for example, from the exploratory, missionary, and commercial expansion of European states into the hitherto unknown areas of America, Asia, and Africa after 1500. In these cases long periods of war eventuated in conquest and forms of imperial organization in which the group with superior military technique dominated until the subject people had acquired enough of that technique successfully to revolt.

When independent groups, utilizing similar military techniques, have rapidly come into closer economic or social contact with each other, periodic wars of serious proportions separated by relatively long intervals of peace have usually occurred. The rapid expansion of international communication, travel, and trade has tended to increase the amount of war. Intergroup political organization, though often attempted in such circumstances, has seldom proceeded with sufficient rapidity to adjust the problems arising from such contacts peacefully. Among the Greek city-states and the medieval feudal principalities the development of international political organization, attempted in the Amphyctionic Council and the Holy Roman Empire, lagged so far behind the development of economic and cultural contacts that increasingly severe wars destroyed the civilizations. Among the Hellenistic states the lag was less. Rome developed a superior military technique, absorbed all these states in an empire, and maintained political stability for several centuries. The destructive wars, before this universal political organization was achieved, may, however, have sowed the seeds of later decay. In the modern period the British Empire developed a superior naval technique and maintained a precarious peace during much of the eighteenth and nineteenth centuries, but in the twentieth century this system proved inadequate to adjust the problems arising from the closer economic and social interdependence of nations stimulated by modern inventions.

While increasing economic and social contacts tend toward wars of union, decreasing contacts tend toward wars of separation. When politically associated groups, utilizing similar military techniques, have diminished their economic and social contacts, because of the development of technological, ideological, or other barriers, they have become involved in wars of revolt, unless political decentralization has kept pace with the growth of economic and cultural autonomy. The failure of the Roman Empire to decentralize politically with sufficient rapidity may have contributed to the revolts in Armenia, Mesopotamia, Palestine, Mauretania, Palmyra, Egypt, Britain, and elsewhere after the end of the first century. The failure of the British Empire to decentralize led to the American Revolution, an experience which was long avoided with respect to the other British colonies by application of the decentralizing dominion-status policy.

The various wars of independence of Spanish, Portuguese, and Turkish dependencies and of the Confederate States of America in the nineteenth century and the violent breakup of the Hapsburg and Romanoff empires in the twentieth century illustrate the same principle. The revolts of Japan, Italy, and Germany from the public law of the world after 1931 may be in part attributed to overrapid centralization under the League of Nations. These propositions may be otherwise stated: that sporadic war is likely whenever, among groups using similar military techniques, the forces of social and economic change outstrip the capacity of recognized peaceful procedures to effect an adjustment between the standards implied by such changes and those established by existing law. Procedures for continually keeping international and constitutional organization in accord with social and economic changes are as necessary as procedures for better enforcement of existing law.

The maintenance of a rate of political and legal centralization or decentralization in exact proportion to the rate of economic and social integration or disintegration has hitherto been the price of peace. This co-ordination may, of course, be maintained not only through the adjustment of law and organization to social and economic change but also through the control of opinion and economy by political and legal authority.

b) Political Change.—Balance-of-power policies, practiced by groups of states utilizing similar military techniques, have tended toward polarization of all states about the two most powerful of the group, leading to serious wars involving all of them. Such a polarization has usually resulted when alliances, counteralliances, and armament races have been utilized to maintain the balance of power. These practices have tended not only to group all the states by alliance in one or the other of two groups but also to create a conviction of the inevitability of war between these groups. This trend is illustrated in the history of the ancient Greek city-states, the Hellenistic states, the medieval Italian city-states, and the modern European states and, in the twentieth century, by trends of world history before the two world wars and during the "cold war."

Related to the tendency of a balance-of-power system to generate periodic general wars has been its tendency to make each civilization the cockpit of the next. The balance of power having reached a state of polarization within a given civilization, each faction tries to draw in states from the outside. As a result, when economic and social contacts have sufficiently progressed, a larger balance of power, dominated by states of a different civilization, has developed around the original area. The states of the original area, even though utilizing more advanced military techniques, remain divided by historic animosities and are unable to defend their civilization as a unit. Consequently, the civilization is overwhelmed. The ancient civilizations of Syria and Palestine became the cockpit of the surrounding monarchies of Egypt, Mesopotamia, Anatolia, and Persia. The ancient Greek civilizations of the Aegean, Greece, and Sicily became the cockpit for wars among the Hellenistic states of Macedonia, Rome, and Carthage. The area of the ancient Roman Empire after its decadence and division became the cockpit for crusading wars of Islam and Christendom in the Middle Ages. The area covered by the highly developed Italian city-states of the late Middle Ages became the cockpit for wars between France, Spain, Austria, and Great Britain in the sixteenth century. The disintegrating Holy Roman Empire was the cockpit for wars of all Europe in the seventeenth century. Europe, still intent upon its balance of power, was the cockpit for wars involving the United States, Japan, Russia, China, and the British Empire. With a world balance of power established among these states, this process can no longer continue without interplanetary wars.

c) *Military Change.*—Wars among a group of states which have utilized a common military technique without radical change over a long period of time have tended to end in stalemate or mutual attrition. Without change in rules, weapons, or tactics the strategic defensive has tended to gain over the strategic offensive, and wars have tended to end only by mutual attrition. They have become rarer and worse. As a corollary to this tendency, among such states the gravity of war has tended to be inversely related to its frequency. As a civilization has advanced, its wars have tended to become absolutely and relatively more destructive and less frequent. At its height there may be a period of comparative tranquillity. As a civilization has declined, it has sometimes had more frequent but less destructive wars initiated by groups revolting from within or attacking from without but utilizing inferior military techniques. Under such circumstances, however, the attackers have gradually acquired the improved techniques, and the tendency toward attrition has developed again, usually wrecking the civilization. The nuclear stalemate which has developed since World War II, based not on mutual defense capability but on mutual deterrence, may have a different result because the destructive effects of war are more obvious.

Closely related to this tendency of the severity of war to rise and fall in long waves during the life of a civilization has been the tendency for very severe war periods to be followed by movements for peace. A strong pacifistic sentiment arose in Greece during and after the Peloponnesian War, but the movements for federation were inadequate. The desire for peace after the severe imperial and civil wars at the end of the Roman republic created the conditions for the successful organization of the empire. In the Middle Ages the destructiveness of the raids of nomads from the steppes and the Vikings from the sea created a strong desire for peace, utilized by the church in such institutions as the truce of God and the peace of God. In the late Middle Ages the hardships of the Crusades and the wars of dynastic rivalry led to the pacifism of humanists and of reformist sects and to many proposals for world-organization in the sixteenth and seventeenth centuries. The devastating Napoleonic Wars led to the Holy Alliance and to numerous peace organizations after 1815. The mid-nineteenth-century wars of nationalism led to a powerful movement for abitration and the codification of international law after 1870. World War I led to the League of Nations, World War II to the United Nations, and the "cold war," with a nuclear arms race, to movements for peace and disarmament of unprecedented magnitude.

This natural reaction toward pacifism after very severe wars has tended to widen the gap between wars of that type, as also has the necessity for a measure of economic recovery before further hostilities are practicable, especially in the modern period of highly capitalized war. Anthropologists have pointed out that even primitive peoples, whose military equipment is very simple, may fight wars of steadily increasing gravity until there is a "war to end war" which, because of its extensive destructiveness of life, is followed by a considerable period of peace.

d) Cultural Change.—With the progress of a civilization the justification for resort to war has tended to become more abstract and more objective. As the civilization has become economically and culturally integrated, the subjective desire of a small group has appeared to constitute a less and less adequate reason for resort to violence. More and more the interest of the civilization as a whole, objectively manifested in principles of law, has been invoked. From being justified as a protection of "natural rights" interpreted by the fighting group itself, war has progressively been justified as a "duel" or "trial by battle" to vindicate honor or to establish rights in pursuance of the general interest that disputes and feuds be definitively settled; as an instrument of policy authorized by legitimate authority to improve the welfare of the community; and finally as a sanction for enforcing peace and justice within the civilization as a whole.

But whatever the theory or rationalization, in practice, war has been resorted to in response to the subjective interpretation of their interests by the entities actually possessing political power. Usually the earlier stages in the development of a civilization have been marked by the integration of smaller into larger units. Political power has tended to expand, so the evolution of legal justifications for war has been parallel to the realities of politics. War has in fact and in law been initiated in the interest of expanding communities. But in the later stages of a civilization disintegration has taken place. Effective political units have become smaller, though the theoretical political unit has become as large as the civilization itself. The trends of legal pretexts and of political objectives have thus been in opposite directions.

This tendency of the *pretexts* of war to depart farther and farther from the *reasons* for war, as a civilization declines from its maximum, is paralleled by the usual inability of a civilization to develop a political organization in pace with the integration of its economy and culture. Both of these tendencies, illustrated in the later Middle Ages and the Renaissance, contributed to the perpetuation of war and the destruction of the civilization. During those periods wars undertaken in the name of Christian solidarity or for the promotion of justice were usually really intended solely for princely aggrandizement or plunder.

These tendencies, together with that resulting from the development of a given military technique (see Sec. *c*), have given a normal sequence to the character of war during the life of a civilization. Civilizations have usually begun with a period of imperial and balance-of-power wars characterized in these terms in law and in fact. These wars have tended to become increasingly destructive, after which there has sometimes been a period of tranquillity followed by wars of internal revolt and defense from external invasion. Both sides have usually tried to justify resort to such wars in the name of the political and legal authority of the civilization. Actually, however, political authorities interested neither in the internal stability nor in the external defense of that civilization have initiated such wars. Such authorities have in fact been initiating a new civilization, though often asserting loyalty to the principles of the one they are destroying. These wars also have increased in gravity, ending with the complete disintegration of the civilization.

If organic evolution as a whole is envisaged, the initiating causes of war have tended better to accord with its theoretical justification. Animal warfare, instituted in response to the hereditary drives of the individual animal, has functioned primarily in the interests of that animal, but because of natural selection, it has tended to serve also the species of which the individual animal is a member, although the service is unperceived by the initiating animal. Primitive warfare, undertaken at the dictates of the group mores, has served at first the primary fighting group—the clan or the village—but with the integration of a tribe or even a tribal federation or kingdom as the fighting group, war has served that larger unit. Historic warfare, undertaken in response to the group's conception of its interest, has served at first the military chief. With political integration, however, it has served the kingdom, the empire, or even the civilization as a whole until the latter has disintegrated. Modern war, which has been undertaken in response to the authority of national law, has served at first the ambitious prince or faction, later the national state or the alliance. The idea of making it serve primarily the world-community has been developed in theory but not yet in practice. War has only to a limited degree become the police activity of the United Nations serving the world-community.

The failures to achieve co-ordination between the motives of war and the needs of the continually expanding social group have resulted in the eventual extermination of most animal species, most primitive peoples, and most civilizations, but the process of evolution has approached nearer to achieving such a co-ordination with each successive attempt. If contemporary efforts to reduce war to the position of a servant of the world-community fail, there will be similar efforts in the future if remnants of mankind survive.

2 . WAR AND STABILITY

Catastrophe, conquest, corruption, and conversion may, singly or in combination, operate to destroy a social equilibrium and to terminate a civilization. These four processes are important in accounting, respectively, for physical, biological, sociological, and ideological change. These processes lie in the realm of contingency rather than determinism. They resist prediction but, related to conceptions of static, dynamic, oscillating, and adaptive stability, they may explain the history of warfare. Animal warfare, while contributing to the dynamic stability of life during geologic periods, has tended to maintain a condition of static equilibrium during the life of particular biological species, communities, and societies. Fundamental changes in the equilibriums of organic groups and periods of rapid evolution have usually arisen from catastrophes such as mountain formations separating races, widespread glaciations, land elevations

or submergencies leading to large-scale migrations, and extermina-
tions. It has been suggested that the last European glaciation was
primarily responsible for the emergence of modern human types
from the earlier ape men and for the transition from animal to primitive warfare.

Primitive warfare has assisted in preserving a condition of dynamic stability among primitive peoples. The great traditions and
inventions—the use of language, ideas, tools, fire, agriculture, loyalty
to customs, social subordination—slowly cumulated and diffused
through the stimulation and contacts of war. The total human population increased and distributed itself over the entire earth. Although
catastrophes such as flooding of the Nile or Mesopotamian valleys
may have stimulated the survivors in the area, contributing to social
evolution and the emergence of civilizations, it seems more probable
that inventions in the fields of writing, agriculture, government, and
military technique and the wars accompanying these changes were
the immediate cause of the transition from primitive culture to historic civilizations. These inventions not only intensified intergroup
contacts but also increased the value of land to be attacked or defended, the size and co-ordination of political groups, and the efficiency of instruments of conquest. War for conquest and political
unification initiated civilization and spread it.

Historic warfare has contributed toward the oscillating stability
which has characterized the course of civilization during the last
five or six thousand years. Civilizations have risen and fallen, and in
their fall the human race has often lost traditions and inventions of
great value; sometimes permanently, sometimes not beyond hope of
recovery centuries or millenniums later. Yet the reaction of humanity has always been adequate to invent or rediscover the instruments, institutions, and ideas necessary to build a new civilization.
Catastrophes—desiccations, epidemics, famines—have sometimes
contributed to the downward sweep of these great cycles. Wars have
also contributed. Military conquests and migrations have expanded
civilizations. Wars of attrition have destroyed civilizations. But
probably more important than either catastrophe or conquest in
causing the disintegration of historic civilizations has been the
cumulative and corrupting effect of gradual internal population and
institutional changes. Population has outgrown the food supply; ex-

cessive inbreeding has deteriorated the stock; wealth and influence differentials have developed, leading to conflict and revolution; and institutions under the influence of tradition have grown inflexible and incapable of making the necessary adjustments.

In the more recent changes of civilization, conversion through conscious propaganda has perhaps been more important than catastrophe, conquest, or corruption. People with a religion or an ideology have consciously sought to modify public opinion, to change institutions, and to reshape society in the direction of an ideal. The early Christians, as pointed out by Gibbon, may have contributed to the fall of Rome, along with the epidemics of the second century, the exhaustion of Italian soil, the barbarian invasions, the decay of Roman institutions, and the decline of population. Certainly the most recent great transition, that from the medieval to the modern world-civilization, while due in part to the epidemics of the fourteenth and fifteenth centuries, to the attrition of later medieval wars, to the discoveries and military conquests in the sixteenth century, and to the decay of medieval secular and ecclesiastical institutions, was also due in part to the conscious propaganda of the philosophy of science and liberalism by societies and writers assisted by the art of printing. In recent times general literacy, the press, the radio, television, and the cinema have greatly increased the importance of education, propaganda, and conversion as agencies of change.

During the last four centuries of world contact, all of these influences toward change have been operative, but on the whole the influence of natural catastrophe has diminished with the progress of medicine and technology, while the influence of war and conquest has increased. War has been an important instrument in building world interdependence and world-civilization, but changes in its techniques, coupled with the very intimacy of world economic and cultural interdependence, threaten to make it an agency to destroy what it has built. Though the absolute influence of war has increased, its relative influence has probably declined. Change has proceeded more rapidly than ever before, and its most important agencies have probably been the corruption of old and the construction of new institutions, the abandonment of old and conversion to new faiths, and the accelerating march of science and invention.

3. MODERNISM AND STABILITY

Each of the historic civilizations eventually reached a point at which some of its members recognized that the security of each was dependent on the stability of the whole. Attempts were often made to organize the entire civilization into a political community within which each could preserve its identity. These attempts failed to produce adequate institutions usually because of influences external to the civilization or because of inadequate internal communications. After World War I, for the first time in human history, with means of instantaneous communication available throughout most of the world, the attempt was made to organize the whole world politically.

On the eve of World War II, a special committee of the League of Nations drew attention to the increasing interdependence of all the constituent parts of the world and to the problems which had arisen because states tried to adapt themselves to this situation by independent action. "Indeed, to attempt such isolation is one of the first natural reactions to the more frequent and intenser impact of these world forces. But it reflects rather a blind instinct to ward off these impacts than a desire of the constituent parts of a changing world to adapt themselves to what in the long run must prove the irresistible dynamism of these changes; and there can be no development without adaptation."

This statement suggests that the historic cause of war has been the "blind instinct" of a group to preserve its identity by isolation from the "irresistible dynamism" of increasingly frequent and intense contacts. The blind instinct of civilized communities has been the faith, handed down from the communities past, constituting its unity and establishing the values by which its members guide their lives. The functioning of faiths has in the past depended upon general belief in their eternal validity; consequently, whatever has appeared to impair the integrity of the faith has been resisted by the community.

The inevitable dynamism of increasing world contacts has been the consequence of the development and diffusion of science and technology, which have continuously modified the human significance of time, space, and matter, have continuously elevated the horizon of men, and have continuously disclosed new ways of living unknown to the historic faiths. The cumulative growth of science and invention has offered men the opportunity to rise above the limitations of earlier faiths. On the other hand, the political, social, and religious traditions, the continued existence of the community, itself, have persistently demanded that they keep within those limits.

Expanding contacts have therefore been the cause both of progress and of war—of progress because human contacts are the condition of science, invention, and change; of war because change has always been resisted by human institutions, customs, and faiths. Among primitive people war became serious when borrowing or invention broke the control of custom. Among civilizations the slow advance of science, though suggesting new policies, could not at first modify those sanctioned by traditional beliefs and supported by powerful institutions. Thus in each civilization the disparity between policies based on what has been and those based on what might be grew, until the gap was closed by long periods of violence in which the civilization often collapsed.

In modern civilization the cumulative and accelerating growth in the achievements and prestige of science has made the obsolescence of traditional beliefs more rapid than ever before, while the power behind the advocates of both the future and the past has become greater. Science, seeking to eliminate human catastrophes and ready to be converted to new ideas, has been in conflict with faith, seeking to prevent the corruption of ancient formulations and institutions and prepared to conquer a wider area in which they might flourish. Modernism has sought to develop a higher frame of reference in which both science and faith might be subsumed. It has envisaged society as a process by which institutions and beliefs are continuously adjusted to the most accurate forecasts which science can offer of the future. Modernism has hoped to eliminate human catastrophes and conquests by social and scientific procedures for continuously testing the present value of ideas and beliefs. It has, however, recognized that such procedures can be effective only if humanity becomes less reluctant to accept the new and to abandon the old than it has been in the past.

PART TWO

CONDITIONS
MAKING
FOR WAR

CHAPTER V

CAUSATION AND WAR

1. MEANING OF "CAUSE"

The phrase "causes of war" has been used in many senses. Writers have declared the cause of World War I to have been the Russian or the German mobilization; the Austrian ultimatum; the Sarajevo assassination; the aims and ambitions of the Kaiser, Poincaré, Izvolsky, Berchtold, or someone else; the desire of France to recover Alsace-Lorraine or of Austria to dominate the Balkans; the European system of alliances; the activities of the munition-makers, the international bankers, or the diplomats; the lack of an adequate European political order; armament rivalries; colonial rivalries; commercial policies; the sentiment of nationality; the concept of sovereignty; the struggle for existence; the tendency of nations to expand; the unequal distribution of population, of resources, or of levels of living; the law of diminishing returns; the value of war as an instrument of national solidarity or as an instrument of national policy; ethnocentrism or group egotism; the failure of the human spirit; and many others.

To some a cause of war is an event, condition, act, or personality involved only in a particular war; to others it is a general proposition applicable to many wars. To some it is a class of human motives, ideals, or values; to others it is a class of impersonal forces, conditions, processes, patterns, or relations. To some it is the entrance or injection of a disturbing factor into a stable situation; to others it is the lack of essential conditions of stability in the situation itself or the human failure to realize potentialities. These differences of opinion reflect different meanings of the word "cause." Social scientists, historians, and politicians often ascribe different meanings to causation, and so they have different views about the causes of war.

a) Scientific Causes of War.—Scientists, in searching for the causes of phenomena, assume that the universal and the particular are aspects of one reality. They attempt to classify, combine, or analyze particular events into general concepts or ideas which represent measurable, controllable, repeatable, and observable phenomena capable of being treated as variables or constants in a formula. Although scientists realize that there are events in any field of study which have not yet been included in classes which can be precisely defined or measured, they are reluctant to believe that any factors are permanently "vague" and "imponderable"—a belief frequently held by practical men, historians, and poets.

The scientifically minded have attempted to describe the normal functioning of the forces, interests, controls, and motives involved in international relations and to formulate abstract propositions relating, respectively, to the balance of power, to international law, to international organization, and to public opinion. While they have sometimes included war as a periodic recurrence in such normal functioning, to some extent predictable by statistical or mathematical analysis, they have more often attributed war to the high degree of unmeasurability, uncontrollability, incompleteness, or uncertainty of the factors which they have studied. They have considered the occasional occurrence of war probable, although imperfectly predictable, if a number of sovereign states in close contact with one another are each guided by a view of self-interest largely controlled from within. They have considered peace a function of the situation as a whole. Consequently, they have believed it always in danger if the major decisions of states are not guided by a correct appraisal of the total situation. They have, therefore, identified the causes of war with the conditions which have made such an appraisal difficult and so unlikely.

Analyzing these conditions, the scientifically minded have attributed war (1) to the difficulty of maintaining stable equilibrium among the uncertain and fluctuating political and military forces within the state system; (2) to the difficulty of utilizing the sources and sanctions of international law so as to make it an effective instrument for determining the changing interests of states, the changing values of humanity, and the just settlement of international disputes; (3) to the difficulty of so organizing political power that it can maintain order in a universal society, not threatened by other societies external to itself; and (4) to the difficulty of making peace a more important symbol in world public opinion than particular symbols which may locally, temporarily, or generally favor war. In short, scientific investigators, giving due consideration to both the historic inertia and the inventive genius of mankind, have tended to attribute war to immaturities in social knowledge and control, as one might attribute epidemics to insufficient medical knowledge or to inadequate public health services. The basic cause of war, in their opinion, is the failure of mankind to establish conditions of peace. War, they think, is inevitable in a jungle world; peace is an artificial construction.

b) Historical Causes of War.—Historians assume that the future is a development of the past which includes, however, forward-looking intentions and aspirations. They attempt to classify events into ideas which represent commonly observed processes of change and development. Because of the common experience of small incidents releasing stored forces—the match and the fuse—they frequently distinguish the occasion of war from its causes. Because people think they are familiar with human nature, with economic and political interests, with social, political, and legal processes and organizations, and with religious and ideological commitments, historians have often classified the causes of war under these headings.

This method may be illustrated by the causes of the Franco-Prussian War set forth in Ploetz's *Manual of Universal History.* These are divided into "immediate causes," "special causes," and "general causes." The first were said to be certain events which shortly preceded the war, including the election of the prince of Hohenzollern to the throne of Spain, the French demand that the

Prussian king should never again permit the candidacy of the prince for the Spanish crown, and the Ems telegram from Bismarck announcing the king's refusal. The special causes were said to be the internal troubles of the French government, the controversy concerning French compensation for the Prussian aggrandizement of 1866, and the news of new German infantry weapons threatening the superiority of the French chassepot. The general causes were stated to be the French idea of natural frontiers as including the left bank of the Rhine and the long struggle of the German nation for unification, together with the French anxiety over it.

Historians have thus sought to demonstrate causes by drawing from a detailed knowledge of the antecedents of a particular war events, circumstances, and conditions which can be related to the war by practical, political, and juristic commonplaces about human motives, impulses, and intentions. When they have written of the causes of war in a more general way, they have meant simply a classification of the causes of the particular wars in a given period of history. Thus certain of the causes of the Franco-Prussian War have been described by such words as "aggressive policies," "changes in military techniques," "domestic difficulties," "unsettled controversies," "dynastic claims," "aspirations for national unification," "historic rivalries," and "insulting communications." Even broader generalizations have been made classifying the causes of war in the Western world as political, juristic, ideological, and psychological.

When generalization has reached this stage, the result is not unlike that achieved by the scientific approach, for such words as "policy," "law," "ideology," or "attitude" represent concepts which, though limited by the historian to a historic epoch, are universals which may be manifested in varying degrees in all times and places. They are, in fact, variables susceptible, in theory, to mathematical treatment, however difficult it may be practically to measure their variations.

c) *Practical Causes of War.*—Practical politicians, publicists, and jurists assume that changes result from free wills operating in an environment. They attempt to classify events according to the motives and purposes from which they seem to proceed. Their assumptions have thus resembled those of the historians, though they have formulated their problems toward practical ends and have often excluded events and impersonal forces which the historian frequently considers. Because men like to rationalize their actions, publicists have often distinguished the pretexts from the causes of war. Because they recognize that no free will ever really acts without antecedents, and therefore the origin of a series of causal events has to be determined arbitrarily, they have distinguished proximate from remote causes. Although they have sometimes attributed wars to the failure of society to adopt particular reforms or to modify certain conditions, they have usually distinguished causes attributable to a responsible person from impersonal conditions and potential reforms. In the same way physicians more frequently attribute a patient's illness to a germ or to his failure to take proper precautions than to his susceptibility because of heredity or a run-down condition or to the failure of society to provide public health measures.

Practical men have, then, usually thought of war as a manifestation of human nature with its complex of ambitions, desires, purposes, animosities, aspirations, and irrationalities. They have insisted that the degree of consciousness or responsibility to be attributed to such manifestations is an important factor in devising measures for dealing with the problem. Classification of human motives from this point of view is familiar in law and economics. Publicists have often distinguished necessary, rational, customary, and capricious acts in the causation of war. They suggest that wars arise in the following situations: (1) Men and governments find themselves in situations where they believe they must fight or cease to exist, and so they fight from necessity. (2) Men or governments want something—wealth, power, social solidarity—and, if the device of war is known to them and other means have failed, they use war as a rational means to get what they want. (3) Men and governments have a custom of fighting when their culture requires fighting in the presence of certain stimuli, and so in appropriate situations they fight. (4) Men and governments feel like fighting because they are pugnacious, bored, or the victims of frustrations or complexes, and accordingly they fight spontaneously for relief or relaxation.

Thus among each class of writers, whether the effort has been to construct a formula relating measurable factors, to narrate a comprehensible process of change, or to describe the reactions by which the generally recognized human motives affect state behavior, the process of generalizing from concrete events has developed similar categories. The historian, however, has usually kept closest to the events, and the scientists have been most bold in generalization, often resting to a considerable extent on the shoulders of the historian and the publicist. (1) Scientists, historians, and publicists have each generalized about material forces in the state system, though they have referred to them, respectively, as the balance of power, political factors, and necessity. (2) So also has each generalized about rational influences under the names of international law, national interests, and reason. (3) They have generalized concerning social institutions under the heads of international organization, ideology, and culture or custom, respectively. (4) The reactions of personality have, finally, been generalized by the three classes of writers under the names of public opinion, psychological or economic factors, and caprice or emotion.

Whether evidence is sought in the study of wars themselves or in the study of competent generalizations about war, the same classification of the causes of war is suggested. War springs from politico-technological, juro-rational, socio-ideological, and psycho-economic conditions. Each of these classes implies certain assumptions about the meaning of war.

2. ASSUMPTIONS AND CAUSES

The causes of war depend not only upon the meaning of the term "cause" but also upon the meaning of the term "war." The definition of war here accepted recognizes that war has technological, legal, sociological, and psychological meanings, each implying distinctive assumptions. Furthermore, the occurrence of war, whatever its meaning, may be explained on the assumption that it is determined by conditions or on the assumption that it is an act of free choice. The causes of war appear differently according as analysis is directed toward one or the other of these assumptions concerning the nature of war and of the universe in which it occurs.

a) Technological Point of View.—From the technological point of view war is a violent encounter of powers, each of which is conceived as a physical system with expansive tendencies. All the powers together are thought of as a larger physical system or balance of power. War occurs whenever the tension, arising from pressures and resistances on a given frontier, passes beyond the tolerance point and invasion occurs.

Each belligerent power is conceived as a military hierarchy, the units of which are the individual soldiers and workers who, through discipline, respond automatically to the word of command, elaborating military materials and supplies in farm, mine, and factory, transporting them to the front, and launching them against the enemy. Each power thus resembles a single great machine, the efficiency of which in a given war can be calculated in terms of its own power in men, materials, morale, manufacturing capacity, population, and resources; in terms of the resistances of the enemy; and in terms of the distances in miles and natural obstacles separating them.

As the efficiency of all actual or potential belligerents might be measured in the same terms, the group of powers can be viewed as an equilibrium of power analogous to the equilibrium of the heavenly bodies. If the efficiency of the military machines were as calculable as the masses and distances of heavenly bodies and if the assumption that governments are motivated only by power considerations were as true as the assumption that heavenly bodies are motivated only by inertia and gravitation, then the occurrence of wars might be predicted as accurately as that of eclipses. As it is, the difficulties of prediction can hardly be overestimated.

Many people still view world politics as a balancing of power. If everyone adopted this point of view with rigorous completeness, war would present no problem of law or morals. Power as manifested in military policy, organization, weapons, and operations would be the sole influence in international relations. With respect to these relations, every individual would be a potential soldier, every power a potential army, and the world as a whole an active or potential struggle for power. Hostilities would end only with conquest or reestablishment of the balance of power and would recur whenever the equilibrium became seriously disturbed.

b) *Legal Point of View.*—The assumption that governments are motivated only by power considerations has in reality been far from the facts. Governments rationalize their decisions to initiate war or to resist war initiated by someone else in the name of the state and in terms of law. States go to war by means of a constitutional procedure in order to defend themselves, to resist injustice, to fulfil a duty, to enforce a right, to vindicate national honor, or to implement policy. Wars, whether of defense or offense, are not in fact unreflective behavior as suggested by physical analogies. They are deliberate decisions in accord with the state's law.

War may therefore be conceived as the consequence of the diversity of legal systems. Each state, because it claims legal sovereignty, assumes that its law must prevail wherever it extends, even on the high seas, over foreign territory, or over aliens. Practical conditions of power, however, have made it necessary for each state to recognize limitations of its jurisdiction. These limitations under the systematizing tendency of legal thought have developed into international law. When recognized by governments, this body of law tends to acquire an independent jural authority. It provides not only the pretexts or rationalizations for war but also the reasons for war. Its rules, in so far as they are based on generally accepted customs and morals and can be interpreted to support a desired policy, are useful to governments as propaganda symbols for the public both at home and abroad. In so far as they are based on a realistic consideration of the aim and nature of the state under present conditions, they explain the reason for state policy and action.

The object of a war, whether economic, political, religious, or dynastic, must rest on a systematization of ideas, or law in the broadest sense, which gives that object a value. Values do not grow out of events but out of ideas. The land utilized by another tribe must be thought of as valuable for grazing cattle and cattle must be thought of as valuable in the economy of a tribe, or that tribe cannot consciously decide to drive the other tribes off in the interests of its cattle. A modern state must regard the achievement or maintenance of independence or sovereignty as valuable, or it cannot consciously go to war to achieve this object. The definition and organization of these values by the state constitute its interests and its law to be enforced internally by police and externally by war. A larger synthesis, including the values of all states, constitutes the goal of international law. In so far as this synthesis is possible and is actually expressed in the rules of international law, the policy of each state comes to be the maintenance and development of international law.

War, like a duel, is the consequence of a situation in which legal sanctions are unable to maintain an accepted system of law. Violent self-help is in principle incompatible with the idea of law. Law implies rules and principles which must be observed with the result of justice and order in the community. The conceptions of absolute sovereignty and of war as a permissible procedure have interfered with the evolution of legislative and sanctioning procedures adequate to realize effective law in international relations. A conception of sovereignty compatible with the outlawry of war is, however, possible.

If everyone adopted the legal point of view and pursued it to its logical conclusion, war as legitimate violence between equals would disappear. All acts of violence would become either crime, defense, or policing as they are in developed systems of municipal law. Law would be the sole influence in international relations, and with respect to those relations, every individual would be subject to the municipal law of some state, every state would be subject to international law, and the world would be a society of nations in which all conflicts would be soluble in accordance with the law.

c) *Sociological Point of View.*—Men and groups do not act only, or even in large measure, to achieve conscious objectives. No more does civilization proceed only by the logical development of ideas. Legal systems are only the conscious aspect of group cultures. Subconscious or unconscious attitudes and behavior patterns constitute their more important aspect. Culture as a whole profoundly influences the application of the law to particular cases and gradually changes the laws themselves. Behind the state is the nation. The latter implies a group whose members feel themselves a unit because of common culture, customs, practices, and responses, and react spontaneously as a unit against encroachments. Wars may therefore be considered consequences of the contacts of diverse cultures.

As systems of municipal law, when in contact with one another, develop and are influenced by international law, so national cultures, when in contact with one another, develop and are influenced by world culture. War is a conflict of cultures, but it is also a breach in a higher culture. In the modern world this is little less true of international than of civil wars. Conflicts between communities which have no culture in common, such as may occur in the organization of newly discovered colonial territory, should not properly be called "wars." The pioneer does not make war on wild beasts which obstruct his plans of development. He exterminates or tames them. The same is sometimes true of the civilized man's attitude toward savage tribes. Mutual recognition by the opponents that they have something in common is an essential element of the concept "war."

From the sociological point of view the propaganda of symbols

Conditions Making for War

of internationalism and nationalism are illustrations of the general process of group integration and differentiation. War has been the predominant method for integrating political groups. The identification of cultural nationalism with legal sovereignty has concentrated political and military power in national governments and has augmented the severity of wars.

If everyone adopted the sociological point of view, wars would be perceived as forms of social conflict occuring spontaneously from group behavior patterns or from the effort of leaders to preserve these patterns by intensifying loyalty to the symbols of the group or extending the influence of preferred symbols into new areas. Military activity would be considered an incident in the continuous conflict of propagandas upon the course of which depends the values, interests, cultures, activities, and eventually the conditions of human life.

To the sociologist the nation is but one of many possible political groups, and hostilities are but one of many forms which intergroup conflict may take. The form both of groups and of conflicts depends eventually upon the types of symbolic construction which are accepted as important. Changes in symbolic constructions have occurred in the past, marking the rise and fall of institutions and of civilizations. Human ingenuity may do much by juristic interpretation, education, propaganda, politics, and administration to effect such changes according to conscious design and without violence. "The world," "humanity," "mankind," and "the United Nations" might become more important symbols than France, Japan, and the United States, just as the latter, during a century, became a more important symbol than Massachusetts, Virginia, and New York; but sociological factors make it difficult for individuals to identify themselves with a universal group, which, by definition, cannot be opposed by a group outside itself.

d) Psychological Point of View.—Cultures are but abstractions of common psychological elements in aggregates of human beings. Wars are ultimately clashes not of armies, laws, or even cultures but of masses of individuals, each of whom is a distinct personality whose behavior, while affected by the command of a superior officer, by laws, and by significant symbols, is also affected by individual heredity and individual experience. Upon these individual elements rests the power of social, legal, and political superstructures.

The fact that opinions rather than conditions induce political action, the ease with which opinion can be manipulated by special interests, and the presence of irrational drives of adventure, persecution, escape, and cruelty account for the usual irrationality of war and for the relatively slight correlation of its occurrence with any definable population or economic changes. The tendency for individuals to concentrate their loyalties upon a concrete group and to concentrate their aggressive dispositions upon an external group makes it possible that an incident in the relations of the two groups will acquire a symbolic significance and stimulate mass reactions which may produce war. Mass reactions, dividing the private and public consciences of individuals, have also been important in creating solid groups capable of securing internal peace. Attempts to prevent war by increasing the autonomy of the personality and its responsibility to choose among groups competing for its allegiance in each crisis present dangers to domestic peace. A high order of general intelligence is required in a liberal society if nations are to be kept from becoming so strong as to threaten international anarchy, without becoming so weak as to threaten domestic anarchy. Such conditions might flourish with the "rational man" ideal of human personality and the democratic ideal of political organization. Under present conditions both of these are contingent upon a reasonably secure organization of the world against violence.

If everyone adopted the psychological point of view, war would exist whenever, as in the "cold war" following World War II, there was a widespread attitude of hostility within a population directed against another population which reciprocated this attitude. As the human race is biologically a unit, all hostilities could, therefore, be regarded as revolts against human solidarity.

Attitudes, while sociologically interpreted as functions of the group, derived from its culture and symbols, are psychologically interpreted as wishes of the individual, derived from his heredity and experience. Incompatible desires to dominate are frequently at the root of hostile attitudes. War results from progressive intensification of hostile attitudes and behaviors in two populations through the reciprocal stimulus of the anxiety of each once their relations are interpreted as those of rivalry.

Impressed by the tremendous variability in the conditioning of human responses, most psychologists perceive possibilities of adequate outlet for the hereditary drives in forms other than war. The division of humanity into races, classes, nations, etc., influences human behavior because of social meanings, not because of any specific hereditary drives.

There are sociological reasons why the human race, lacking an "out-group," has been a less important social unit than many of its subdivisions, but there are no psychological reasons why, under suitable sociological conditions, conflicts of attitude within neighboring populations should not be solved without violence.

e) Deterministic and Voluntaristic Points of View.—Each of the meanings of war considered may be viewed deterministically or voluntaristically.

The deterministic point of view, which has often been accepted in scientific analysis, holds that every event can be explained by natural laws manifesting the essential continuity and homogeneity of the universe in which it occurs. With a formula expressing the relationship of such laws and with complete knowledge of the state of the universe at any moment, it would be possible to predict what will happen in any part of the universe at any future time. Since all parts of the universe are interrelated, no predictions can be absolutely certain in the absence of such omniscience.

A scientific analysis of war proceeds from a deterministic point of view, but social scientists follow the usual practice of historians and publicists in allowing some room for contingency and choice. War has been defined as "a legal condition which equally permits two or more hostile groups to carry on a conflict by armed force." This suggests that the requirements of law, the conditions of hostility, the process of conflict, and the technique of arms set the scene for war, leaving to free choice only a minor role. With much oversimplification and with decreasing accuracy as civilization develops, it may be said that wars usually result (1) technologically, because of the need of political power confronted by rivals continually to increase itself in order to survive, (2) legally, because of the tendency of a system of law to assume that the state is completely sovereign, (3) sociologically, because of the utility of external war as a means of integrating societies in time of emergency, and (4) psychologically, because persons cannot satisfy the human disposition to dominate except through identification with a sovereign group.

The voluntaristic point of view holds that occurrences may be caused by the policy of the initiating entity. It assumes a pluralistic universe with many free agents. Instead of the whole determining the behavior of lesser entities, such entities by their self-determined behavior influence the whole. Writers on war have been inclined to adopt this point of view. Historians have often insisted that war is the great contingency of history. Embarkation upon war is an act of free will, and its consequences change the course of history. While no one denies that antecedent conditions, circumstances, tendencies, and generalizations of experience exercise some influence, practical writers and jurists often treat the initiation of war as in large measure an act of choice by at least one of the parties.

From the voluntaristic point of view war might be defined as the utilization by a group of violent means to remove political obstructions in the path of group policy. War is simply policy when speed is deemed necessary and political obstructions will not yield to persuasion. From this point of view the causes of war consist of the particular ends, interests, or political objectives of the various states, of the policies of other states which they believe obstruct the achievement of these ends, and of the circumstances which make violence seem the most expedient procedure to the initiating government.

Explanations of war from the deterministic and voluntaristic points of view differ in degree rather than in kind, and they tend to approach each other as the knowledge and intelligence of the entity which initiates war approach zero or infinity. Determinism is a function either of matter or of God. Man, being superior to unconscious matter and lower than the angels, can exercise uncertain choices. If a government had no knowledge at all of the external world, its reactions would be entirely determined by the natural law defining the behavior of entities of its type in contact with an environment of the type within which it exists. It would have no more freedom than would a particle of matter obeying the laws of gravitation and inertia. On the other hand, if a government had perfect knowledge of the universe in which it exists, it would be able to frame policies and adopt methods which were certain to succeed without disabling costs. Proposals which did not conform to these conditions would not be accepted.

Wars arise unpredictably because governments know something but not everything. Their image of other governments, of the world situation, and even of themselves, upon which their policies, decisions, and actions are based, is always in some measure distorted.

CHAPTER VI

THE BALANCE OF POWER

1. MEANING OF THE BALANCE OF POWER

Among the hypotheses suggested to explain the recurrence of war is the difficulty of maintaining a stable equilibrium among the uncertain and fluctuating political and military forces within the system of states. The phrase "balance of power" has sometimes designated the achievement and sometimes the effort to achieve that difficult task. In the static sense a balance of power is a condition which accounts for the continued coexistence of independent governments in contact with one another. In the dynamic sense "balance of power" characterizes the policies adopted by governments to maintain that condition.

The term "balance of power" implies that changes in relative political power can be observed and measured. In the rough calculations of world politics transfer of territory has been the most important evidence of changes in political power, just as in business changes in wealth have been the important evidence of changes in economic power. This is partly because territory, with its potentialities in relation to population, taxation, resources, and strategy, usually adds to military power but even more because the value of territory has been accepted in the international mores, and consequently the fact of acquisition gives evidence of the power to acquire not only territory but anything else, whereas the fact of cession gives evidence to the contrary.

The term is based on the assumption that governments have a tendency to struggle both for increase of power and for self-preservation. Only if the latter tendency checks the first will all the governments continue to be independent. Whenever one increases its relative power, its capacity to increase it further will be enhanced. As a consequence, any departure from equilibrium tends to initiate an accelerating process of conquest.

Evidence that a static balance of power has ceased to exist is at hand when certain governments begin to disappear or to lose territory and others to increase in territory, a process which may continue until only one government survives with the others inside it, as illustrated by the Macedonian and Roman empires of the ancient world.

Oppenheim assumed that the power of international law and organization will always be less than the military power of sovereign states, and consequently only if national military forces are in stable equilibrium can the other two exist. On this assumption discussions of the balance of power have usually ignored considerations of law, social solidarity, and public opinion except as they bore upon the military power, immediate or potential, of the states involved in the system.

Recent analyses of the concept of political power, however, cast doubt upon this assumption. Charles E. Merriam writes:

> The power does not lie in the guns, or the ships, or the walls of stone, or the lines of steel. Important as these are, the real political power lies in a definite common pattern of impulse. If the soldiers choose to disobey or even shoot their officers, if the guns are turned against the government, if the citizenry connives at disobedience of the law, and makes of it even a virtue, then authority is impotent and may drag its bearer down to doom.

Whether direction of military forces gives an individual or institution more "power" than does title to legal prerogatives or control of social symbols or influence upon public opinion depends upon historical circumstances and upon the time interval considered. Although in some historical periods international stability has depended upon a balance of military forces, at other times factors of a wholly different type may have been more important. While it may be true that military unbalance has in all historic epochs constituted an *immediate* threat to international stability, at certain periods, perhaps in most, other factors have been more important *in the long run.*

David Hume wrote: "As force is always on the side of the governed, the governors have nothing to support them but opinion." Alexander Hamilton said: "Opinion, whether well or ill founded is the governing principle of human affairs." Abraham Lincoln said: "He who moulds public sentiment goes deeper than he who enacts statutes or pronounces decisions." H. D. Lasswell writes to similar effect: "The ascendancy of the ruling few, the political élite, depends upon the acceptance by the masses of a common body of symbols and practices." "I repeat," wrote Machiavelli, "it is necessary for a prince to have the people friendly, otherwise he has no security."

Any conception of stability, whether in civics, biology, sociology, or psychology, rests on some kind of equilibrium, but the nature of the factors in equilibrium may vary greatly. As used in international politics, however, the term "balance of power" assumes that power is measured primarily by armaments and military potential.

In the dynamic sense, the term refers not to a condition of blind forces—as, for instance, the balance of inertia and gravitation which keeps members of the solar universe revolving in fixed relation to one another—but to a policy actively pursued by the member-governments of a political system to preserve equilibrium. The balance of power is not something that just happens but something that is actively willed and maintained. Thus policies of rearmament and disarmament, annexation and cession of territory, alliance and counteralliance, intervention and nonintervention, are frequently said to be intended to preserve the balance of power. Canning said he called the new world into existence to redress the balance of the old. Several treaties of the eighteenth and nineteenth centuries declared in their preambles that they were made to preserve the balance of power. The British Army General Act authorized forces to be raised by the Crown to preserve the balance of power.

Balance-of-power policies are sometimes pursued by single states, sometimes by groups of states, and sometimes by all the states in concert or in combination. Some states have been said to make the balance of power the goal of their policy more than others. In some periods of history states have been influenced by the balance of power more than in others.

It is important to emphasize, however, that, whenever maintenance of the balance of power becomes a guide to the policy of a government, that government is on the threshold of conceding that the stability of the community of states is an interest superior to its domestic interests. Doubtless it concedes this only because it believes that stability is a *sine qua non* of its own survival. The concession is, however, an enlightenment of self-interest which approaches altruism or submergence of the self in a larger whole. In the dynamic usage of the term "balance of power" there are already rudiments of a situation in which law, organization, and opinion may become more important than military power.

Balance of power in the static sense, that of the physical analogy, can apply literally only when states struggle for self-preservation and aggrandizement directly and immediately without conscious effort to maintain the balance of power. The moment a government consciously frames its policies in view of the stability of the larger whole, it has ceased to behave like "power" in the physical sense. Maintaining a balance of power is, in the words of seventeenth-century political scientists, the first step in formulating the social contract among nations.

Vattel so considered it. "Europe forms a political system in which the nations inhabiting this part of the world are bound together by their relations and various interests into a single body. It is no longer, as in former times, a confused heap of detached parts, each of which had but little concern for the lot of the others, and rarely troubled itself over what did not immediately affect it. The constant attention of sovereigns to all that goes on, the custom of resident ministers, the continual negotiations that take place, make of modern Europe a sort of Republic, whose members—each independent, but all bound together by a common interest—unite for the maintenance of order and the preservation of liberty. This is what has given rise to the well-known principle of the balance of power, by which is meant an arrangement of affairs so that no state shall be in a position to have absolute mastery and dominate over the others."

The emphasis when the term "balance of power" is used is always upon the static sense of the word. Governments insist that the state is independent, that it acts only in self-interest, and that self-interest concerns only survival and augmentation of power. The balance of power is a form of thought which grew out of the post-Renaissance interest in physics and astronomy and may be contrasted to the ways of thinking on politics later inaugurated by Benthamite jurisprudence, Darwinian biology, and Freudian psychoanalysis. While balance-of-power politics may lead to group consciousness, international society, and international law and while a stable balance of power may have been an essential condition for international law during the past centuries, yet, in the future, effective international organization may prove to be an essential condition for either a stable balance of power or international law.

2. THE STABILITY OF THE BALANCE

From the point of view of the balance of power, the probability of peace or war at a given moment depends upon the degree of stability of that balance. An investigation of the conditions of such an equilibrium depends upon certain assumptions concerning the motives and capacities of states, the measurability of their power and separation, and the intelligence of statesmen.

First, balance-of-power diplomacy assumes that every sovereign state tends to impose its will on every other, choosing first that one least capable of resisting; that every state tends to resist the imposition upon itself, or upon any other state in the system, of another will; and that an attack will occur whenever the pressure of imposition exceeds capacity to resist, unless other states are willing, for their own preservation, to assist the menaced power in order to prevent the dangerous aggrandizement of the other.

Second, balance-of-power diplomacy assumes that the capacity of a state to resist or to attack, at any moment and at any point on its frontier, is a function of the relative power of the two states separated by the frontier and of the degree of their separation. This assumption implies a complete mobility of the state's military power within its territory, making possible a rapid mobilization on any frontier and a continual alertness to the dangers of attack.

The third assumption, very difficult to realize in practice, asserts that the power of each sovereign and the degree of its separation from every other sovereign can be measured. Although "political power" in a broad sense includes legal, cultural, and psychological factors, from the point of view of the balance of power it has usually been confined to actual and potential military power. Actual military power includes land, naval, and air armament. This includes personnel, matériel, organization and morale of the armed forces. It also includes railroads, motor vehicles, civil aircraft, and other means of communication and conveyance which, though used in normal times for civilian purposes, are immediately available for military purposes. Potential military power consists of available population, raw materials, industrial skill, and industrial plant capable of producing military power. The morale of the civil population and the capability of gaining allies have often proved more important elements of power than arms-in-being or economic potential but are so difficult to measure that they have frequently been ignored in calculations of relative power.

It is no less difficult to measure the degree of military separation of two states from each other. This conception involves estimates of the importance of distance in miles, of the character of the barriers occupying this space, such as seas, lakes, rivers, deserts, mountains, and the length of the frontiers which are in proximity or adjacency to each other. With the development of instruments of sea and air transportation, countries separated by wide oceans may be militarily nearer than adjacent countries with very high mountains on the frontier, and with the development of missiles, the separation of all countries has been practically reduced to zero.

Finally, it is assumed that statesmen pursue a balance-of-power policy and do so intelligently—that they measure the factors involved in the balance of power accurately and guide their behavior by these calculations. This assumption is particularly difficult to realize in democracies because public opinion is likely to be more interested in domestic than in foreign affairs and to be influenced in the latter by considerations, such as nationality, justice, or traditional friendships and enmities, which may be inconsistent with maintenance of the balance of power. The latter often requires shifts in political relationships, threatening gestures, or even war, which public opinion is likely to regard as perfidious.

Analysis of the relationships between the variable factors in the balance of power seems to warrant the following conclusions, provided all states act in accord with the assumptions of that system:

First, stability will increase and the probability of war will decrease in proportion as the number of states in the system increases. Obviously a tendency to localize relations would be equivalent to reducing the system, in any particular instance, to a small number of states and so would make against stability. So also the grouping of states in permanent alliances which are committed to act together would tend to reduce the number of independent entities in the system and so would decrease stability. As a consequence, on the assumptions of the balance of power, policies of rigid neutrality and of permanent alliance both make for instability. With all power concentrated in two alliances, stability is reduced to a minimum and war can be expected unless factors other than those assumed by the balance of power intervene.

Second, stability will increase as the parity in the power of states increases. If there were only two states, there would be great instability unless they were very nearly equal in power or their frontiers were widely separated or difficult to pass. The same would be true if all the states had become polarized in two rival alliances. Even with a large number of states acting independently, comparative equality of power would tend to augment the capacity of each to defend itself and so to increase stability. With the great disparity in weaponry which existed in the period of the discoveries when only European powers had firearms and in the period following World War II when only a few states had nuclear weapons, the states with inferior weapons tended to lose their independence through conquest or unequal alliances, and instability continued through rivalry among the small number of power centers which remained.

Third, stability will be promoted by a moderate separation of states from one another. If every state were separated from every other by impassable barriers, there would be complete interstate stability but there would not be an international system. States would have no more relations with one another than does the earth with Mars. If, on the other hand, states of different power faced each other on certain frontiers, then great separation of states would make for instability because other states would be unable immediately to help the weaker state if attacked. If, however, states were so little separated that they had to rely primarily on the assistance of others for security, their independence would be curtailed, and the first assumption of the balance-of-power system would no longer prevail. That system would give way either to empire or to collective security. Thus stability under a balance of power is promoted by artificial devices, such as disarmed zones or strong fortifications, which increase the separation of especially vulnerable frontiers. If all frontiers are vulnerable to sudden attack, resulting in constant anxiety, a stable balance-of-power system is impossible, a condition inevitable in the missile age.

Fourth, stability will be promoted by certainty as to the states which enter into the equilibrium. Only with such certainty is accurate calculation possible. If there is a possibility of outside states' intervening sporadically on one side or the other with motives other than those assumed in balance-of-power politics, the situation becomes unstable. Thus the entry of such states as France, Spain, and Austria into the Italian balance of power during the Renaissance created instability in that equilibrium. In the same way the entry in the twentieth century of the United States and Japan into the European equilibrium has rendered it less stable. In the long run, however, as an increase in the number of states renders an equilibrium more stable, so the complete incorporation of non-European states into the system, creating a world equilibrium, should in itself eventually make for stability. This has not happened because other factors have made for instability.

3 . THE BALANCE OF POWER IN HISTORY

While other factors have had an influence, the concept of the balance of power provided the most general explanation for the oscillations of peace and war in Europe from the Thirty Years' War to the end of World War II. Most European wars during that period and all serious ones became balance-of-power wars if they did not begin as such. Frederick the Great wrote:

Foreign politics embraces all the system of Europe, labors to consolidate the safety of the state and to extend as much as is possible by customary and permitted means the number of its possessions, the power and consideration of the Prince. . . . Christian Europe is like a republic of sovereigns which is divided into two great parties. England and France have for a century given the impulse to all movements. When a warlike Prince wishes to undertake anything, if both powers are in agreement to keep the peace, they will offer their mediation to him and compel him to accept it. Once it is established, the political system prevents all great robberies and makes war unfruitful, unless it be urged with greater resources and extraordinary luck.

Not only was this conception explanatory but its wide acceptance by statesmen tended toward its continued realization in practice. Statesmen in general directed foreign policy toward preserving or augmenting the relative power of the state. As a means to the first all recognized the expediency of joining forces to prevent the aggrandizement of others, and as a means to the second all recognized the expediency of taking advantage of the quarrels of others to aggrandize themselves. "Curb the strongest" and "divide and rule"—these were the two incompatible shibboleths of the game of world politics.

It is partly because of this inherent contradiction in the assumptions of the balance of power that it did not give permanent stability. If states had been interested only in self-preservation and in the maintenance by each of its relative power, stability might have been preserved for long periods, although under such conditions general changes in technologies, ideas, laws, economies, and policies would have eventually shattered it. Each of the powers, however, especially the great powers, was interested not only in preserving but also in augmenting its relative power; consequently, there was never wholehearted devotion to the balance-of-power principle among them. Each statesman considered the balance of power good for others but not for himself. Each tried to get out of the system in order to "hold the balance" and to establish a hegemony, perhaps an empire, over all the others. Eventually the overgreat power found itself encircled but did not give up without war.

England alone among the European states was able to "hold" the balance for a long time, but only because of its relative invulnerability to attack and its persistent recollection of the Hundred Years' War. Because the navy was sufficient for defense, Britain did not require a large land army which would have menaced others, and because of the failure of the long effort to conquer France, it did not attempt to aggrandize itself on the Continent. The fact that overseas enterprise in commerce and colonies offered abundant opportunity made it easier for Britain to pursue a peaceful policy in regard to Europe. To a limited extent after the Armada and to a large extent during the century after Waterloo, England dominated the extra-European world with naval and commercial power and held the balance in Europe.

While there were great changes in navies during the century after Trafalgar, it happened that all of them—steam navigation, screw propeller, iron hull, armor plate, rifled naval guns—at first added to British predominance, although at times British opinion was seized by panic before realizing the effect of these inventions. The long-run influence of these inventions was, however, to weaken sea power operating far from its base. These inventions, together with the relative decline of British commerce and finance, weakened British power overseas. The invention of the airplane greatly increased the vulnerability of the British Isles themselves. As a result Britain could no longer hold the balance of power. It was forced to join one of the great Continental alliances in 1903 and has not since been able to create such a stable equilibrium in Europe that it could safely remain outside.

The predominance of the balance of power in the practice of statesmen for three centuries, however, should not obscure the fact that throughout world history periods dominated by balance-of-power policies have not been the rule. The balance of power scarcely existed anywhere as a conscious principle of international politics before 1500, and even its unformulated functioning can hardly be studied except among the Italian states of the two centuries preceding, among the Hellenistic states of the Mediterranean in the first three centuries B.C., among the Greek city-states for three centuries before that, among the Chinese city-states of the Ch'un Ch'iu period (700–480 B.C.), and perhaps in the "times of trouble" of Indian, Babylonian, and Egyptian civilizations.

In the long periods of the Roman Empire and the medieval church, factors other than the balance of power were of major importance in controlling the action of statesmen and in giving political form to the civilization. Even in the nineteenth and early twentieth centuries, though balance-of-power politics were undoubtedly important, many historians consider that other factors, ideological and economic, assumed a greater importance.

With the advent of nuclear explosives and intercontinental missiles, the power situation tended to bipolarity because of the great disparity in power between nuclear and non-nuclear states and the complete vulnerability of all to missile attack. This unstable condition developed a "cold war" but did not eventuate in a hot war because all the nuclear powers recognized that the initiation of nuclear war would be suicidal and that conventional war between them would probably escalate. Nevertheless, the "cold war" continued because each side thought threats of nuclear attack were a useful instrument of policy. The balance of terror, prone to accident, pre-emption, irrational decision, or the escalation of civil strife or border hostilities, has not produced stability; consequently, states have increasingly sought disarmament and stability by means other than a military balance of power.

With this history in mind, it is possible to summarize the reasons for the collapse of balance-of-power systems and their succession by longer or shorter periods of war, disorder, or imperial domination. Collapse may occur because of unpredictable events such as the rise of a conquering genius establishing tyrannical rule; the invention of a new weapon or military formation facilitating aggression; the emergence of a new religion or ideology arousing popular enthusiasm leading to civil strife and international crusading; or the intervention of powerful states on the periphery of the balance-of-power system.

There are also tendencies inherent in a balance-of-power system which eventually destroy it. There are trends toward reduction of the number of participating states by the voluntary union or the conquest of smaller states; toward polarization of the balance around the two most powerful states; toward the development of democracies and constitutionalism in some states; and toward a general weakening of confidence in balance-of-power policies as a means of security.

A stable balance of power, by stimulating international trade, communications, and cultural diffusion, develops conditions favorable to constitutionalism, democracy, international law, and international organization, which tend to unify the civilization and to create a preference for welfare over power, thereby weakening the disposition of governments to give primary attention to power.

These attitudes, however, may not be universal. Their prevalence offers an opportunity to the few who prefer power to welfare, adventure to security. Law without effective force cannot curb that minority. States relying on law which has ceased to be supported by an effective balance of power but is not yet supported by organized collective power may be destroyed by conquest. As efforts to federalize the states of a civilization have usually failed, universal empire or anarchy has usually followed balance-of-power periods.

CHAPTER VII

POLICIES AND POWER

It is the assumption of a balance-of-power system that the preservation of the relative power position of the state and, if possible, the improvement of that position constitute the major interests of the state, to which its interests in the economic welfare and cultural advancement of its population are subordinate. A state's interests are determined by those of the politically influential. Consequently, this assumption is justified only if the opinion of the politically important public generally demands security first, aggrandizement second, and other advantages, economic and cultural, in lesser degree; if national traditions have developed from the experience of the foreign office in meeting these major demands in the light of the state's peculiar geographic, cultural, economic, and political conditions; and if law will be respected only in so far as it serves these primary interests. These conditions have generally been realized by the great powers with the result that balance-of-power politics has dominated relations in modern history.

If states are to pursue balance-of-power policies, statesmen must have in mind the evidences of disequilibrium and the procedures for restoring balance. Evidences of disequilibrium have been found, on the one hand, in the movements of the indexes of political power and, on the other, in manifestations of aggressive intention. The latter include declarations of policy looking toward expansion and increased armament and legislative or executive acts, annexing territory, consummating alliances, enlarging military programs, making threats or ultimatums, and initiating hostilities.

Different indexes of power have been deemed significant at different periods of history. During the seventeenth and eighteenth centuries territorial change in Europe was the main index. Population was mainly agricultural and illiterate. Any European area annexed by a state added approximately equal increments of recruits and taxes per acre. It was thought equilibrium would be adversely affected by every territorial acquisition in proportion to its size with adjustments for great differences in population density. It was difficult to estimate the power value of colonial acquisitions overseas, and such acquisitions entered into balance-of-power calculations surprisingly little.

J. R. Seeley's remark that "we [Britain] seem as it were to have conquered and peopled half the world in a fit of absence of mind" suggests that the rest of the world was even more absent-minded on the value of colonies. Seeley, who regarded colonies as a major factor in the balance of power, was anxious to show that their value was not entirely disregarded in the eighteenth century. Mercantilist economists usually regarded colonies as adding to the state's power, but this was not universally true. Adam Smith, laissez faire economists, and especially the "Manchester school" considered colonies a political disadvantage.

With the rise of industrialism and nationalism, however, economic resources, industrial plant, and manufactured armament became more important. After the population, first of western and then of eastern Europe and finally of Asia and Africa, had become infected with the virus of nationalism, acquisition of a territory with a considerable minority population might weaken rather than strengthen the state. Thus in the latter part of the nineteenth century territorial acquisitions became a less important index for the measurement of disturbances to the balance of power. Instead, armament budgets, changes in military and naval legislation, and accumulations of mili-

tary and naval materials, size of standing armies, and trained reserves tended to be the measure of power. The development of a
new military invention, the proposal of an enlarged military or naval
budget, or a military reorganization law by one of the great powers
would usually start a flurry in all the others. In the period from 1870
to World War I high politics consisted mainly in the reaction of the
European great powers to such events. Equilibrium was maintained
with increasing difficulty. Contentions arose during the era of colonial expansion after 1880, naval and military armaments of each
country piled up in response to increases of the others, and the
powers became organized into two great rival alliances.

The procedure followed in order to rectify departures from equilibrium has usually had a relation to the disturbing phenomenon.
The answer to enlarged armament programs has usually been increased armament by others. The answer to an alliance has usually
been a counteralliance. To territorial aggrandizement the answer
has sometimes been preventive war to compel renunciation of the
annexed territory, sometimes agreement upon compensatory an-
nexations. For aggression the answer has been resistance by the
victim and assistance, benevolent neutrality, or collective intervention by others. Sometimes, however, international arrangements
designed to effect a general stabilization of the balance have been
attempted, such as guaranties of the territory and independence of
certain states, armament limitations, commitments to periodic consultation and conference, and collective security systems to assure
orderly procedures for settlement and change.

What has been the influence either in disturbing or in restoring
the balance of power (1) of territorial changes, (2) of alliances and
guaranties, (3) of neutrality and the localization of war, and (4) of
rearmament and disarmament?

1. TERRITORIAL CHANGES

Changes in the political map have always been disturbing to the
balance of power. Such changes and demands for them have been
the main problem with which power politics has dealt during the last
four centuries. The problem has also been faced in the partition of
the American continents during the sixteenth, seventeenth, and
eighteenth centuries, in the partition of Africa and the Pacific

islands in the nineteenth century, in the influence of the western territories upon the rivalry of the American North and South before the Civil War, and in the general concern of the Latin-American states over the struggles concerning undetermined boundaries in Tacna and Arica, the Gran Chaco, Leticia, and elsewhere.

Because territorial acquisition is usually thought to increase the acquiring state's position in the balance-of-power system, states very rarely cede territory voluntarily. Whatever the apparent justice of demands for change based on economic, racial, cultural, linguistic, geographic, or other circumstances, statesmen usually argue that preservation of the state's integrity is an obligation superior to justice for others. Governments, therefore, are reluctant to yield even in apparently small matters, especially when territory is involved.

Efforts to compel an acquiring state to renounce its gains, as in the case of Louis XIV's claim to the Spanish succession, or to provide compensatory territory for that state's principal rival in the balance of power, as in the gradual dissolution of the Ottoman Empire during the nineteenth century, have usually involved hostilities. Russian renunciation of its gains from Turkey in 1878 by the General Conference at Berlin, the partition of Africa in the late nineteenth century, and the partition of Poland in the late eighteenth century were nominally peaceful.

It is to be expected that territory will continue as an important index of power and that the balance of power will continue to be disturbed by claims for territorial change. Since there is no single principle, whether it be "nationality," "economic necessities," or "natural frontiers," application of which will fully satisfy the sentiment of justice in all territorial controversies, since historical claims long dormant may rapidly rise to importance if political conditions seem favorable, and since new conditions precipitate new demands, it is unlikely that the problem of a just territorial distribution can ever be solved permanently or be assured a peaceful solution in the future under a balance-of-power system.

2. ALLIANCES AND GUARANTIES

Alliances and regional coalitions among the weak to defend themselves from the strong have been the typical method for preserving the balance of power. Such a combination may take the form (*a*) of an *ad hoc* alliance to meet a particular crisis or to wage a particular war; (*b*) of a permanent guaranty to a particular state or territory in a strategic position, often as a buffer between two powerful states; (*c*) of a permanent regional bloc, coalition, confederation, or federation co-ordinating the foreign policy of several states; or (*d*) of a general system of collective security.

a) *Alliances.*—The first of these devices, the *ad hoc* alliance, is probably most favorable to the perpetuation of a balance-of-power system. Such alliances do not reduce the number of independent participants in the system but leave each state free to add its weight against the state threatening to destroy the balance at any time. They have been the usual devices employed in modern European history. Alliances have usually been concluded for two or three years or for the duration of a war, and when they have been for longer they have often not been honored. Expediency, as dictated by balance-of-power politics, has, in fact, usually outweighed respect for alliance obligations.

Alliances and confederations intended to be permanent have seldom proved reliable unless carried to the point of federation, transferring much of legal sovereignty and the conduct of external affairs to the central organs. Such a development has seldom been possible unless geographic and cultural factors have conspired to unite the group. Alliances purely for defense have broken up if the state against which they are directed ceases to be menacing. Otherwise they have usually been utilized by one of the parties as an opportunity for aggression against an outside state and have led to war. Even if not so utilized, they have tended toward a polarization of the balance-of-power system, and this has usually eventuated in general war.

It appears, therefore, that a balance-of-power system is more stable if permanent alliances are avoided, if all states remain free to determine their action until a crisis actually approaches, and if in a crisis the states not directly menaced by aggression attempt to break up dangerous combinations rather than to make counteralliances. These precepts are, however, difficult to follow. The British government, with the experience of the pre–World War I alliances in mind, sought to apply this insight to the crisis precipitated by Hitler's occupation of the Rhineland in 1936. It attempted to break up the axis by appeasing Mussolini at the expense of Ethiopia and

Spain. But so long as threats and *démarche* based on threats succeeded, the partners in aggression were not inclined to separate. Peace was temporarily maintained, but confidence in the League was destroyed, and appeasement had encouraged new demands. Again appeasement was tried, this time for the benefit of Hitler at the expense of Czechoslovakia. Again it failed, and in 1939 Great Britain turned to the policy of counteralliance, and general war soon followed.

This experience suggests that satisfied states, in applying balance-of-power policies, are likely to be confronted by the alternatives of appeasing or threatening the unsatisfied states. Appeasement will encourage aggression until it reaches a point threatening the independence of all, but threats against the unsatisfied may unite them and leave no alternative but counteralliance and augmentation of the tendency toward polarization of the balance of power. Either will lead to general war, which will imperil the security of all. Thus, under modern conditions, balance-of-power policies defeat their own ends. They operate not only against peace but also against the security of states.

b) *Guaranties.*—Guaranties of the status quo in buffer areas have been common and are intended to stabilize the balance of power by increasing the separation of overpowerful states from their neighbors. The cases of Switzerland (1815), Belgium (1839), Luxembourg (1867), the Åland Islands (1921), and the Pacific Islands under the Washington Treaty of 1922 are illustrations. The danger of such guaranties lies in the uncertainty of their observance. The guarantors are often the only states that would be likely to violate the territory, and when a guarantor becomes itself an aggressor, the others are likely to act in accord with the dictates of power politics of the moment rather than to observe the obligation of the guaranty. It is, in fact, doubtful in law just what the obligation of guarantors is if one of them violates its obligation. Thus such guaranties have frequently been expressly renewed as crises arise, and under conditions of balance-of-power politics such renewals seem expedient.

c) *Regional Arrangements.*—Leagues, confederations, and "regional understandings like the Monroe Doctrine" envisage collective security within limited areas. They have usually been unreliable arrangements in which the members, because of defensive emergency or because of geographic, historic, or cultural bonds, have accepted the leadership of one or have united their policies by agreement with full reservation of sovereignty. They have often moved toward closer imperial or federal union or have dissolved through internal controversy.

Regional arrangements have sometimes bound together potentially hostile states in a common regional guaranty as at Locarno; sometimes they have consolidated a geographical group like the American countries, through acceptance of common policies toward the outside world; sometimes they have united states for defense against a particular danger as in the case of the Little Entente against Hungary, the Arab League against Israel, and the NATO and Warsaw pacts against each other. Such arrangements have sometimes resulted in a federation in which the conduct of foreign relations has been vested in a single body. Often, as in the cases of the Little Entente and the Scandinavian and Baltic states, the spirit of national independence has so retarded union that the members could be invaded one at a time. Again, as in the cases of Locarno and the Nine-Power Treaty, the members have failed to meet their responsibilities in an emergency, and the arrangement has become obsolescent. Finally, as in the German Confederation of 1815, internal controversy has sometimes resulted in formal dissolution. The regional arrangements established since World War II, although tolerated by the United Nations Charter as collective defense organizations, have tended to bipolarize power and to weaken collective security through the United Nations.

d) Collective Security.—Universal alliances or systems of collective security were vaguely envisaged in the diplomacy of Wolsey and Henry IV, were hesitatingly initiated in the treaties of Westphalia and Utrecht, were actually attempted in the post-Napoleonic "confederation of Europe," the nineteenth-century "concert of Europe," and the "confederation of the Hague Conferences," and were provided with permanent institutions in the League of Nations and the United Nations. They have, in the past, been dependent on a stable balance of power. None of them succeeded in subordinating the balance of power to their juridical and ideological postulates. Consequently, they were not able to survive serious disturbances of the balance of power.

The relations of the balance of power to collective security have been at the same time complementary and antagonistic. They have been complementary in that both recognize stability of the international community as a major interest of each nation, in that collective security has been able to develop only during periods of a stable balance of power, and in that a stable balance of power has not been able to exist without at least a modicum of international organization. They have been antagonistic in that the policies necessary to restore the balance of power when seriously threatened have often been inconsistent with the obligations of collective security.

The fundamental assumptions of the two systems are different. The balance of power depends upon national policies of ganging up against the overpowerful, whereas collective security depends on an international obligation to collaborate against any aggressor. A government cannot at the same time behave according to the Machiavellian assumptions of the balance of power and the Wilsonian assumptions of international organization. During the modern period, while the balance-of-power system has on the whole dominated, there have been periods of increasing length, particularly during the nineteenth and twentieth centuries, when that system has been modified by the functioning of collective security under international organization.

3 . NEUTRALITY

The idea of neutrality has been exemplified (*a*) in *ad hoc* policies of non-participation in war, (*b*) in the guaranteed neutralization of states or areas, (*c*) in general rules or principles tending toward the localization of war, (*d*) in collective organizations to enforce rights of neutrals and to prevent wars from spreading, and (*e*) in permanent policies of abstention from alliances.

a) The *Policy of Neutrality* emphasized particularly by the United States and to a lesser degree by Great Britain among the great powers but characteristic also of many lesser powers, especially Switzerland, the Netherlands, and the Scandinavian powers in Europe, has not always been hostile to the balance of power. Neutrality is, in fact, the policy which all states, particularly those with maritime commercial interests, have tried to achieve in the balance-of-power system. To be able to remain neutral is to hold the balance of power. Whether taking the characteristic American form of profiting by other people's wars, the characteristic British form of divide (the continent of Europe) and rule (elsewhere), or the character-istic Scandinavian form of peace at almost any price, neutrality has assumed a balance of power, and the neutral has shaped its policy accordingly.

Small neutrals usually try to keep out of war, but their ability to do so depends on the interests of the belligerents. Great powers are torn between remaining neutral or entering the war on one side or the other.

Any one of these policies may promote the balance of power, even the bandwagon policy, in case the stronger in a given war is a relatively weak state whose strengthening is necessary to hold a more powerful neighbor in check. When, however, the great powers have been involved, the underdog policy has generally been thought to conform to balance-of-power politics and has generally been fol-lowed by uncommitted great powers. The discriminatory policy required by the League of Nations, and later by the United Nations, unless a special exception was made, would usually have a similar result, on the assumption that the weaker state will seldom have initiated an illegal attack upon a powerful neighbor. Nations have usually assumed that the underdog has justice on its side. In prin-ciple, however, there is a vast difference between these policies. The underdog policy tends toward the perpetuation of the balance of power, the discriminatory policy tends toward collective security, the bandwagon policy tends toward absorption of all in a universal empire, and the isolationist policy may encourage aggression.

The United States and the Latin-American countries, because of their geographical position, have been particularly prone to develop policies of neutrality into a shibboleth of isolation. In the case of the United States, however, particularly since it became a great power, isolation, as an implication of neutrality, has been more marked in word than in deed. The United States has, in fact, mani-fested interest in the course of world events and has usually entered European wars when balance-of-power considerations called for such action, although usually without complete consciousness of the reasons for its action. Popular discontent with passivity in the face of humiliations and belligerent propaganda has been a factor in the growth of war-mindedness, added to concern over the disturbance to the balance of power and legal claims, tending to draw the United States into general European wars. Its increasing power position and increasing commitments in world politics led the United States after World War II to abandon its traditional policy of neutrality.

b) *Guaranteed Neutralization* has often accompanied guaranty of a territorial status quo, and, like the latter, may create buffer states or areas stabilizing the balance of power. Such arrangements, however, have proved unreliable, unless the guaranteed states were prepared adequately to defend their frontiers and unless the guarantors renewed the pledge in each crisis.

c) *Status of Neutrality.*—General rules of international law establishing neutrality as a status that prescribes rights and obligations have been a phase in the transition from the balance-of-power to international organization in modern civilization. This development tends toward collective neutrality and international organization. Immediately, it may make the balance of power less stable by encouraging aggression. If it can be anticipated that any war will remain localized, powerful states will not hesitate, guided by balance-of-power principles, to attack their small neighbors. Small states have continued to exist only because of the expectation, according to the balance-of-power principle, that they would be helped by great neighbors if attacked. In so far as international law, by formalizing neutrality, has created an expectation against such help, the balance of power has become less stable.

The legal institution of neutrality has not, in fact, had much influence upon the operation of the balance of power among the great European states. All of them have usually entered wars in which at least one great power was a belligerent on each side and which therefore threatened the balance of power, if the war lasted as long as two years. The status of neutrality may have assisted the smaller states, which have been the beneficiaries rather than the actors in the balance of power, to keep out of war. The rules of neutrality may increase the assurance of the great belligerents that they would lose more than they would gain by encroaching on a neutral. On the other hand, it may have sometimes lulled neutrals into a false sense of security, causing them to neglect more substantial defenses. Since the smaller states could in any case contribute little military force beyond that necessary for their own defense, their abstention from war has not greatly affected the stability of the balance, and their neutrality has sometimes made it possible for them to contribute to peace by mediating between the belligerents.

The status of neutrality reached its climax in the nineteenth century with the especial support of Great Britain and the United States, both of which, because of geographic invulnerability, were indifferent to the world-community and, because of commercial and shipping interests, favored the localization of war and freedom of the seas. Its roots, however, are to be found in the writings of eighteenth-century publicists and in practices which reach back to the later Middle Ages; and its rules were codified in general conventions of the nineteenth and early twentieth centuries. The experience of World War I and the development of international organization tended to undermine these rules in the 1920's. In the 1930's, interests in the dynamic states dependent upon aggression, interests in the United States committed to isolation, and the failures of collective security tended temporarily to revive the idea of neutrality.

A movement arose in the United States to make of neutrality a more positive policy of isolation by departing from the earlier doctrine of freedom of the seas. This followed unsuccessful attempts to implement the Pact of Paris and to assist League of Nations sanctions by providing for discriminatory embargoes against aggressors. Acts of 1935 and 1936, inspired by an elaborate investigation of the influence on war of arms-traders and financiers, embargoed the export of arms, ammunition, and instruments of war and prohibited the extension of loans and credits to all belligerents.

In 1937 the policy of permitting belligerent trade on the cash-and-carry basis was adopted, and after war in Europe had begun in 1939, the arms embargo was repealed. This constituted an obvious discrimination in favor of powers controlling the seas, and further discrimination favorable to Great Britain was manifest in the passage of the Lend-Lease Act in March, 1941, on the theory that Germany was engaging in hostilities in breach of the Pact of Paris and so was not entitled to the benefits of neutrality. Thus, neutrality had merged into collective security.

In spite of the growth of the legal status of neutrality during the nineteenth century, the policy of nonbelligerent states was determined less by rules of international law than by expediency and public opinion. Within great powers public opinion, affected by interested propaganda, sentimental preferences, juridical ideas, and balance-of-power considerations, usually rapidly became unneutral, and help short of war was given to the favored belligerent, often eventuating in war itself.

d) Collective Neutrality was envisaged in the armed neutralities of 1780 and 1800, in various proposals for a league of neutrals during World War I, in provisions of the Argentine antiwar treaty, and in proposals emerging from conferences of the American powers and of the Oslo powers since 1936. This system may tend toward international organization. Neutrals are bound to be adversely affected by war, so a league of neutrals tends to be a league against war, though its immediate object may be to assure the profits while avoiding the risks of neutral trade with belligerents, to keep hostilities out of specified regions, or to prevent or frustrate aggression. If, however, a league is confined to neutrals, it can have no influence in preventing hostilities, and its influence in stopping them is limited. If directed toward the protection of neutral trading rights only, such a league is not likely to be effective unless the neutrals are prepared to enter the war to defend their rights. If directed toward keeping war out of a region, its effectiveness will depend upon the geographical situation as well as the willingness of the neutrals to use force.

The solidarity manifested by the American countries in meetings at Panama, Havana, and Rio de Janeiro from 1939 to 1942 was remarkable. It seems likely that if there is sufficient solidarity among neutrals to create a league, they will hardly stop at this ineffective step but will move on toward a league of nations not only to limit but to prevent war. The American states did this. After entering the war, they developed the Organization of American States as a regional arrangement within the United Nations.

The northern European "neutrals" tried to develop a compromise between collective security and collective neutrality after the failure of sanctions in the Ethiopian case. This "neoneutrality" proposed to abandon impartiality and passivity as the essence of neutrality and to emphasize the determination to remain out of the "collective psychosis" of war. As means to this end, neutrality was to require active efforts against war, perhaps including commercial embargoes against one or both belligerents. Branding of one as the aggressor was to be avoided as likely to exacerbate the hostilities, though discrimination against the belligerent unreasonably continuing war was suggested. The difference between neoneutrality and collective security seemed to be in large measure terminological; but the reversion of the Oslo powers to the terminology of neutrality weakened collective security, and neither conception saved them from invasion in 1940.

e) Non-alignment.—Somewhat similar to "neoneutrality" was the policy espoused by India and pursued after World War II by many Asian and African states, which were known as the non-aligned states because they belonged to neither the Communist nor the Western group, led respectively by the U.S.S.R. and the United States. It differs from other forms of neutrality in that it refers to policy in time of peace rather than of war. It aims at preventing war by refusing to participate in great alliances in order to remain impartial and to be in a position to mediate between them. The non-aligned states are, however, members of the United Nations and are ready to carry out Charter obligations to co-operate against a state found to be an aggressor.

The position of the non-aligned is not defined by law, as is that of neutralization or neutrality in time of war, but is a political position similar to Washington's and Jefferson's policy of avoiding "entangling alliances." It differs, however, in looking not toward isolation from world politics or "neutralism" but toward active participation for peace through the United Nations.

4. ARMAMENT AND DISARMAMENT

In the nineteenth century, with the industrialization and capitalization of war, armaments became the normal measure of state power. Consequently, rearmament and disarmament assumed a role of major importance in the balance of power. Armament increases in one state have usually been motivated primarily by anxiety as to actual or prospective armament increases or manifestation of aggressive policies in neighboring states. Thus the balance of power, always influenced by the technology of war, has become peculiarly dependent upon it during the nineteenth century.

a) The Influence of Military Invention.—The history of the art of war has been dominated by the effort of the strategists to devise new weapons, new maneuvers, and new organizations with which to win a rapid victory. This effort is opposed by the tendency of war to reach a stalemate in which victory can be won only by years of mutual attrition, so expensive to the victor that war ceases to be an efficient instrument of policy. The race has been continuous between improvements in offensive and defensive weapons, formations, and tactical combinations. On a tactical level the offensive or defensive quality of a unit may be estimated by considering its utility in an attack upon an enemy unit like itself or in an attack upon some other concrete enemy objective, such as territory, commerce, or morale.

The offensive power of surface naval vessels against other such vessels has increased in the modern period. The range and penetrability of naval artillery and torpedoes have increased more rapidly than the resisting power of ships' armor, until today a naval battle usually results in elimination of the inferior force. The use of the submarine and airplane in naval engagements has further increased the power of the tactical offensive.

The prime object of naval war is, however, the control of commerce. The offensive against the enemy fleet is for the purpose of defending our commerce and rendering his vulnerable. With respect to war on commerce the tactical offensive has probably also gained. Before the nineteenth century an armed merchant vessel had a good chance of escaping or successfully defending itself against an enemy privateer or frigate. Resistance by a merchant vessel to a cruiser, however, became hopeless in the late nineteenth century. The state with the superior surface force could destroy convoys and control all maritime commerce of the enemy.

The utilization of submarines, mines, and airplanes in commercial war has further increased tactical offensive power against commerce. Even the state with the inferior surface navy can destroy much commerce in waters near its bases. Compared with the Napoleonic period, the hazards to the maritime commerce of the belligerents and the neutrals have become much greater. The belligerent weaker in surface war vessels can be entirely blockaded, but even the belligerent stronger in surface navy is in grave danger of that fate. Superiority of the tactical offensive in sea war tends to reduce warfare to attrition. The belligerent with the greater economic resources and civilian morale will win, though only after both have been ruined. In naval war, progress in the relative power of the tactical offensive increases the rate of mutual attrition.

Air war as an independent service has the objective of destroying enemy naval forces, shipping, bases, troop concentrations, munition

depots, transportation centers, and war factories. The air attack upon the enemy air force is to give the attacker freedom of the air,
as the naval attack upon the enemy naval force is to give freedom
of the seas. The invention of aviation gave an immediate advantage
to the tactical offensive, but during World War I the defensive, by
development of pursuit planes and antiaircraft guns, gained against
the offensive bomber and attack plane. In World War II, however,
the aviation offensive gained such an advantage over the defensive
that the only defense became the fear of reprisals. In spite of this
deterrent, mutual destruction from the air of both land and sea
objectives became increasingly serious and would be intolerably so
with general use of nuclear weapons. In air war, as in sea war,
superiority of the tactical offensive tends to reduce war as a whole
to attrition. But in air war the rate of mutual attrition is far more
rapid.

The prime object of land war is the occupation of enemy territory. Capture or destruction of his armies and fortifications is a
means to this end. If the infantry, which has always been considered
the backbone of land forces, is considered alone, the power of the
defense has, on the whole, gained since the fifteenth century. A
smaller force with rifles, machine guns, and intrenchment spades
can today effectively resist a much larger force similarly equipped.

With respect to attack on prepared positions on land, it is difficult
to detect a trend. Medieval castles were almost invulnerable to
direct attack until gunpowder was invented. The advantage which
artillery gave the offensive in siege operations in the sixteenth and
seventeenth centuries was, however, lost by the superior methods of
fortification invented in the eighteenth century. The offensive gained
an advantage with new forms of heavy mobile artillery in the nineteenth century, but the stalemate of World War I created the impression that the defense again had an advantage; the German
invasion in World War II, however, indicated the offensive superiority of highly mechanized armies in the field. The thoroughly integrated force, combining planes, tanks, motorcycle contingents, infantry, and light artillery, had a tremendous advantage over all field
defenses and minor fortifications. In land warfare, differing from
sea and air warfare, increase in the relative power of the tactical
offensive tends to avoid the war of attrition and to terminate hostilities by rapid occupation of the territory of the state with inferior
land forces.

It is clear that no study of the relative defensive or offensive power of particular weapons, of particular tactical movements, or of particular branches of the service can indicate the relative advantage of the offensive or the defensive in war as a whole at a given stage of technology. A tremendous tactical advantage of the offensive may not compensate for less obvious strategic, political, and economic advantages of the defensive, such as capacity to resist blockade by organization of industry, agriculture, and the use of substitutes; the lesser human and material costs of defensive as compared with offensive operations; and the capacity for passive resistance and guerrilla tactics even in occupied territories.

In the broadest sense it is difficult to judge the relative power of the offensive and defensive except by a historical audit to determine whether on the whole, in a given state of military technology, military violence had or had not proved a useful instrument of legal and political change. Satisfied powers favor the status quo. They do not resort to arms except in defense. During periods when dissatisfied powers have, on the whole, gained their ends by a resort to arms, it may be assumed, on the level of grand strategy, that the power of the offensive has been greater. During periods when they have not been able to do so, it may be assumed that the power of the grand strategic defensive has been greater.

A general superiority of the defensive in war may result in stability or in destruction of the civilization according as this superiority is or is not known in advance and acted upon. Superiority of the offensive, on the other hand, will result in changes desired by those dissatisfied powers best prepared for war. Since by assumption those powers place a premium on the use of arms, it is clear that superiority of the grand strategic offensive tends to augment the warlikeness of a civilization.

The progress of social organization has combined with progress in the art of war to make successful aggression more difficult. This progress, however, has rendered the civilization more vulnerable to destruction through internal or external use of a wholly new military technique by the advocates of change. This development has contributed to the eventual destruction of most civilizations.

b) *Political Aspects of Disarmament.*—The natural tendency during the rise of a civilization has been in the direction of a stable balance of power. The policy of disarmament has been intended to reinforce this tendency, but it has been confronted by the policy of national strategists whose object is to break the deadlock and to acquire for their own country temporary monopoly of a new strategy or technique with which to dominate. There has, therefore, been a conflict of aim between disarmament conferences, on the one hand, and national military departments, on the other. One has sought to stabilize the balance of power and to assure that any resort to arms will result in at least a temporary stalemate. The other has sought to break the balance of power and to assure speedy victory to its own arms or at least to create the conviction among others that the risk is too great to justify resistance to an aggressive policy.

It is, of course, true that financial as well as political considerations have often constituted an important motivation in disarmament efforts. Disarmament movements have been common after great wars when countries were nearly bankrupt and wished to save money. After the Napoleonic Wars such a movement was led by Tsar Alexander of Russia. When armament rivalry was becoming very intense, toward the end of the nineteenth century, another tsar of Russia, influenced by the pacifist writings of Ivan Bloch and Bertha von Suttner and advised by his minister of finance that his exchequer could not stand the strain of maintaining competition with Germany in making rapid-fire field artillery, called the first Hague Conference in 1899. After World War I and World War II the same motivations were evident.

It has been said that disarmament cannot affect the frequency of war, because people will fight with fists or with clubs if they are denied superior weapons. It is true that wars may develop between disarmed people, but that does not prove that they might not be less frequent or less destructive. Mark Twain reports that, as a second in a French duel, he was to suggest the weapons to be used. His first suggestion was axes. The opposing second thought these might cause bloodshed and, anyway, were barred by the French code. Twain then suggested, successively, gatling guns, rifles, shotguns, and revolvers. All were objected to, and he proposed brickbats at three-quarters of a mile. This was satisfactory except for the danger to passers-by. Finally they agreed on comparatively small pistols at a comparatively great distance, and the duel went on to the mutual satisfaction of the duelists. The story indicates that the type of weapons may affect the probability of hostilities. If armaments are of such a character that both countries are sure to destroy each other, there is less likely to be war than if they are of such a character that each country feels it has a chance to win with comparatively slight expense. For this reason, nuclear weapons created a "peace of mutual terror" after World War II.

It has also been suggested that disarmament arrangements are of no value because they will be violated. Nations at war, it is assumed, will pay little attention to bits of paper. Doubtless if two countries go to war they will start to build armaments as rapidly as they can without attention to any treaties which may exist. However, "production lags" may prevent such activity from changing the military position for a considerable time. These lags vary among different types of armament, but the increasing mechanization of war tends to increase them. Although the United States had been preparing for a year prior to entry into World War I in April, 1917, and although after that date it stepped up all military production processes to the utmost, it was not until the spring of 1918 that American military equipment other than explosives began to get to the front in France. The disarmament treaty might even strike at the means of producing armaments. Instead of limiting the quantity of rifles or missiles, it might limit the number and size of factories for the production of these instruments. Such a treaty would make the "production lag" even longer but could not meet the insuperable difficulty that factories for production of non-military articles can also produce war equipment. There is also a lag in developing the personnel of armies. It takes a considerable time to train effective pilots. If the treaty does not allow military organizations to function or reserves to be trained in time of peace, months must elapse after the war breaks out before adequate military organizations can be put in the field.

The sanctioning value of "production lag" depends upon the efficiency of the peacetime international inspection. The treaty must provide for an impartial body to visit periodically all the countries bound and thus to assure that any violation will immediately become known. This problem was mainly responsible for the difficulty in reaching disarmament agreements after World War II. The agreement of 1963 to eliminate the nuclear testing in environments other than underground was possible because such tests could be ascertained from outside the territory of the testing state.

It has also been said that states will not reduce armaments unless they are given an equivalent in political guaranties of security. Under the pressure of taxpayers, governments, it is supposed, maintain armaments at no greater level than they consider necessary for security or, if they are dissatisfied with the status quo, at no greater level than they consider necessary to effect the changes desired. They will not, therefore, agree to disarm until assured of a substitute method of security or of change. There is certainly evidence to support this contention. Successful disarmament treaties have always been accompanied by political arrangements which were believed by the parties to augment their political security or to settle their outstanding political problems. The two have gone hand in hand, and considering the conditions of successful negotiation, it is unlikely that agreement will ever be reached on the technical problems of disarmament unless the parties have lessened the tensions by political settlements or by general acceptance of international procedures creating confidence that such settlements can be effected peacefully.

It is, however, clear that the armament required by one country for security is a function of the armament of other countries, though statesmen have more easily perceived the influence of foreign increases upon their own needs than the influence of their own measures upon foreign needs. Theoretically, therefore, it is possible to conceive a self-executing treaty which would stabilize the balance of power and reduce the probability of war, although it dealt with nothing but the armament programs of the states and a system of inspection.

c) *Armament-building Holidays* have been of value in diminishing tensions. This is the easiest type of disarmament treaty to negotiate and is illustrated in the Argentine-Chilean Treaty of 1902, the Washington Treaty of 1921, and the London Treaty of 1930. The psychological effect of such treaties, however, is not likely to endure for a long time. Usually after four or five years, changing conditions will convince some of the parties that the existing armament status quo is no longer equitable.

d) *Quantitative Disarmament* implies a general reduction of armaments to a specified level. Such a reduction might increase the frequency of war. One factor tending to reduce the frequency of war has been the probability that a war will result in mutual destruction unacceptable to either side. If the scale of armaments of all belligerents is very large, such a result is more likely than if the scale of armaments is small.

A quantitative reduction of armaments inevitably affects the *relative size* of armaments in different countries. Proposals for quantitative reduction have usually attempted first to solve the problem of ratios. The relative strength at the moment the convention goes into effect may be accepted, as was done at the Washington Conference. A ratio may be defined on the basis of some theoretical consideration, such as the relative populations of the states, their areas, their coast lines, or similar consideration thought to measure defensive needs. Agreement on ratios is exceedingly difficult to achieve and to maintain. Japan denounced the Washington treaties and the London treaties in 1934 because it was denied equality with the United States and Great Britain. Political and prestige considerations always render acceptance of any ratio less than equality difficult for any state, whereas defensive as well as prestige considerations make it difficult for states that have a relative superiority to abandon it. Even if the existing status quo is the basis of the ratio, a reduction of armaments will almost certainly mean an actual change in the balance of power because it will augment the importance of the non-military resources of the states. If navies are reduced, the larger merchant marine will count for more. If stocks of arms and munitions are reduced, the larger iron and chemical industry will count for more. If effectives are reduced, the larger population will count for more.

If agreement is reached on ratios, the problem of measuring armaments remains. Should only armaments be counted, or should total military power, including resources, industrial plant, and population, be estimated? France suggested during the Geneva discussions that it should have more actual armament than Germany to compensate for Germany's advantage in population and industry. Because of the difficulties of measurement and ratios, it has been suggested that disarmament might proceed by permitting each state to have equality not in armament but in security. Each would state the program in each type of equipment and personnel it deemed essential for maintaining internal order and for defending its frontiers. These programs would then be incorporated in a treaty. This procedure, however, neglects the dependence of the defense component upon the armament of others. No state could tell what was essential until it had seen the programs of all the others. Thus the problem of ratios, though it may be concealed by treating armament categories separately, can hardly be avoided.

Armament agreement may, therefore, influence the balance of power. By properly arranging ratios and categories, it may be possible to promote the prospects of a stalemate in case military operations develop and thus to reduce the prospects of war. During the disarmament discussions in 1932, it was accepted that France and its allies still had such a superiority in arms that they could overrun Germany, in spite of probable German lapses from the requirements of the Treaty of Versailles. Germany wanted equality, by which was meant not only equality between its armaments and those of France but equality between the armaments of itself and its allies, on the one hand, and France and its allies, on the other. France feared that Germany, anxious for a war of revenge, would take the field, even though the prospects for victory were no more than even. The French argument, therefore, denied that peace could be promoted by disarmament and with proper logic France asked rather for a strengthening of collective security.

The advocates of disarmament have replied that so long as the military situation was such that France could easily win a "preventive war," Germany would not cease to militarize itself in the name of "defensive necessity"; but this aggressive attitude of Germany, being a consequence of the military disequilibrium, would disappear if genuine equilibrium were achieved.

Subsequent events hardly supported this hypothesis. Germany did rearm in 1935 and ended the Rhineland demilitarization in 1936, thereby achieving "equality." But the tensions of Europe increased. German rearmament and aggressions continued until in 1938 Great Britain and France inaugurated vast but insufficient programs of rearmament to restore "equality." These events suggest that in practice quantitative equality will not in itself necessarily assure a stable balance of power.

e) *Qualitative Disarmament,* as the conception developed at the Geneva Disarmament Conference of 1932, meant the elimination of certain types of military instruments and methods deemed to be particularly valuable for aggression. Its object is to increase the possession of defensive weapons and to decrease the possession of offensive weapons to such an extent that each country will approximate a perfect defense against any probable attack. Invasion will then be physically impossible. The conception that the object of disarmament is to prevent the possibility of territorial invasion was especially emphasized by the American delegation at the Geneva conference.

Is such universal perfection of defenses possible to achieve? The answer depends not only on the characteristics of armament but also on the characteristics of the things to be defended. The defense of territory, the defense of overseas commerce, the defense of nationals abroad, and the defense of expansive foreign policies may require very different equipment. A particular nation's interpretation of defense depends upon its economic, political, and psychological circumstances as well as upon existing international law. It was generally assumed in the Geneva discussions, however, that the defense of the territory to which the state was entitled under existing law was intended.

It has been questioned whether a valid distinction can be made between defensive and offensive weapons. Although the shield would ordinarily be spoken of as defensive and the sword as offensive, it is clear that even in this simple case the distinction is relative. The shield increases the offensive effectiveness of the sword, and the sword can be used to parry as well as to cut or thrust. Among the materials which may be examined to ascertain weapons regarded as especially offensive are the provisions of unilateral disarmament treaties, the discussions of disarmament conferences, and the analyses of military writers. The major disarmament conference of 1932 produced a vague formula and incomplete agreement on its application. A majority agreed that long-term professional armies, heavy mobile artillery, heavy tanks, capital ships, aircraft carriers, submarines, bombing airplanes, poison gases, and bacteria were predominantly offensive weapons. Effectiveness of the instrument in facilitating the invasion of territory and the destruction of civilians seems to have been the main criterion.

Disarmament discussions after World War II first concentrated on nuclear weapons and the means of delivering them with the objective of preventing vast destruction. Later proposals for general and complete disarmament covered all weapons except those necessary for internal policing, with the hope of eliminating the possibility of aggression.

Military writers have studied the problem functionally and analytically. Functionally they have distinguished the offense and the defense at various levels. At the levels of law, policy, and grand strategy the offensive consists in the intention to change the legal status quo by force; the defensive, to preserve it. At the level of strategy and tactics the offensive consists in a movement toward the enemy; the defensive, in waiting for the enemy to attack a position. Clearly, both the offensive and the defensive at the political and legal level will at times and places be strategically on the offensive and at other times and places strategically on the defensive. At the level of weapons and organizations, those instruments most useful for the strategical and tactical offensive may be called offensive or aggressive armament. Military writers recognize that all weapons may be used either offensively or defensively. Even fortifications, though primarily defensive, can provide a screen for offensive movements. Weapons to be most valuable in the tactical offensive must, however, be capable of movement toward the enemy, rapidly and over varied terrains. A political offensive cannot be advanced by a purely defensive strategy, nor can a strategic offensive be advanced by a purely defensive tactic. Since some weapons are more useful than others in the tactical offensive, it is clear that, according to military theory, a regulation of weapons may have an influence on the capacity of the political offensive to advance itself by resort to arms.

Military analysts have considered that an offensive weapon consists in the combination of four elements: mobility, protection, striking power, and holding power. Striking power in itself does not make an offensive weapon. A gun firmly implanted in a fort can defend the fort, can defend a certain surrounding area, and can defend an advancing force for a limited distance, but it cannot move and conquer the enemy. A gun carried by a means of transportation which is at the same time highly protected is a powerful offensive weapon. A machine gun in a trench is a defensive weapon, but a tank equipped with machine guns is a powerful offensive weapon, as is an airplane carrier or a bombing aircraft. An intercontinental missile with a nuclear warhead is the most powerful offensive weapon ever devised. All of these weapons have mobility, protection, and striking power combined, but they lack holding power. Infantry has greater holding power than other arms and thus continues to be an indispensable element for the offensive if the object of war is territorial occupation rather than destruction and terrorization with the hope of inducing surrender.

The problem of qualitative disarmament, on land, sea, air, or outer space, involves complex technical questions as well as political and psychological questions. But in their mastery lies the most important avenue for achieving greater stability through disarmament.

f) Rules of War may recognize disarmament not of materials but of methods. Such rules have existed even among savage tribes, but the modern system which was eventually codified in The Hague Conventions of 1899 and 1907 developed from medieval chivalry, sixteenth-century honor, seventeenth-century military discipline, eighteenth-century commercial treaties, and nineteenth-century humanitarianism.

Rules of warfare of the type abundant in the seventeenth and eighteenth centuries, designed to promote the safety, honor, and prosperity of rulers and high officers, tended to make war a game rather than a destruction, easy to start and easy to end; but for that reason such rules have tended to disappear with the nationalization and democratization of armies.

Rules of the type abundant in the eighteenth and nineteenth centuries, designed to moderate the hardships of war for noncombatants and neutrals in so far as military necessity permits, tended to confine hostilities to the armed forces, to prevent wars of attrition, to localize wars, to favor aggressors, and to make wars short and frequent. As the proportion of the population contributing directly or indirectly to the making of the policy and the military effort of the enemy have increased, economic and propaganda measures have gained in relative importance. Attacks upon civilians and neutrals have increased under the plea that traditional rules must be applied in the light of "military necessity" as developed under changing technical conditions.

In general, a far-reaching regulation of war, confining its destruction to definite military objectives, has tended to reduce the bitterness and destructiveness of war, to make both resort to war and restoration of peace more easy, and, consequently, to bring about a state of affairs where wars are short, inexpensive, but frequent. Such a modification of war is looked upon with favor by many military writers who believe that the totalitarian war, originating at the time of Napoleon in conscription, propaganda, and a multiplication of war objectives and developed since that time by mechanization of military transportation and national industrial mobilization, has been a misfortune. They believe the situation might be improved by reverting to the more gentlemanly and limited type of war characteristic of the eighteenth century.

This program seeks to reverse the natural trend of war toward utilization of all means available to bring about complete submission of the enemy. Rules of war have habitually proved of little practical significance when they have failed to give sufficient heed to "military necessity"—when they have attempted to prohibit methods and weapons which, in the existing state of military and political technique and with due consideration to the possibilities of reprisal by the enemy and of entry into the war by neutrals, promise military results. The experience with conventional regulation of submarine warfare, aerial bombardment, and poison gas in the prewar and interwar periods gives little reason for believing that such efforts to regulate warfare will be effective in the future. Efforts to prohibit the first use of nuclear weapons have not yet been successful although a limited ban on nuclear testing was widely accepted in 1963.

The sanctions of rules of war have been inadequate between peoples of similar civilization, and observance of such rules has been almost wholly lacking in wars between peoples of very different civilizations. Among the Greek city-states, for example, rules recognized in hostilities between one another were considered inapplicable in war with barbarians. The Western nations manifested few scruples in hostilities against American Indians, Australian aborigines, Asiatic tribesmen, and African natives and at times even claimed that the normal rules are not applicable to hostilities against such recognized political entities as China, the Sudan, Syria, and Abyssinia. The British argued in the Hague Conference of 1899 against the adoption of a rule prohibiting the use of dumdum bullets on the grounds that a bullet which not merely penetrated a man but stopped him was necessary when dealing with the fanatical tribes of the Sudan and the northwest frontier of India. More recently it has been suggested that conventional limitations on aerial bombardment should not apply in hostilities against primitive tribes. With the rise of extreme forms of nationalism, passionately adhering to revolutionary doctrines, all external political groups come to be considered as inferior civilizations to be denied the benefit of rules of war whenever expedient. The regulation of war, even though sanctioned primarily by the military self-interest of the belligerents, implies also recognition by all the belligerents of their common membership in a higher community or family of nations.

g) Moral Disarmament.—Discussion of material disarmament has usually led to a consideration of "moral disarmament." By this is meant limitation or qualification of the will to fight as a prerequisite of limitation of the instruments of fighting. This discussion has included consideration of the regulation of international propaganda and of the political problem of security, on the one hand, and the possible revision of the international status quo, on the other. Moral disarmament from the standpoint of those countries satisfied with their present possessions means genuine belief that they will be able to retain them without resort to arms. But from the standpoint of those that are anxious to modify the territorial status quo it means genuine belief that they will be able to acquire what they want by peaceful procedures. The problem of moral disarmament is psychological rather than technological, but it is related to the problem of material disarmament because statistics of the latter provide evidence of the former and because progress in the former contributes to progress in the latter. When armament budgets, personnel, and matériel are rising at an accelerating rate, it may be assumed that international tensions are increasing and that states are morally, as well as materially, rearming. Armament races, evidenced by such statistics, constitute a form of international relations closely related to war and often ending in war itself.

Efforts since World War II to establish aerial or other inspectors to assure against surprise attacks have aimed at relieving international tensions, an object of moral disarmament. The establishment of direct communication between the White House and the Kremlin in 1963 with the object of avoiding mistakes or misunderstanding had a similar objective.

CHAPTER VIII

CONSTITUTIONS AND POLITICS

Governments must think first of retaining power. Even though they perceive that because of the balance of power a given foreign policy is certain to fail, nevertheless they may pursue that policy if they are convinced that national law, national tradition, or national public opinion is firmly committed to it. Governments tend to place domestic requirements ahead of international requirements because their impact upon the existence of the government is more immediate. They may be obliged to attempt the impossible in order to retain office.

They may underestimate the relentless efficiency of the external power situation and trust overmuch to the potency of a firm will and the supineness of other states. In spite of a relatively stable international system a government may initiate war either because of the real or apparently irresistible pressure of internal forces or because of its own doubt of the strength of external resistances.

1. GOVERNMENT, STATE, AND SOCIETY

In the simplest sense of the term, the government is the group of men who decide how the state shall function at a given moment. Clearly the constitutional structure which determines in a given state the type of men in the government and the considerations which limit their freedom and influence their decisions affect the probability of that state's getting into war. A war does not start unless some government either initiates it deliberately or blunders into it. An analysis of the relationship of constitutions to war is, therefore, important.

The state differs from other social entities by its possession of sovereignty or the capacity to make and enforce law within the society. This capacity implies an ultimate control over the life of the individual and of other social entities. The state, or society in its political aspect, may therefore be identified by its claim to a monopoly of human killing and protection from killing.

The state claims the privilege of killing people for such crimes as treason, sedition, and murder and in such activities as wars, reprisals, and pacifications. The state also tries to prevent any other person or organization from killing within its jurisdiction by enforcing municipal laws against homicides, insurrections, and invasions and from killing its nationals abroad by diplomatic protection and intervention. Since this monopoly in killing is conceived as a characteristic of the state in the abstract, the recognition by each state of other states implies recognition of the equal right of every state to exercise the monopoly within its jurisdiction. This jurisdiction, however, is not easy to define because of the migratory character of nationals and armies and the frequent instruction of armies to kill foreigners abroad and to protect nationals abroad from being killed. It is the inadequately achieved task of international law to demarcate the jurisdiction of states, internally and externally, so that conflict may be avoided.

The government exercises the state's authority internally to coordinate the various elements constituting the national society and externally to adjust that society to ever changing conditions. The manner and efficiency with which a government performs this difficult task depend upon the patterns of behavior implicit in the structure and relationships of the organizations and institutions which in the broadest sense constitute the society's constitution. The state's constitution consists of that part of the society's constitution formulated in public law. The latter may be called the "political constitution" and may be distinguished from the remainder of the society's constitution called the "social constitution."

Different states have varied in warlikeness at the same time, and the same state has varied in warlikeness at different times. Can these variations be related to variations in national constitutions?

2. CONSTITUTIONS AND FOREIGN POLICY

Detailed studies have indicated that states with widely different constitutions have tended to react similarly under similar external pressures. Foreign policies have been influenced more by the external situation, especially the political and economic activities of other nations, than by the society's internal constitution. The latter, however, has not been without influence. Although states must in the long run adapt their constitutions to external pressures which cannot be changed, yet they may through wise policies to some extent adapt the external environment to the existing constitution. Democracies pressed by emergency may have to become dictatorships, but by foresight they may organize a world safe for democracy.

The constitution, as well as the foreign policy, of a state results from the interaction of internal and external conditions, but the constitution need not and usually does not respond to a changing world situation as rapidly as does the foreign policy. As a consequence the organs of government responsible for a state's foreign policy are under continuous tension, especially in time of rapid external change. The internal constitution urges a policy founded on national traditions and domestic public opinion, while external conditions and events, as disclosed by information from the diplomatic service, urges a policy of immediate adaptation to shifts in the balance of power by preparations for defense or utilization of favorable opportunities.

This tension has been dramatized in the United States because the Constitution, based upon a system of checks and balances, emphasizes the conflict between the President, in continuous contact with external conditions, and the Congress, influenced mainly by internal opinion. Joseph Alsop and Robert Kintner wrote in 1940:

> The President and Secretary of State together propose, and the Senate, speaking with the voice of American public opinion, in the long run disposes. Yet neither the President nor the Secretary nor the Senate really *makes* American foreign policy. The cables make it. Senators, who do not read the cables, may be isolationists. But men who see the cables coming in, week by week and month by month, are either enlightened or afflicted with a professional deformation, as you may choose to call it. These long mimeographed sheets, with their heavy, secretive stamp, too insistently proclaim this country to be one member only in the community of nations; too grimly suggest that what threatens the community threatens us. Recent history does not record a President in office or a Secretary of State who believed the United States could safely be indifferent to the fate of the rest of the world.

Such belief, however, does not necessarily determine policy. Senators with ears to the ground and eyes to the past may ignore the requirements of foreign policy which seem obvious to the executive, and disaster may occur. The same conflict exists in all countries, though it is usually manifested in the privacy of cabinet meetings or inner councils.

The lag of the domestic constitution behind changing international conditions may be an important factor in the fluctuations of war and peace. In times of general expectation of peace, politics tend to become democratic. Governments tend to become agents for the execution of national public opinion rather than leaders in forming it. Public opinion, springing from sources other than the government, tends to dominate policy, and that opinion, in so far as it bears upon foreign affairs, reflects the attitude of the average man and of interest groups. The average man inclines to be suspicious of the foreigner, to be more interested in domestic than in foreign affairs, to be educated in nationalism limiting his interests to the national horizon, and to be more ready to resent than to understand the complaints of foreign governments. Interest groups seek protection of their special interests, and this is more likely to result in government action when protection is sought against the foreigner than when it is sought against some other domestic group. The export trades and foreign investments tend to expand, frequently giving rise to friction within the area of expansion and to demands for diplomatic or military protection.

As a consequence, in time of peace governments tend to pursue policies, springing from the domestic public opinion, which neglect the balance of power and which have the dual effect of increasing the vulnerability of the state to economic and military attack and of increasing the number of controversies with foreign states. States become materially more interdependent and morally more aloof. In course of time certain states may pass a threshold either of vulnerability or of irritation or of both, leading to constitutional changes. The government of these states will assume a leadership devoted primarily to integrating domestic opinion behind an aggressive foreign policy. The unpreparedness of other states will provide the opportunity to utilize foreign propaganda, diplomatic threats, and military coercion with effect.

As a result of this process, illustrated in the 1930's, tensions may rise in all states, military preparations may become general, diplomatic grievances may become intense and eventually minor and then major wars will be fought. These may so exhaust all participants that a period of peace will follow.

The alternation of domestic constitutions from autocracy to democracy may therefore be related both as cause and as effect to the alternations of peace and war. In history, however, the expectation of war has prevailed in most times and places. Consequently, autocracy, at least in the handling of foreign affairs, has been the prevailing constitutional form.

3. THE POLITICAL CONSTITUTION AND WAR

Among factors which appear to influence the warlikeness of a state are the degrees of constitutionalism, federalism, division of powers, and democracy established in its political constitution.

a) *Constitutionalism* implies that the scope of all political power is limited by law. It is distinguished from absolutism, which implies that political power is hierarchically organized under a supreme authority superior to the law. Law is interpreted as the commands of that supreme authority.

The early advocates of state sovereignty, like Bodin and Althusius, considered that the power of the state as a whole was limited by international law and constitutional principles. The tendency of democratic theory, however, as developed both by the romanticists and by the utilitarians, was toward the absolute sovereignty of the state, though toward limitations of the authority of the government by the political constitution. The state as a corporate body was absolute, though the monarch was limited.

Modern despots have combined the democratic absolutism of the nineteenth-century state with the divine-right absolutism of seventeenth-century monarchs. They have put the despot above the law, not, however, from hereditary title but from a self-discovered capacity to mold the national will.

It is clear that constitutionalism is more favorable to peace than is absolutism. By envisaging law in the abstract as superior to organization, constitutionalism tends toward a universalizing of law and thus facilitates a harmonizing of international law and municipal law through application of the former in national courts. It avoids the assumption of ineradicable conflicts between the legal sovereignties of different states. Absolutism is faced by the dilemma of assuming, as did Dante, a single sovereign empire to make law for the world, thus reducing national states to mere administrative circumscriptions, or of assuming, as did Machiavelli, a number of sovereign states each with unlimited authority to make law, thus eliminating international law altogether and reducing international relations to relations of power.

b) Federalism.—The influence of geographic centralization of government upon peace and war is similar to that of totalitarianism or the expansion of the functions of the state. The highly centralized government tends to prepare for and wage war more efficiently than the decentralized federation, and it is likely to be under greater necessity for doing so, in order to divert attention from dissatisfactions, certain to arise in local areas, because of the very intensity of centralization.

The difficulties which federations, especially those formed by the union of sovereign states, have sometimes encountered in concluding and carrying out international engagements have caused diplomatic friction but have had little direct importance in causing war. These difficulties may, however, have hampered the participation of federal states in international organization.

States are continually undergoing a process of centralization or decentralization according as the centripetal or centrifugal forces are stronger. If either centralization or decentralization of government proceeds out of pace with the integration or dissolution of the society and the culture, it is likely to lead to civil war, whether of self-determination against central interference or of sanctions against local nullifications of the constitution. Furthermore, while the process of decentralization may stimulate attacks from outside, because of the impression of weakening, rapid centralization may alarm other states and lead to preventive wars, if indeed the political strengthening which it gives to the state does not induce it to embark upon aggressions.

   c) *Separation of Powers.*—Whether centralized or federal, the national government may have a functional union or a functional separation of governmental powers. It would appear that the system of separation of powers, maintained by checks and balances, augments the tendencies of liberalism and federalism. Even when governments have a considerable separation of powers for domestic purposes, it is common for the control of foreign relations to be centralized in a single authority. Such a policy was advocated by the early prophets of the separation of powers, such as Locke and Montesquieu. The United States has been peculiar in extending the system of checks of balances into the conduct of foreign relations, and the perpetual antagonism between the Senate and the President has rendered a persistent and efficient foreign policy extremely difficult. This system has doubtless influenced the isolationist tendencies of American foreign policy and the unwillingness of the United States to enter actively into the balance of power or into international organization. When the isolationist policy was based upon a high degree of strategic and economic invulnerability, the effort to make the most of those conditions may have been wise policy. But with a rapidly shrinking world, both strategically and economically, it may be doubted whether the check-and-balance system with its extreme inefficiency has proved adequate to the conduct of foreign affairs. It has augmented the reluctance of the United States to enter war, but it has also decreased the possibility of the United States' taking constructive measures to prevent war. It, however, has not decreased the vulnerability of the United States to attack or the vulnerability of its population to war fever induced by the propaganda of special interests and by false images. Although functional centralization, at least in foreign and military affairs, is a prerequisite for the effective preparation and waging of modern war and may be the price of survival in a jungle world, it also increases the warlikeness of the state.

d) Democracy.—What is the influence of democracy upon war and peace? By democracy in the political sense is meant the general conviction that the source of governmental authority and of the duty of obedience should be the freely manifested consent of the governed population and the realization of this conviction through appropriate institutions. Democracy is distinguished from aristocracy, oligarchy, and autocracy, which assert that governmental authority and the duty of obedience flow from the superior ability, status, or title of an individual or small group of individuals. Recognizing that freedom and equality, while both elements of consent, if pushed to the extreme become incompatible, democracy has usually insisted that both must be exercised under law which develops with changing conditions of culture and technology. Although suspicious of both totalitarianism and centralization, democracy has usually recognized that changing conditions may require an increase in the functions of the state and in the centralization of its government. It has tried to combine individual liberty with social progress.

It was a favorite theme of the allied powers during World War I that democracy tends toward peace. The masses of the people who have to do the fighting, it has been said, never want war, and if they control the state they will not consent to war. This theory has been carried to its logical extreme by proposals for a referendum on war.

Statistics can hardly be invoked to show that democracies have been less often involved in war than autocracies. France was almost as belligerent while it was a republic as while it was a monarchy or empire. Great Britain is high in the list of belligerent countries, though it has for the longest time approximated democracy in its form of government if not in its social attitudes. More convincing statistical correlations can be found by comparing the trend toward democracy in periods of general peace and away from democracy in periods of general war. This correlation, however, may prove that peace produces democracy rather than that democracy produces peace.

It seems probable that although democracies have frequently

been involved in war, this has usually been because they were attacked by non-democratic governments. Yet democracies have displayed some aggressive characteristics. Former Secretary of State Root wrote:

> Governments do not make war nowadays unless assured of general and hearty support among their people, but it sometimes happens that governments are driven into war against their will by the pressure of strong popular feeling. It is not uncommon to see two governments striving in the most conciliatory and patient way to settle some matter of difference peaceably while a large part of the people in both countries maintain an uncompromising and belligerent attitude, insisting upon the supreme and utmost views of their own right in a way which, if it were to control national actions, would render peaceable settlement impossible.

Probably there are tendencies toward both peace and war in democracies as there are in autocracies—tendencies which approximately neutralize each other and, under present conditions, render the probabilities of war for states under either form of government about equal. Perhaps it would not be far from the truth to say that democracies, although in principle opposed to war, are, in practice, often opposed to the organization of peace; whereas autocracies, though in principle unwilling to abandon war as an instrument of policy, in practice often achieve their ends without actual breach of the peace.

Democracies normally require that important decisions be made only after wide participation of the public and deliberate procedures which assure respect for law and freedom of criticism before and after the decision is made. They are, therefore, ill adapted to the successful use of threats and violence as instruments of foreign policy. Autocracies, on the other hand, are accustomed to ruling by authority at home and are able to make rapid decisions which will appear to be accepted because adverse opinion is suppressed. Consequently, in the game of power diplomacy, democracies pitted against autocracies are at a disadvantage. They cannot make effective threats unless they really mean war; they can seldom convince either themselves or the potential enemy that they really do mean war; and they are always vulnerable to the dissensions of internal oppositions, capable of stimulation by the potential enemy, whatever decision is made. Thus it is not surprising that democracies have usually desired to abandon war as an instrument of policy, whereas autocracies have desired to retain it.

Yet if war occurs, democracies fight effectively, display equal endurance, and survive the shocks of disaster and defeat even better than autocracies. In World War I it was the autocracies rather than the democracies that suffered violent revolution, even those on the victorious side. Democracies are likely to be more prosperous in times of peace because their economy is likely to aim at welfare rather than at military invulnerability, and in wars of attrition their superior economies give them an advantage. They have therefore survived, even under conditions of power diplomacy, but at the expense of temporary dictatorships for the conduct of war and other emergencies. Such dictatorships may prove difficult to shake off after the emergency, especially if it is protracted and soon followed by another. The modern world appears to be threatened by such a succession of emergencies, arising from the overrapid growth of technological interdependence and the lag of political adaptation.

To sum up, it appears that absolutistic states with geographically and functionally centralized governments under autocratic leadership are likely to be most belligerent, whereas constitutional states with geographically and functionally federalized governments under democratic leadership are likely to be most peaceful. The types of government tending toward warlikeness are also those tending toward efficient operation of the balance-of-power system, whereas the types of government making for peace tend in the long run toward an international system based upon law and organization. Governments of the peaceful type tend to develop within a stable balance of power, but such governments have succeeded neither in organizing the world for peace nor in maintaining the equilibrium of power. Peaceful governments have created conditions favorable to the rise of warlike governments. There have, therefore, been historic successions from periods dominated by peaceful to those dominated by warlike governments.

4. THE SOCIAL CONSTITUTION AND WAR

Among factors in its social constitution which appear to influence the warlikeness of a state are its age, cultural composition, economy, progressiveness, and integration.

a) Age.—There is some evidence that the warlikeness of a state alters with its age. Holland, Sweden, and Denmark, for instance, were all much more belligerent in the seventeenth century than they have been in the nineteenth and twentieth. Even France and Austria, the most belligerent of the powers during most of the modern period, declined somewhat in belligerency in the nineteenth century. Russia, Prussia, and Italy, on the other hand, have increased in belligerency. These states were also the latest comers into the general power complex of Europe.

It has been suggested that states have a life history like that of individuals. In youth the population increases rapidly, and consequently, there is a larger proportion of young people and a smaller proportion of old people. This induces an adventurous and warlike tendency. After a time the population becomes stabilized, the proportion of the young becomes smaller, the culture is more concerned with economy and welfare, less with adventure and expansion, and there is less inclination to go to war. In advancing age the proportion of the old in the population becomes greater, the state's position in the balance of power becomes stabilized, and its willingness to risk this position by war becomes progressively less.

Revolution brings youth to the front and usually involves civil strife, paving the way to autocratic rule and foreign war, followed by restoration of many aspects of the *ancien régime* and the rule of older men who water down the ideology of the revolution if they do not abandon it. Thus belligerency may be measured by the time since the last revolution.

b) *Cultural Composition.*—Cultural heterogeneity within a state tends to involve it in wars of two types: civil revolts of cultural minorities to resist oppression or to establish national independence and imperialistic wars to expand empire or to divert attention from domestic troubles. As the large states, especially those with overseas empires, have tended to be the least homogeneous, this is a factor accounting for the greater warlikeness of the "great powers." Wars to suppress colonial revolt and to expand empire in backward areas, although numerous, have seldom involved national aspirations or the balance of power and have usually remained "small wars."

If there is great cultural heterogeneity within the home territory of the state, as was notably true of the Hapsburg Empire, wars of self-determination or diversion may occur. The cure for incipient civil war is said to be foreign war. Instances appear in the history of all the great powers where this device has been considered or utilized, notably by the Hapsburg Empire in 1914. Such wars often spread because they usually involve the balance of power.

Another method for avoiding the dangers of a heterogeneous culture has been the propaganda of nationalism. The effort to advance cultural homogeneity has, however, been even more productive of war than the existence of cultural heterogeneity. Nationalism has produced more serious wars than imperialism. The characteristics of national cultures differ, however, in respect to warlikeness. These qualitative differences are probably more important than the degree of uniformity of the culture throughout the state's territory.

c) Economy.—The system of economic production or resource utilization has had an important influence upon warlikeness. States with economies based on agriculture, though less warlike than those based on animal pasturage, have generally been more warlike than those based on commerce and industry. This tendency has been pointed out by economists, sociologists, and historians and was to be observed in the contrast between the agricultural West and the more commercial East in the United States during the Napoleonic period; in that between the agricultural South and the industrial North in the United States before the Civil War; in that between the agricultural east and the industrial west of Germany before World War I; and in that between the agricultural east and the industrial west of Europe in the nineteenth century.

This difference appears to rest on inherent conditions of the two economies. Self-sufficient agriculture in modern civilization originated in feudalism and in the settlement of nomadic conquerors or pioneers with the spirit of adventure and self-reliance in which each man defends his home with his arms; the spirit of feudalism has continued in the former case and has tended to develop in the latter, with increasing inequality in the possession of land. The landowners defend their estates with their own forces, exalt the military virtues, engage in hunting for food and sport, and maintain familiarity with the weapons of war. Furthermore, in an agricultural civilization, land is the major commodity of value, and land is something that can be acquired by war. The growth of population, which is usually more rapid in rural than in urban areas, makes evident a continuous need for more land if the rising generation is to have an equal number of acres.

On the other hand, industrial activities tend toward urbanization

and demilitarization of the leaders and the society and to the exaltation of business shrewdness, which in trade and industry is a more
efficient instrument than war. It is, of course, true that highly industrialized states must import foodstuffs and raw materials, must
export manufactured goods to pay for them, and may profit by opportunities to invest capital and to utilize technical and managerial
ability abroad. In the absence of excessive trade barriers and excessive government control of trade, these requirements and opportunities can, however, be more profitably secured by peaceful bargaining than by conquest. The rise of industrialism, capitalism, and freedom of enterprise in Europe was probably both the cause and the
effect of that continent's century of relative peace between the
battles of Waterloo and the Marne.

Feudalism, by subordinating economy to polity, increased the
warlikeness of earlier agricultural economies. Excessive government
planning has often tended to subordinate economic welfare to political power and to increase the warlikeness of states with an industrial economy. Socialistic economies have produced the most warlike states of history.

d) Progressiveness.—Modern war appears to have had an adverse
influence on social progress. Military preparedness and war have
made for rigidity, unadaptiveness, and traditionalism. Progressive
and dynamic states, however, which continually strive to arouse the
society as a whole to a consciousness of national values and to adapt
social institutions and activities in order to realize these values, are
more warlike than traditional and static states, which leave the definition and realization of social values to the interplay of the ideas
and propagandas of individuals and private organizations. The effort
of government rapidly to change society tends to produce internal
dissensions, and to eliminate these dissensions, governments often
resort to regimentation of opinion and the creation of scapegoats.
The psychological mechanisms of repression, displacement, and
projection are made to serve the purposes of government. In so far as
they are successful, the state becomes socialistic in economy and
aggressive in foreign policy.

e) Integration.—Youth, cultural homogeneity, national economic
planning, and dynamism are characteristics which often make for
war and are all characteristics upon which totalitarian societies have
prided themselves. It is therefore not surprising that totalitarianism

has made for war. The liberal society, which confines the functions of government to the maintenance of law and order and which recognizes the autonomy of national minorities, churches, economic enterprises, and educational, research, and publicity organizations, may be expected to command a less united loyalty among its population and a less perfect administrative machine with which to mobilize its resources for war than will the totalitarian society whose people are taught to believe that the state is the supreme value. The leaders of liberal states have, therefore, taken a longer time to prepare to fight, even after that decision has been reached, and, in the early stages of fighting, have been less efficient. Furthermore, the liberal society, because of its liberalism, presents an opportunity for propagandas of disintegration and, because of its unpreparedness, presents a tempting target for attack by aggressive neighbors. These relationships were illustrated by the aggressions of the totalitarian states after 1931 and suggest that either excessive or inadequate social integration within a state presents dangers for peace. Furthermore, the degree of integration may be less important in this respect than the method by which integration is achieved.

Changes in the social constitution of states may in part account for the changes in the warlikeness of a civilization during its history. The youthfulness and dynamism of states make for warlikeness in the heroic age. The increasing homogeneity and integration of states in the time of trouble sustain their warlikeness in spite of increasing age. In the period of the universal state, however, the increasing age of states as well as the developing uniformity and integration of the civilization as a whole makes for peace. The peacefulness of aged states in the period of decline renders the civilization as a whole vulnerable to attack from outside, and defensive wars become frequent.

5. VULNERABILITY AND WAR

The internal constitution of states exercises less influence upon their foreign policy than do the external conditions with which they are faced. The state must adjust to conditions even at the expense of its theories, or it may cease to exist. Among these conditions are the relative power of the state and its military and economic vulnerability.

   a) *Relative Power.*—There seems to have been a positive correlation between the warlikeness of a state and its relative power. The "great powers" in all periods of history have been the most frequently at war, and the small states have been the most peaceful. The great powers not only have engaged in balance-of-power wars among themselves but also have engaged in frequent small wars and military expeditions against lesser states and semicivilized communities. Some countries large in area and population, such as India and China, did not participate as subjects in the balance of power during the modern period. They were not, however, "great powers" in the political sense. France and the Hapsburg Empire were, through most of that period, the greatest of the powers, about which the balance of power revolved, and they were most frequently at war. Next in rank of warlikeness during the period from 1480 to 1940 were Great Britain, Russia, Prussia, Spain, and Turkey; the smaller powers—Holland, Sweden, Denmark, and Norway—were comparatively rarely at war.

This relationship is partly due to the fact that the more powerful the state, the more it thinks that it can win a given war, although statistics show that, because of the functioning of the balance of power, the state best armed at the beginning of a war has been defeated in a majority of cases. A weak state, if it fights, is likely to be opposed by a more powerful state, and consequently, it is not likely to be at war unless attacked. When such attacks have occurred, the small state has sometimes failed to survive; but frequently, it has been protected by the functioning of the balance of power, and in general wars, both belligerents may find it to their advantage to maintain the neutrality of small neighbors in order to protect that portion of their frontiers by a less costly means than military defense. The more important reason for the excessive belligerency of great powers, however, lies in the structure of the balance of power, which practically assures that all great powers will enter wars which threaten the balance in order to preserve it, a responsibility which the smaller states do not have.

b) *Strategic Vulnerability* tends to involve the state affected in war. States with widely scattered territories are more difficult to defend at all points than are states with compact territory, and at the same time they are under pressure to expand at more points in order to achieve more satisfactory strategic boundaries. The states with the most concentrated territory, such as Switzerland, Sweden, and Norway, have been least at war in the modern period.

Natural barriers to invasion probably make the state so protected less likely to be involved in war, though the influence of invulnerability at home in creating civil strife and an aggressive spirit cannot be overlooked. Japan, England, and the United States, separated by oceans or straits from the other great powers, were less at war than the states of continental Europe during the modern period, particularly in its early portion. The trend of modern invention has made natural barriers a decreasing source of security, although even in earlier times Hannibal and Napoleon crossed the Alps. Wide oceans did not preserve the empires of the Aztecs and the Incas and did not protect Japan, India, and China from military attacks by the Western powers. It should also be noticed that natural geographic frontiers may be more favorable to the state on one side than to that on the other. The frontier itself has sometimes become a bone of contention, as has, for example, the Tyrolean frontier between Italy and Austria and the Rhine frontier between Germany and France.

There is a tendency for communities so isolated or protected that they fear no hostile neighbors to break up even when they do not suffer from serious cultural heterogeneity. The population of certain of the small Polynesian Islands has split into two quarreling groups. Great Britain was divided by civil war following its withdrawal from Europe after the Hundred Years' War. During its seclusion in the seventeenth and eighteenth centuries Japan was broken into numerous feudal baronies continually fighting one another. China during the period when it was comparatively isolated was frequently the victim of civil war. Even the United States, which after the Napoleonic Wars was in comparative isolation, broke into two halves and had the bloody Civil War. Thus the immunity from foreign war, arising from the effect of strategic invulnerability, may increase the danger of civil war. This danger tends to be compensated by an aggressive spirit, prone to indulge in foreign war as a diversion from domestic ills.

c) The *Economic Vulnerability* of a state requires it to maintain economic contacts with foreign territory in times of both peace and neutrality. Among primitive peoples warlikeness was found to be closely correlated with the number of contacts; among civilized peoples the relation is less clear. Economic contacts are certain to involve occasional friction and may lead to war in the absence of international machinery of adjustment, but civilization implies superior capacity to make such adjustments. It has, in fact, been suggested that the shrinking of the world through rapid transport and communication and the increase of economic interdependence among the states have made for peace. This may be true in the long run, but the first effect of increasing economic and cultural contact among states has been to augment the probability of war.

Economic self-sufficiency has therefore been urged in the interest of peace. This policy, however, also has within it the seeds of war, because a state which has been engaged in extensive trade, in seeking to increase its economic invulnerability by trade barriers, is certain to injure others thereby deprived of markets. Efforts at economic invulnerability or autarchy may also cause domestic discontent because of the lowering of standards of living and the disorganization of many economic enterprises, thus creating a demand for territorial expansion. A vicious circle of autarchy, conquest, and deteriorating economies is thus set up. The expansiveness of Germany, Italy, and Japan in the interwar period arose in part from their efforts toward economic self-sufficiency as measures of military defense. The hardships which the peoples of certain countries had suffered from blockade during World War I and the world economic disorganization afterward made such policies popular though they were pursued at great cost.

Economic self-sufficiency, even when not developed out of a condition of wider trade, may be unfavorable to peace. Among dynamic cultures it tends to create a spirit of conquest, as illustrated by the efforts of the relatively self-sufficient England to conquer France and of Japan to conquer Korea in the late Middle Ages. Perhaps because of the very invulnerability of the population to economic attack, self-sufficiency tends to create overconfidence in the state's invincibility. Economic and cultural self-sufficiency also tend to produce a divergence between the cultural and economic standards and the military methods of the isolated state and its neighbors. When contacts do occur, there is likely to be both the opportunity and the urge for conquest from one side or the other because of these differentials. On the other hand, countries in continual contact will copy each other's military advances, and consequently, if not too disparate in size, a speedy victory for either side becomes unlikely.

Self-sufficiency, if achieved only at the price of destroying a complicated world trading system, undoubtedly makes for war, as illustrated by the anarchy of the fifth and twentieth centuries following the breakup of the Mediterranean trading system established during the *Pax Romana* and of the world trading system established under the *Pax Britannica*. The international problems of the economically vulnerable states are difficult; but if guided by commercial and financial minds, such states are more likely to recognize the economic advantages of conciliating or adjudicating controversies than are states guided by the military and land-centered minds usually influential in self-sufficient states. Free trade, encouraging economic interdependence of states and commercial rather than military values among people, was considered the key to peace by economic liberals such as Richard Cobden and Cordell Hull.

The conditions of modern civilization have certainly tended to become progressively less favorable to an international political system based exclusively upon the balance of power. The advent of democracy and constitutionalism has made it extremely difficult for governments to take action devoted solely to rectifying the balance of power. The development of nationalism, liberalism, and interdependence flowing from international commerce and communication has diverted much influential opinion from problems of power to problems of welfare. New military inventions, the rise of industrialism, the rise of literacy, and the sentiment of nationality have augmented the importance of economic, diplomatic, and propaganda activities as instruments of war and of policy and have made calculation of the relative power of states less easy. The development of international law and the network of treaties and of international organizations have created moral and customary barriers to free action on the basis of power politics. The increasing destructiveness of hostilities and the rapidity with which they may spread have created hesitancy to resort to war even when necessary to restore equilibrium. The economic objectives of states have become less capable of advancement by war than was the case in a less interdependent and less industrialized world.

These factors make against the dominance of balance-of-power politics in international affairs and tend to develop international law, international organization, and world public opinion as new guides for foreign policy. The latter institutions, however, have not as yet developed sufficiently to give a sense of security to the satisfied and confidence in the possibility of change to the dissatisfied. As a consequence, the balance of power has continued as the basis of international relations. So long as it is the basis, states will attempt to conform their constitutions and policies to its exigencies, and war will be regarded as the final arbiter. Diplomacy, economic pressure, propaganda, litigation, consultation, and investigation may be utilized for obtaining particular objectives, but the problem of which state is most powerful will continue to dominate until the radical change wrought by the missile with a nuclear warhead is more widely appreciated.

CHAPTER IX

LAW AND VIOLENCE

1. LAW, WAR, AND PEACE

Among the hypotheses suggested to explain the recurrence of war
was the inadequacy of the sources and sanctions of international law
continually to keep that law an effective analysis of the changing
interests of states and the changing values of humanity. Although
certain branches of law have as their end the definition and regulation of permissible violence (laws of war and of military occupation) and the organization of collective violence (military law and
martial law) and all systems of law tolerate certain kinds of violence
under certain circumstances (right of self-defense and police action), the normal end of law—the maintenance of order and justice—
is hostile to violence. When Cicero wrote, *Inter arma silent legis*,
he emphasized this generally accepted antithesis between law and
violence. Political philosophers have emphasized the same antithesis
when they have posited the social contract, establishing law and
society, as the process of man's emancipation from the state of

nature, which, if not a perpetual *bellum omnium contra omnes,* was at least a condition in which each man judged his own case, and violence was frequent. Violence has been considered synonymous with disorder and injustice, both of which are eliminated in the ideal legal community. The concept of war has included both law and violence. The same is true of the concept of peace, which, according to Augustine, is "tranquillity in order." But order has never been maintained in practice without occasional violence against the evil doer. Peace may, then, be defined as the condition of a community in which order and justice prevail, internally among its members and externally in its relations with other communities.

It is the function of law to produce this condition—of municipal law to maintain internal peace in each state and of international law to maintain external peace among all states. As crime, rebellion, and insurrection are evidences of the imperfection of municipal law, so interventions, reprisals, and wars are evidences of the imperfection of international law. This proposition is not denied by the existence of abnormal law to regulate these conditions. Just as remedial medicine is necessary to rectify the imperfections of preventive medicine, so abnormal law is necessary to remedy the imperfections of normal law, or, if not to remedy them, at least to ameliorate their resulting evils.

2. WAR AND THE DUEL

The legal position of war was discussed by jurists and philosophers of the classic civilizations of Greece and Rome and of the Western Christian civilization of the Middle Ages as well as by writers of other civilizations. Attempts were made to answer such questions as: Who can wage war, and against whom? When, where, and under what circumstances is resort to war justifiable? How should war be begun and conducted? What attitudes may non-participants take toward a war?

At different times and places jurists and philosophers have likened war to an act of self-defense, to the execution of a judgment, to a political measure, to a crime, and to a duel or judicial combat. Many Renaissance writers discussed the legal propriety of war, and all these analogies were used. It appears, however, that the dominant idea of war at that time was that of a duel between princes.

The words *bellum* and *duellum* have the same origin (from the word *duo*, "two"), and in the Middle Ages the two were often treated together, as by Legnano. The dominant medieval opinion, however, treated war as a proper measure of sovereign authority for promoting justice and remedying wrong. War could be just only on one side, and that side was normally the one acting under superior authority of God, the pope, or the emperor. Although war between equals was discussed, equals were necessarily, under the prevailing theory of a united Christendom, subject to some superior authority. One belligerent, if not both, would presumably be disobeying this authority; if not his direct command, at least the divine law or the law of nature which he sanctioned.

The doctrines of the equality of sovereignties and the absolutism of monarchs had, however, been developing in the later Middle Ages, and by the Renaissance it had reached such a stage that political and juristic writers (who, with the wider development of literacy and the press, ceased to be exclusively ecclesiastics) took cognizance of it and presented war as a combat between equal princes. Even churchmen like Victoria and Molina, who clung to the medieval tradition that war could be just only on one side, modified this tradition in fact by the doctrine of "invincible ignorance." This doctrine held that if the side in the wrong remained ignorant of the unjustness of its cause after due study, the war should be treated as just on both sides. Churchmen also began to consider honor a cause of war. Lay jurists like Gentili and Grotius found these circumlocutions unnecessary and simply said that in doubtful cases "neither can be called unjust." Consistent with the analogy of war to the duel, neutrality, which had been inconsistent with the medieval conception of society, began to take root, and lay writers like Machiavelli and Hobbes, as well as reformers like Luther, perceived war as the natural consequence of controversy between equals subject to no common authority and so in a state of nature.

More significant of the relation of war to the duel than this logical similarity was the assumption that war was a personal affair of the prince. He alone could initiate war (except perhaps in defense). And though he ought to consult the grandees of the state, he had discretion to reject their advice. Although the medieval customs by which princes had sometimes actually settled international controversies by a personal duel (and usually instituted war by sending a defiance by herald in the manner of a challenge) had fallen into abeyance in the Renaissance, these practices showed that modern war and the duel were one and the same in origin, though the two institutions had diverged. Only persons of a certain legal capacity could fight duels. In the Middle Ages the king was only *primus inter pares,* but in the Renaissance the rise of monarchy placed him in a class by himself. Gentlemen and nobles continued to fight duels, but only kings could fight wars. The fact that duels were usually fought personally, although substitutes might be used, and that wars were usually fought with armies accentuated the developing difference between the two institutions, especially as armies became more formidable in size.

The rise of the corporate theory of the state, with its accompaniments of constitutionalism, nationality, and democracy, led to a conception of war as a means to political or economic ends or as a spontaneous manifestation of cultural or biological urges and obscured its genetic relationship to the duel. Nevertheless, their homology makes the history of the duel still instructive in explaining war. Many of the curious conventions of the duel flow from psychological factors which are present also in war.

Historians of the duel recognize three forms—the state duel, the judical combat, and the duel of honor. In the first a champion fights in behalf of the state. It is thus a war in miniature. In the judicial combat or trial by battle the duel becomes a prescribed procedure under state authority to prove guilt or innocence. In the duel of honor gentlemen defend their honor by a fight under conditions prescribed by practice and convention.

These forms of the duel are related to one another; in fact, they developed, with some overlapping, in the sequence named, and they are all related to war. War was a state duel in that the army fought as the representative of the prince. It was a trial by combat in that it decided the justice of the cause under the regulation of international law. It was a duel of honor in that "national honor" was and continues one of its main causes.

Duels are fought in defense of reputation, prestige, or honor. They do not directly concern facts or material injuries. Thus the insult which "gave the lie" (the accusation of falsehood being the accepted slight upon honor) did not necessarily repeat the statement said to be false, nor did it necessarily mention the person accused. The insulter might say, "So-and-so has lied," or he might say, "Whoever said so-and-so lied." There was no argument about the truth of this allegation. The fact that the allegation had been made was an insult or a stain on honor, and if the person thus insulted did not issue a challenge he would cease to be a gentleman, a circumstance which might carry with it grave disadvantages. The fact that the issue was not on a question of fact but about words, which each claimed sullied his honor, meant that each could be defending his honor. The problem which troubled the medieval writers on war—How could both sides be acting in defense of justice?—could not arise. In a duel of honor there was no issue which could be submitted to any form of adjudication. The only defense against an insult was a willingness to risk one's life in order to prove one's honor.

In situations where the administration of justice was inadequate, an individual's freedom from harassment depended in no small measure upon his reputation for avenging insult. Once honor was gone, reputation was gone; no one would fear to commit trespasses against the dishonored, who would rapidly sink in the world.

The duel of honor was, therefore, in reality a mode of defending material interests when there was no established code of religion, morality, law, or custom adequate to mobilize social authority. Reputation, prestige, and honor were, under such conditions, the practical road to security and advancement.

The conditions favorable to the duel existed with the breakup of the traditional social controls in the late Middle Ages. Similar conditions have led to fights as a protection from bullying among small boys, to warfare among primitive tribes, to rapid gunplay among cowpunchers of the early American West, to duels among medieval monarchs, to feuds, vendettas, lynch law, and vigilantism, and to wars among modern states, especially when governed by despotic regimes contemptuous of international law.

There is a tendency for the duel of honor to develop similar conventions in all these diverse circumstances. Each of the parties is motivated by two strong but antagonistic drives—a desire to preserve a reputation for courage in order that no one will risk a trespass and a desire to preserve life and limb. To reconcile these opposing motives, there is a tendency for conventions to develop which will make it possible for each party to say the other is the coward without himself actually fighting. In the Italian duel of the sixteenth

century experts discussed such points as: What form of words constituted a *mentita,* or insult? What form of words constituted a
challenge?

The analogy of such practices to the diplomatic parleying of today
is obvious. Each of the states wishes to keep its reputation for fighting, and through that reputation to acquire territory or hold what it
has without actually fighting. Each desires to impress the world with
its willingness to fight if an attack is made, but at the same time each
tries to avoid making an actual challenge which might precipitate
the fight. The process is illustrated in the exchange of insults between Germany and Poland in 1939 and in the effort of nuclear
powers to make their threats of "massive retaliation" so credible that
there will be no war.

Private dueling was gradually eliminated by the rise of the bourgeois temperament, which preferred litigation in court to fighting,
recognized the acquisition of wealth as the appropriate means to influence and prestige, and regarded killing as immoral, and by the
contemporaneous rise of more efficient government, providing adequate courts and police. In the same way, reliance upon national
honor, prestige, and military reputation as instruments of national
policy might gradually subside, if statesmen developed a trading
spirit, humanistic morals, and efficient international institutions.
Until such attitudes and institutions are established, perhaps because of universal fear of nuclear war and more abundant international communication, states will place a high value upon military
reputation as an essential means for preserving national existence
and will find it difficult to maintain that reputation without occasionally risking war.

The duel has at times been a legal institution. But it was characteristic of the duel of honor that it flourished most when it was in
principle illegal. In this respect also war and threats of war today
resemble the duel. The duel of honor appears at a stage in the
development of a legal community in which principle is ahead of
institutional realization. This often occurs during periods in which
ancient institutions have crumbled or in which people have carried
advanced conceptions into a backward environment. The modern
state system, with legal conceptions borrowed from advanced systems of municipal law beyond the possibility of realization in the
backward state of international organization, presents a parallel

situation. Men and nations in such circumstances seek prestige to assure respect for rights which the community lacks capacity to protect. The struggle for prestige is an advance in law above the struggle for power. The struggle for legal rights marks a further advance but is dependent upon a more completely organized society. In international relations, however, the distinction among rights, prestige, and power is blurred. The sovereign rights of the state, as a corporate body, were maintained by its prestige and by the power of its government to enforce law internally and to defend its frontiers. The war for honor, the war for rights, and the war for power were all covered by the phrase "reason of state."

3 . WAR AND ETHICS

The breakdown of medieval law and religion and the rise of powerful monarchs created a situation in which both gentlemen and princes maintained their positions by defending honor with the sword. The literature and ideas of the Middle Ages were, however, carried on in the writings of Victoria and other ecclesiastical jurists and continued to influence the position of war in the developing international law. In fact, to the medieval tradition of "just war" was added the pacifistic attitude characteristic of the Stoics and the early Christians which had been revived in the study of classical sources and early Christian literature by Erasmus and other Renaissance writers.

Both of these traditions, centering attention upon human or Christian ideals of individual welfare, tended to ignore political interests, princely prerogatives, prestige, and honor. They classified war from its outstanding manifestation, the maiming, slaughter, and impoverishment of human beings, and they appraised it ethically according to the sixth commandment and the Sermon on the Mount. To this way of thinking, there was no distinction between state ethics and private ethics. There was just one community—Christendom—which to the Christian was potentially universal. International law was therefore identical with private law, both resting upon "natural law," which reflected human nature. Nature, Grotius pointed out, is the mother of natural law, whose child is the obligation of promises which begot civil society. Consequently, "nature may be considered the great grandmother of municipal law."

The Stoics, early Christians, and Renaissance humanists jumped one of these generations and decided that war was contrary to human nature, thus paving the way to non-resistant pacifism.

The Catholic tradition, initiated by Augustine in the fourth century, qualified this position by a more realistic consideration of the need of police in actual human societies and the need of defending Christendom from its external enemies. It asserted that war was permissible to promote peace, that is, order and justice, provided the war was initiated by a proper authority and provided that authority had found peaceful procedures inadequate in the situation and had assured itself that the injustices arising from the war would not be greater than the injustices which the war was to remedy. Further elaboration made it clear that war would not promote peace unless there was a "just cause," usually limited to defense from aggression, remedy of a wrong, and punishment of a crime, and unless this cause constituted the actual motive, not a mere pretext, of the initiating authority. This thesis was supported by biblical exegesis to show that the New Testament tolerated just war and permitted soldiers and citizens to give the ruler the benefit of the doubt in respect to a particular war. This carefully balanced theory of war figured in the classical writings on international law, continued as the official theory of the Catholic church, and influenced modern international law, different as are its assumptions from those of the theory which assimilated war to the duel of honor.

The Catholic theory was adapted to the religion-dominated medieval Christendom, which lacked strong political organization and often degenerated into feudal anarchy. This theory was, however, difficult to apply in the post-Renaissance world of powerful princes, claiming sovereign authority to organize their states internally on national lines. With the realization of a world economic and cultural community in the nineteenth and twentieth centuries, the Catholic theory attracted more attention. Jurists, however, tended to develop municipal-law analogies rather than to revert to the medieval theory when considering the problem of checking resort to war. Positive law and ethics had become too much separated to be easily drawn together, although the Catholic theory of just war and Renaissance pacifism were a continuous reminder to international lawyers that law and ethics can never be wholly separated. Both derive from human needs and interests rather than from the accidents of sovereignty. Appreciating that the law is eventually for man, not man for the law, Pope John XXIII in his encyclical of May, 1963, ignored the age-old Catholic tradition and questioned whether any war could be "just" in the nuclear age.

4. WAR AND PRIVATE-LAW ANALOGIES

Modern international law is a primitive system of law. It lacks the wealth of sources, the precision of propositions, and the efficiency of procedures which characterize the municipal law of modern states. Its advocates, usually schooled in some system of municipal law, both because of habits of thought and because of the opportunity offered, tend to develop their subject by analogy to the rules of those more mature systems. Among the classical writers Roman law was an important source, but more recently jurists have drawn from contemporary systems and particularly from those rules or principles found to be common to most of them. This practice was indulged in by the "naturalists," for whom international law was fundamentally a law for individuals and the state was only an instrument for the benefit of its citizens. The "positivists" also used such analogies, although they considered international law as law only between states, which were no longer sovereign princes but sovereign corporations with complex constitutions.

Many international lawyers questioned the analogy between the individual, who could be physically brought to court, jailed, or, if need be, executed, and the state, to which none of these treatments could be applied. There have, consequently, been many warnings about the careless application of private-law analogies, but the practice continues. Bilateral treaties are considered analogous to contracts and multilateral treaties to legislation. Protectorates and mandates are considered analogous to the relationships of guardianship, agency, and trust. State domain is likened to real property, states to natural persons, and international unions to corporations. It is not surprising, therefore, that the familiar legal allocations of internal violence to the categories of crime, insurrection, defense, and police should have been utilized in dealing with war. The League of Nations' Covenant, the Pact of Paris, and other similar treaties accepted this analogy, as the author commented in 1930:

What has heretofore been called an act of war became, under the Pact, either a criminal breach of the peace, an act of self-defense, or an act of international police. As the legal consequences of each would be very different, the situation of states engaged in these different acts should no longer be characterized by the common term, war. Similarly what has heretofore been called neutrality becomes the situation of states, not actively engaged in illegal violence or its suppression, but bound, to paraphrase Grotius, "to do nothing to strengthen the side" of the pact-breaker "or which may hinder the movements" of his adversary.

The influence of this analogy was found in the numerous suggestions for revision or elimination of the idea of neutrality which is hardly analogous to any situation recognized in municipal law. The analogy of the nonbelligerent to the witness of a crime was developed in the Budapest Articles of Interpretation of the Pact of Paris and in the Harvard research draft on the rights and duties of states in case of aggression. Suggestions were made that nonbelligerents should be permitted to participate in a primitive form of collective security analogous to the "hue and cry" or the vigilantes, even in the absence of international organization. The analogy has also figured in the continuous effort to define aggression in the United Nations.

Although it is clear that ideas of justice cannot be reconciled with legal toleration of acts of war found, by procedures accepted by all the states involved, to have been in violation of international obligation, it is also clear that the problem of controlling states by international law is very different from the problem of controlling individuals by municipal law. The units are proportionately larger, and coercion may lead to the initiation of war rather than to an effective exercise of police. National sentiment hampers the creation of a unified international police force; punishment of guilty nations by fine, indemnities, or losses of territories are likely to undermine the economic structure of society to the injury of all nations; and the moral responsibility cannot usually be attributed to one nation and almost never to the entire population of a nation.

The fact that the problem of control is different does not mean that it is incapable of solution. Nor does the lack of analogy in respect to the sanctions of a rule necessarily vitiate the analogy in respect to the rule. On such a theory the analogy between the interpretation of treaties and of written instruments of municipal law would have to be denied. The difficulty of enforcing effective sanctions against states has, however, induced many to consider whether individuals and public officials, against whom sanctions could more easily be enforced, should not be subjects of international law. Such a liability, long enforced against captured enemy persons accused of breaking the laws of war, was extended by the war crimes trials after World War II to high government officials responsible for initiating aggressive war. Confederations, if they survive, usually develop a direct relationship between the individual and the central government, as did the United States in the more perfect union of 1789. The United Nations, in principle, is committed to protect human rights, as is the western European community in practice by the establishment of the Court of Human Rights with jurisdiction over violations of human rights by the member states.

Such a development would be a reversion to the ethical theory of the Middle Ages, which tended to reduce the states and their sovereigns from entities of pre-eminent value in themselves to the position of administrative conveniences relating the individual to mankind.

5. WAR AND MODERN INTERNATIONAL LAW

What is the position of war in modern international law? No categorical answer can be given. International law is a dynamic system, and a careful examination of its sources—treaties, customs, general principles, and the authority of jurists and judges—give different answers if examined in successive decades of the twentieth century.

In 1924 the writer examined the changes in the concepts of war since the Middle Ages with the conclusion:

Under present international law "acts of war" are illegal unless committed in time of war or other extraordinary necessity, but the transition from a state of peace to a "state of war" is neither legal nor illegal. A state of war is regarded as an event, the origin of which is outside of international law although that law prescribes rules for its conduct differing from those prevailing in time of peace. The reason for this conception, different from that of antiquity and the Middle Ages, was found in the complexity of the causes of war in the present state of international relations, in the difficulty of locating responsibility in the present regime of constitutional government, and in the prevalence of the scientific habit of attributing occurrences to natural causes rather than to design. It was recalled, however, that the problem of eliminating war has gained in importance while the possibility of solving it through the application of law has improved with the development of jural science. Thus efforts have been made to eliminate war (1) by defining the responsibility for bringing on a state of war, (2) by defining justifiable self-defense, and (3) by providing sanctions for enforcement.

Ten years later the writer examined the concept of aggression, then growing into jural usage, with the following conclusions:

A state which is under an obligation not to resort to force, which is applying force against another state, and which refuses to accept an armistice proposed in accordance with the procedure which it has accepted to implement its no-force obligation, is an aggressor, and may be subjected to preventive, deterrent or remedial measures by other states bound by that obligation. There cannot be an aggressor in the legal sense unless there is an antecedent obligation not to resort to force. Doubtless there are some such obligations in customary international law; thus the pre-war textbooks define limitations upon the resort to intervention and reprisal, upon the use of force during a state of war, and even upon the initiation of a state of war, although during the nineteenth century the latter was considered a moral rather than a legal question. Treaties, however, especially post-war treaties, have imposed extensive obligations not to resort to force, and the conception of aggression has developed mainly in connection with the interpretation and application of these treaties, of which the League of Nations Covenant and the Pact of Paris have been the most widely ratified. . . .

Even if a state violates an obligation not to resort to force, it would still not be an aggressor under the definition proposed unless the law draws some practical consequences therefrom. Several official texts have described aggressive war as a crime, but the definition here proposed does not demand that the consequence of aggression be of the nature of criminal liability. The measures consequent upon aggression may be preventive, deterrent, or remedial rather than punitive, and their application may be discretionary, rather than obligatory with other states, but unless there is some sanction, some legal consequence of the breach, the breaker is not, under this definition, an aggressor.

While it is believed that the test of aggression here proposed conforms to the standards of practicability and justice, it cannot be applied satisfactorily without discretion. While it is as automatic as may be in the varied conditions of international relations, a test applicable with mechanical precision cannot be expected. The body proposing the armistice cannot merely order the parties to stop fighting. It must propose a line of separation, provide a commission for observing the withdrawal of troops behind the line, and act rapidly, always with due consideration to the military problems of transport and terrain, in determining the period necessary for withdrawal. While the line of battle at the time would probably have to be given primary consideration, various tests of aggression should be in mind in formulating the terms of the armistice. What was the respective attitude of the parties toward pacific settlement of the dispute before hostilities began? Who first violated the *de facto* frontier? Which was best prepared with an offensive strategy? Such questions, if easily answered, might be given weight in determining the terms of the armistice. It is believed, however, that the basic tests of aggression must be the attitude and behavior of the parties in response to the armistice after it is presented.

These principles prohibiting aggression, establishing criteria for determining the aggressor, and permitting all states to discriminate against the aggressor were applied in a number of cases, including the Greco-Bulgarian dispute (1925), the Chaco War (1929), and the Manchurian (1931), Ethiopian (1935), and Chinese (1937) hostilities, but in the most serious of these cases sanctions proved inadequate. In the aggressions of Germany in Austria (1939) and Czechoslovakia (1939), no effort was made to apply the League Covenant. The German aggression in Danzig and in Poland was made the occasion for war by Great Britain and France but less on the basis of general principles of law than on the basis of special guaranties given to Poland on balance-of-power principles. With the further German and Italian aggressions in 1940, most states in a condition to exercise independent judgment denounced these states as aggressors, and the United States justified its discriminatory action in favor of Great Britain on this ground.

After Pearl Harbor most states signed the "Declaration by United Nations" condemning the axis powers as aggressors, a position confirmed by the war crimes trials after the war. These judgments and the United Nations Charter, which prohibited the threat or use of force in international relations with certain exceptions, clarified the "new international law," but a precise definition of aggression was still lacking. Referring to the efforts of United Nations organs to provide such a definition, the writer stated in 1956:

These discussions have indicated that different definitions are necessary according to whether the problem is that of preventing breaches of the peace, of restoring peace, of determining international responsibility for breaches of the peace, or for determining the criminal responsibility of individuals for crimes against peace. . . . Debate in the United Nations has concerned the question whether a precise definition of aggression is possible or desirable; whether, if desirable, a definition of aggression should be by concrete enumeration of acts of aggression or by abstract definition; whether aggression should include only illegal threats or uses of armed force or should also include "indirect aggression" by economic pressure or ideological infiltration; whether aggression should be confined to international action or should in some circumstances include civil strife; whether aggression should include only actual aggression or threats of aggression or should include also "potential threats of aggression"; and whether the aggressor was always a state or might also be a government or a band of individuals.

The conclusion was that, as used in the Charter, an act of aggression

is the threat or use of armed force across an internationally recognized frontier, for which a government, *de facto* or *de jure*, is responsible because of act or negligence, unless justified by necessity for individual or collective self-defense, by the authority of the United Nations to maintain international peace and security, or by consent of the state within whose territory armed force is being used.

While hostilities in Iran (1946), Greece (1947), Indonesia (1947), Kashmir (1948), Palestine (1949), Suez (1956), Lebanon (1958), and the Congo (1961) were ended by provisional or policing measures without need to determine the aggressor, the United Nations found North Korea and Communist China aggressors in the Korean hostilities (1950) and the Soviet Union an aggressor in the Hungarian hostilities (1956). The United Nations took no action in the hostilities in southeast Asia, in the Strait of Formosa after 1954, in Algeria after 1955, in Goa in 1961, in the Chino-Indian incident of 1962, or in the Cuba quarantine of 1962 because they were regarded as civil strife, as colonial emancipation, as within the competence of other agencies, as involving non-members, or as susceptible of diplomatic settlement.

The developments from 1920 to 1964 suggest that the customary international law, tolerating and regulating resort to war, which had existed before 1914 had received important modifications during this period by treaty, juristic interpretation, and diplomatic practice, influenced by ethical considerations and private-law analogies. International law had begun to differentiate the conceptions of aggression, defense, and sanction, all of which may involve the use of armed force, from the conception of war and had differentiated the conceptions of peaceful procedures and peaceful change from the conceptions of intervention and aggression.

It is also clear that these new conceptions had not worked themselves into the minds of all jurists, much less of all statesmen. They had not acquired the sanction of custom, their logical ramifications had not been fully developed, nor were there institutions capable of enforcing them. Although international law struggled to improve its sanctions by clarification of its rules, by procedures of adjudication, by education of public opinion, and by focusing world opinion upon threats to its principles, it did not during this period create an expectation that its rules would be observed and enforced. Statesmen were convinced that the state, fortified by military power and prestige, had a superior status to the state fortified by legal powers and rights. The latter were of value but not of sufficient value to supersede the former.

CHAPTER X

SOVEREIGNTY AND WAR

Modern international law took form in the sixteenth century while princes were claiming and in some cases maintaining a monopoly of violence in territories larger than the feudal domains and smaller than Christendom. The political theory was developing that princes could build stable states by using force and fraud. The ethical assumption was being made that the state society was superior to the religious community. The economic doctrine was being applied that commerce should be regulated in the interest of state power. These conditions and doctrines conspired to create sovereignty as a developing fact and an inchoate idea. The distinguishing feature of international law was its assertion of the sole competence of the sovereign state to make war. According to Arnold Brecht, "there is a cause of wars between sovereign states that stands above all others—the fact that there are sovereign states, and a very great many of them." Perhaps it would be no less accurate to attribute war to the fact that there are no sovereign states in the political sense but a great many that want to be.

1. THE CONCEPTION OF SOVEREIGNTY

Sovereignty in the legal sense has been defined as "the status of an entity subject to international law and superior to municipal law." By this definition sovereignty could be ascribed in 1964 to over one hundred twenty of the thousands of political organizations in the world. They were the source of authority for negotiating treaties, for recognizing new conditions, for submitting international disputes to adjudication or conciliation, and for initiating war, as well as for enacting, applying, and enforcing municipal law. The definition does not, however, throw much light upon the characteristics of sovereignty, except to persons familiar with law, both international and municipal. As international law and the various systems of municipal law are not necessarily consistent with each other, the characteristics of a particular sovereign entity may seem very different from one or the other point of view.

As each sovereign entity can modify its own municipal law merely by observing the proper internal procedures, it can give itself whatever rights and powers it pleases under that law. But, viewed from within, municipal law is the only law there is. Rules of international law are not law unless "adopted," and rules of other systems of municipal law are not law unless recognized. From the point of view of municipal law, therefore, each sovereign is omnipotent in the jural universe.

On the other hand, from the standpoint of international law, each sovereign is bound by law, and none can, on its own authority, change it. Furthermore, different sovereigns have different rights under treaties, and some have been more limited than others with respect to their capacities to acquire rights, thus creating variations in status. Furthermore, as international law is continually developing through treaty, customs, and juristic analysis, the sphere within which the normal sovereign entity may act freely is suffering continual modification. Thus, from the international-law point of view, sovereignty is limited by law, and the scope of these limitations has varied in time and place.

In the sixteenth century Bodin defined sovereignty as "the supreme power over citizens and subjects unrestrained by law." Grotius defined it as "that power whose acts . . . may not be made void by the acts of any other human will." Bodin conceived of sovereignty as a relation between a personal ruler and his subjects and gave only casual attention to the relation of such rulers *inter se*. Grotius gave detailed attention to those relationships but thought of them as relationships of individual monarchs. Both were aware of the medieval tradition whereby society was conceived as an organic hierarchy of governing individuals. They modified this conception in the light of changing conditions by giving extraordinary emphasis to one stage in the hierarchy which they denominated "sovereignty."

In the Middle Ages equal importance had been attached to each of the estates, lordships, and ecclesiastical titles existing in the feudal and religious hierarchy from the vassal or the priest up to God, who was the Supreme Lord and ruled on earth through the emperor, either by direct authority or by way of his vicar, the pope. The Renaissance writers emphasized one step in this hierarchy as of supreme importance—that from the sovereign prince to the international order. The authorities in the hierarchy below became subject to the prince, and the princes themselves became subject only to natural law and to the law of nations resting on agreement and custom. The gradual secularization of affairs and of thought reduced the influence of the pope and of divine law with respect to temporal government.

This change was important but scarcely more so than the later change which transferred the prerogatives of the prince to the corporate state. This latter change was initiated by the Swiss and Dutch confederations and stimulated by the American and French revolutions and was generally accepted after World War II.

No less important in changing the meaning of sovereignty has been the growth in the objectivity and the scope of international law. The *jus naturale* and *jus gentium*, which theoretically defined the sphere of princes from the international point of view in the seventeenth century, were maintained by few documents, little practice, and no permanent institutions, though they were maintained by the declining supranational estates of clergy, nobility, and merchants. International law today is a relatively precise body of rules, defined in general and particular treaties, judicial precedents, and four centuries of juristic analyses, with established international institutions, capable of making clear its application in particular cases, although not always successful in preventing violation or in applying remedies.

Not only have the meaning and application of sovereignty changed but its locus in the hierarchy of human government has also changed. When Dante wrote his *De monarchia* in the early fourteenth century, he did not use the word "sovereignty," but he was convinced that there could be only one "monarch" in the world, though it is well to remember that he had only the Christian world in mind. Two centuries later Machiavelli located supreme power, or at least competence to strive for it, in the thousands of princes, dukes, counts, and republics continually waging war with one another. Doubtless, the difference in fact between these two periods was not so great as these descriptions of the locus of supreme power suggest. There were warring baronies in the fourteenth century and aspirations for unity in the sixteenth but there was more ground for attributing sovereignty to the many in the later than in the earlier period. This suggests that the conception of sovereignty has always had some relation to the actual organization of political authority. Although this organization is affected by many factors, administrative, economic, and sociological, it has generally been most closely related to military organization and activity.

The theologians and canonists of the Middle Ages inquired whether the wars of princes and barons were "private wars" or "public wars." They all agreed that the *bellum Romanum* or war against the infidel authorized by the pope and conducted in the Crusades was a public war, but with respect to other wars they differed. According to the theory of the time, a public war could only be authorized by a ruler who had legal characteristics which later would have been designated as sovereignty. Some thought the emperor or the pope alone had these characteristics. Others recognized certain kings as having them, but all the medieval writers assumed that the right to make war was prior to the fact of waging war. Because one was fighting, or even because he was fighting successfully, did not prove that he had the right to fight.

The age of science, initially, reversed this order. Instead of inquiring who can declare a just war, writers began to inquire, "Whom does the army obey?" He whom it obeyed actually made war and actually was a sovereign, whatever might be his title or his morals.

In both periods, then, the war power was associated with sovereignty, but in the Middle Ages the war power flowed from the legal title of the monarch. In the Renaissance legal titles flowed from successful warmaking. The anarchic condition of Machiavelli's world, though not wholly eliminated, suffered attrition during the eighteenth and nineteenth centuries, both because the facts of European political life appeared to accord better with the new international law which developed and because that law appeared to develop a certain capacity to control the behavior of rulers.

In the late sixteenth century the juristic conception of sovereignty could be applied to territorial princes with less doubt than in either the fourteenth or the fifteenth centuries. On the one hand, the papacy had lost prestige and the Empire had lost its shadowy titles to land outside of Germany and northern Italy. On the other, many of the minor princelings had been united by force of arms, so that Bodin could "tidy up Europe" by distinguishing a moderate number of sovereigns who deserved the title according to his juristic definition.

After the Thirty Years' War, the problem of locating sovereignties in Europe was simpler still, because formalities of diplomatic intercourse and treaty-making, not to mention the text-writers, had provided criteria. But already complexities were arising because of the spread of the family of nations and of the suggestion that American, oriental, and African rulers were "sovereigns." It was hard to apply a definition based upon conceptions of European law to communities whose municipal law was of a different type and which had never heard of international law as expounded by Victoria, Gentili, and Grotius. New difficulties developed when principles of natural right were invoked to justify oppressed peoples and nationalities in violent secession. Social and economic changes accompanied political changes. The sociological foundations of sovereignty were one thing in illiterate peasant communities subject to autocratic princes, another in states dominated by literate, trading *bourgeoisie* insisting upon constitutionalism.

However, international law and municipal law accommodated themselves to these changes, and in the nineteenth-century world it was not difficult to identify the sovereign states with power to make municipal law, treaties, and war but subject to international law. The latter regulated the intercourse of states in peace and limited the methods of warfare but imposed no precise limits on the initiation of war.

The world after 1918 tended to recognize a new *jus ad bellum* reminiscent of, but different from, the medieval conception of "just war" and to distinguish "public war" or sanctions authorized by the League of Nations from "private war" or aggression not so authorized. Thus there was a tendency for the locus of jural sovereignty to shift from the national state to the world-community, but the rise of totalitarian states, fascist and communist, the self-determination of colonies, and the development of national sentiments in all states moderated this tendency. The international community was, however, in theory, competent to maintain the peaceful coexistence of equally sovereign states.

The function of sovereignty has also changed during the last four centuries. Bodin valued royal sovereignty because it tended toward peace among the nobility within the relatively large areas subject to the "sovereign" and thus promoted order in a period of transition. Grotius valued it because it regularized international relations and centralized responsibility in the interest of peace and the humanization of war in the European community as a whole. Others have valued sovereignty as a dynamic factor, capable of shattering the status quo for the benefit of political power or popular welfare within a group, or of assuring human progress through competition or co-operation among distinctive groups. Voices have not been wanting who, in the interest of churches, labor unions, or other groups, chafing at the restrictions of sovereignty or in the interest of world peace, have decried the conception of sovereignty as obsolete and harmful and have urged that it be abandoned.

What would be the function of sovereignty if applied to nations in a world organized for peace?

Between the primary communities and the world it might be useful to have a definite breach in the continuity of law and organization. Sovereignty, by distinguishing the sources and sanctions of international law from those of municipal law, makes the state the indispensable mediator between the individual and the international community and assures that the two laws shall not become identical, that neither shall dominate over the other, and that between the two an area of flexible political adjustment shall always remain. This might cushion the pressure of the world-community toward unity and uniformity and permit juridical experimentation and differentiation in sections of the human population on their own responsibility and risk without committing or jeopardizing the whole human race.

Diversification in law, and as a result in ideals and standards of all kinds, might thus be perpetuated, permitting continuous progress.

In the legal sense, national sovereignty, by preserving the dualism of international and municipal law and the independence of systems of municipal law, even at the expense of logical harmony and with some danger of juristic conflict, facilitates national legislative experimentation, international competition, and progressive civilization.

2. POLITICAL AND LEGAL SOVEREIGNTY

The problem of reconciling the legal sovereignty of states with peace is the problem of preventing these logical disharmonies and conflicts, useful if kept within bounds, from degenerating into violence and war.

Those who have emphasized the war-producing characteristics of sovereignty have usually ignored international law and have assumed that sovereignty implies unlimited competence to make political decisions, even decisions to expand territory at the expense of others.

If, however, sovereignty is confined to a legal conception, such political decisions are not inevitable. If sovereignty means freedom under international law, the problem of reconciling sovereignty with peace is merely that of adequately developing and enforcing international law. That problem, however, is difficult to solve because international law has often been deduced from alleged attributes of sovereignty. Political sovereignty has controlled international law not only in practice but also in theory, and as a result, international law has supported doctrines which are inconsistent with a legal system.

Sovereignty has been said to imply that the state is not bound by a judgment or a new rule without its express consent, that it is free to resort to war and to remain neutral during the hostilities of others, and that it is free to govern its territory and to pursue its foreign policies subject only to the duty to make reparation to another state injured by its acts or omissions in violation of international obligations. If adjudication is based on consent of the parties and legislation on consent of all states, an effective judicial or legislative system cannot be developed in the community of nations. If the acquisition and destruction of rights by violence and the impartial treatment of the aggressor and the victim are permissible, an effective executive system is impossible. If all government and policy-making are left to states, subject only to remedial responsibility, an effective administrative system is very difficult to devise. In short, these deductions from sovereignty prevent the development of the institutions essential to an effective system of law.

International law has not been as impotent as this theory suggests. In times of tranquillity international adjudication, legislation, execution, and administration have developed from treaty, custom, general principles of law, judicial precedent, and juristic analysis. The idea that sovereignty is something apart from law has, however, prevented a continuous development of such institutions. If it became generally accepted by the nations and the people of the world that sovereignty under law is a broader and more desirable freedom than sovereignty above the law, international law could develop into an effective system. The germ of such an acceptance by many nations was recorded in the League of Nations Covenant, the Pact of Paris, the "optional clause" of the Statute of the Permanent Court of International Justice, and the United Nations Charter; but the germ has not yet matured.

a) Human Rights.—It has been suggested that international law is confronted by a dilemma between two inconsistent aims—to promote human welfare by protecting minimum human rights and to preserve the independence of distinctive nations by protecting state sovereignty.

Although hitherto international law has not in theory recognized "rights of man" subject to its direct protection and has not often accorded a legal personality to individuals entitling them to direct access to international procedures, it has in fact defined and enforced many such rights. The practice of diplomatic protection of nationals abroad has often resulted in the arbitration of claims, the actual, if not the theoretical, beneficiaries of which are individuals. The states have had the dual interest of maintaining sovereignty within their territory and of protecting their nationals abroad. Among states, each of which has a considerable number of nationals abroad, reciprocity exists. Each is ready to qualify its territorial sovereignty by the duty to accord certain legal rights to resident aliens, provided the others do likewise. Furthermore, among states with a similarity of civilization and governmental organization, there has been no great difficulty in defining the minimum legal rights which international law requires each state to accord to resident aliens. An international standard has been defined by extracting the common elements in the national standards. The humanitarian spirit induced several general treaties before World War I according international protection to individuals particularly liable to abuse by their own government, such as natives in colonial areas, members of racial, linguistic, and religious minorities, and laborers.

There have been difficulties, however, in achieving a universal recognition of human rights through this development of the reciprocal interest of states in protecting their nationals because of the great diversity of cultures and standards. While the United Nations, in pursuance of Charter principles, has accepted a "Universal Declaration of Human Rights" as a goal to be achieved, it has not been able to agree on precise covenants of human rights with procedures for enforcement, although the society of western Europe has done so.

Unless all states respect a minimum of human rights, particularly those assuring the individual access to world opinion and world markets, governments will occasionally prostitute national opinion to illegal ambitions, and large-scale violations of law will follow. A world public opinion is the ultimate sanction of international law, and such an opinion cannot develop unless minimum human rights are respected everywhere. But efforts to compel all states to respect individual rights peculiar to a particular culture or ideology are certain to produce allegations of disrespect for the domestic jurisdiction and sovereignty of states which will result in conflicts dangerous to peace.

b) *International Delinquencies.*—The fact that states are bound by international law implies that they are under a responsibility to make suitable reparation to those injured as a result of their violations of that law. Although many theories of responsibility have been developed by jurists and a mass of concrete rules has been developed by diplomatic practice and international adjudication, it has remained difficult to explain how the state, if its powers flow only from law, can commit an act in violation of law. Would not an illegal act be *ultra vires* and consequently attributable not to the state but to the agent? The explanation lies in the fact that the state is a creature of two laws. From the point of view of international law, *de facto* its powers derive from its own municipal law but *de jure* they derive from international law. Acts authorized by municipal law may violate international law. They are not *ultra vires* by the state's constitution though they are by international law.

In practice, international law has recognized both criminal and civil responsibilities of states. States have been considered responsible to the community of nations as a whole and liable to preventive and deterrent sanctions for aggressions in violation of general antiwar obligations. The term "international crime," however, has usually referred not to acts involving the responsibility of states but to acts involving the responsibility of individuals which jeopardize the procedures and instruments of international relations. Acts of piracy, attacks upon diplomatic officers, libels on foreign sovereigns, counterfeiting of foreign currencies, and breaches of neutral obligation have been considered "offences against the law of nations." This practice has doubtless arisen because of realization that criminal sanctions are by their nature adapted to controlling the behavior of individuals rather than of states.

On the other hand, the civil responsibility of states for injury to the nationals, territory, government, or prestige of other states has been enforced by diplomacy and arbitration in numerous cases.

If international law took the position, as it has tended to do, that, although a state may commit a tort or a breach of contract, it cannot commit a crime, it would be abandoning a large sphere of international relations to lawlessness, unless it at the same time recognized that an individual who in the name of the state resorts to violence in disregard of the state's obligations to the community of nations as a whole is himself criminally responsible to that community. Such recognition of the responsibility of high governmental officials under international law modifies the traditional doctrine that states alone are subjects of international law and has been criticized as tending to break down the solidarity and unity of the state and to open the way for civil disorder by dividing the government from its people.

The doctrine of the legal unity of the state has doubtless been of value in assuring the autonomy of national cultures and the continuance of diverse cultures in the world as well as in assuring peace and order within the state's territory. If, however, agents, officials, or individuals within a state have taken action sufficiently injurious to the family of nations as a whole to be characterized as international crime or aggression, if the government of the state, far from attempting to stop this disorder, is its main propagator, considerations of national unity might well be sacrificed to considerations of international order.

In fact, this is exactly what has happened when a large share of the world has envisaged the behavior of the government of a state as an international crime. The Declaration of the Congress of Vienna on March 13, 1815, declared that "Napoleon Bonaparte has placed himself without the pale of civil and social relations and that as an enemy and a disturber of the tranquillity of the world he has rendered himself liable to public vengeance." By Article 227 of the Treaty of Versailles "the Allied and Associated Powers publicly arraign William II of Hohenzollern, formerly German Emperor, for his supreme offence against international morality and the sanctity of treaties." President Wilson in his address to Congress of April 2, 1917, declared the United States a friend of the German people and an enemy only of the German government. In World War II the British and other governments declared that they were acting only against the Nazi government, and after the war the Nürnberg and other war crimes tribunals found high government officials guilty of crimes against international law.

From a practical point of view the first step in making sanctions effective is to divide the delinquent government from its people, and this would be facilitated by a legal theory which held that if a government has resorted to violence, contrary to the international obligations of the state, it should be considered to have violated not only international law but also the state's constitution, which, owing its authority to recognition by the family of nations, cannot be assumed to permit violations of the fundamental laws of that society. Such an act of the government should not therefore impose responsibility upon the state as such but should render the government itself liable not only to international sanctions but also to such constitutional sanctions as are provided in case of a betrayal of the state's fundamental laws. A government guilty of aggression should be guilty also of treason. With this theory, the sanctions against a delinquent government might be supported not only by public opinion in foreign countries anxious to sustain international law but also by patriotic opinion in the state which has been betrayed by the delinquent government.

Such a theory would be parallel to the common practice of dealing with corporations whose acts have violated criminal law by proceeding not against the corporation as such but against its officers. It also was the theory of the United States in dealing with violence supported by the governments of the southern states in the Civil War. Although it has been held that the federal government has power to take measures against a state as such to enforce the state's federal obligations, in practice it has been considered inexpedient to use this power. When coercion has been resorted to, it has not been against states as such but against governments, individuals, or hostile combinations within the state. The expediency of such practice was clearly recognized in the Federal Constitutional Convention of 1787, and a provision for federal execution against a delinquent state included in an early draft was omitted in the final constitution.

In the present age of close international contacts there is a high degree of unreality in insisting upon the dogma that only states are subjects of international law. Eventual responsibility of the state under international law is not adequate to preserve respect for that law in the modern dynamic and interdependent world. Responsibility must be established more immediately and more concretely if the supremacy of law is to supersede the balance of power. The United Nations Charter and decisions of the International Court of Justice recognize that not only states but also governments, officials, individuals, the United Nations and other international organizations, certain private corporations, and international associations may have a status in international law. Entities of all these types have, on occasion, been accorded some international status. International law has tended to become world law. This development implies that the interests of these non-state entities as well as the interests of sovereign states should have an influence in the development of international law, that they should be competent to invoke appropriate international procedures to protect their rights, and that they should be directly responsible for breach of their duties under that law.

Nevertheless, the diversity of cultures and the strength of national sentiments will delay progress in this direction. For a long time peace will require, as the International Court of Justice has said, that international law recognize that a state enjoys sovereignty except in so far as it has expressly or tacitly accepted obligations of international law. This leaves to each state a wide discretion to develop its own values, institutions, and ideologies within its territory.

3. SOVEREIGNTY AND SECURITY

Security through sovereignty is the system by which the family of nations has in the main been governed since the Middle Ages. This system has broken down because of its incapacity to prevent recurrent war and of the increasingly intolerable character of war with the progress of inventions and industrial production. This system rested upon the corpus of customary international law permitting both war and neutrality, applied by diplomacy and *ad hoc* arbitration, sanctioned by self-help and the balance of power. International law could only be adjusted to changing conditions by the gradual processes of custom, juristic commentary, and treaty-making, but it always lagged behind the demands of dynamic states who sought to extend their rights beyond what the law granted them at the moment. For this purpose they used negotiation and equitable arbitration if possible. Otherwise, threats and the accomplished fact were resorted to with the expectation that most states would remain neutral and the fruits of aggression would be legitimatized by subsequent recognition. This was a system of limited security for the militarily strong and unlimited insecurity for the militarily weak. Law governed the unimportant transactions, force the important. The system prevented a world-state, preserved the independence of some states, won independence for others, and destroyed the independence of many. Sovereignty was loudly proclaimed and exemplified in action but was always in jeopardy.

Collective security through international organization had proposed to increase the definiteness of international law by codification in general treaties and by the accumulation of precedents handed down by the World Court. It proposed to perfect the application of law by compulsory adjudication before that tribunal and to substitute for the sanction of self-help the prevention or stopping of violence by collective action.

Neither the League of Nations nor the United Nations contemplated enforcement of all rules of international law. Sanctions were provided only to prevent or to stop illegal hostilities and, in the United Nations, to enforce World Court judgments. These sanctions were intended to preserve peace rather than to maintain law but even for that purpose the procedural requirements made their application uncertain.

Accompanying these collective devices for defining law and rights and preserving them against violence, no less important arrangements for change were provided. Many conferences were held for improving international law through the conclusion of general treaties of legislative effect. Furthermore, procedures for consultation, conciliation, and recommendation were provided to compromise legal claims, to revise treaties, or to rectify dangerous conditions. These procedures had some analogy to the procedures of police and eminent domain in systems of municipal law, but the authority of the world-community as a whole to subordinate rights of particular states to important general interests was not fully established, and the procedures proved inadequate although the United Nations contributed much to facilitating the self-determination of colonial peoples and general respect for human rights against claims of sovereignty.

None of these collective procedures impaired sovereignty in the legal sense. None of them proposed to subject any state to the municipal law of another, or to modify the international law binding a state, except by the established international procedures. The sources of international law and municipal law were kept distinct. Sanctions were applied only for breaches of the peace to protect the weaker from being subjected to the municipal law of the conqueror.

Intervention in the domestic jurisdiction of states was expressly forbidden by both the Covenant and the Charter. Collective interventions or consultations to preserve peace came nearest to depriving states of rights against their will; but in both instruments the competence of the collective bodies was limited to recommendation or advice. However, they enshrined the principle stated by President Wilson that "the peace of the world is superior in importance to every question of political jurisdiction or boundary" and the principle stated in the Charter that political settlements must be not only peaceful but also just. These principles are not inimical to sovereignty functioning as the custodian of the distinction between international law and municipal law, but they are inimical to sovereignty functioning as the right to make war. Exercise of this right, creating discipline within and fear of the enemy without, has been the most important sociological context in which the legal conception of sovereignty has developed, although within the past century the historical-psychological phenomenon of nationality has been of almost equal importance. Thus, in so far as international law, supported by collective institutions holding governments responsible for aggressions and protecting human rights within the states, prevents war, it will modify the political content of sovereignty, if not its legal form.

Legal sovereignty does not prevent peace through law; military sovereignty does. Can legal sovereignty exist without military sovereignty? The question resembles that long ago answered, "Can individual liberty exist without side arms?"

It is clear that non-sovereign nationalities can acquire independence more easily under collective security than under security through sovereignty. Nationalities have sometimes won independence by war, but peaceful procedures for self-determination, developed through the mandate system of the League of Nations and subsequently under United Nations supervision of trusteeship and non-self-governing territories, have practically eliminated colonial imperialism.

With appreciation of national cultural differences, with pride but not prejudice in national characteristics, with adequate systems of civic education, and with national legal systems independent except in respect to obligations under international law, both legal sovereignty and cultural nationality might function as characteristics of the state in a world organized to maintain a legal equilibrium between national independence and human justice.

It may be that sovereignty will shift its locus in the future as it has in the past. Perhaps it will pass from the nation to the world-community. Perhaps it will pass to regions or continents including several nations. Perhaps it will pass to units smaller than the nation. A world-community of several hundred small and more nearly equal sovereign nations might be more stable than a world of one hundred twenty states, varying in size from the United States to Monaco.

Sovereignty may be redefined in the future as in the past. Perhaps it will serve new functions and cease to serve old functions. It is possible that it will cease to be useful altogether and disappear. It seems clear that the effort of states to gain security, each through its own sovereignty, under present conditions of economic interdependence and military technique, endangers the sovereignty of many and is hostile to the security of all.

CHAPTER XI

LAW AND POLITICS

1. LEGAL COMPETENCE AND POLITICAL POWER

International law has fallen short of being an effective system because the development of its substantive rules has not been closely linked with enforcing and correcting procedures.

Jural law implies that the will of the whole is greater than the will of its parts, that the subject of law is subordinate to the community in whose name the law is made and enforced. International law, however, has been referred to as a law of co-ordination, not of subordination—a law which rests on agreement among sovereign states, none of which is subordinate to anything. This theory appears to deny the existence of international law altogether, for, unless agreements are supported by a duty arising from some other source, they can be repudiated at will. If international law is to be a real law, sovereignty must be subordinate to law and sanctioned by the community of nations. That community must, therefore, be superior in legal competence to sovereign states. Such a jural community cannot, however, function effectively unless it is also a political community with political power superior to that of its subjects. The practice of international relations has not provided evidence sufficient to create general belief in the existence of such a political power above sovereign states.

In the history of states the political power of central organs has usually developed before their jural competence. In the history of the family of nations this order has been reversed. International institutions have been given jural competence by treaties, but their executive powers have depended upon the will of the member-states. They have lacked political power.

The contrast between legal and political power should not be exaggerated. Each contributes to the other. Nevertheless, their sources are different. Political power is a psychological phenomenon which springs eventually from the attitude of individuals toward group symbols, whereas legal power is a juridical phenomenon which springs from the sources of a particular system of law.

In systems of municipal law the customs, maxims, judicial precedents, constitutional compacts, legislative enactments, and jural reasoning which have constituted the main sources of legal competence have also constituted important psychological influences uniting the people. In international law, on the other hand, the treaties and resolutions; diplomatic exchanges; national practices, recognitions, and acquiescences; private-law analogies, juristic treatises, and judicial precedents which have constituted the sources of legal competence have been remote from the daily life of peoples. However influential they might be upon the minds of a few diplomats and international lawyers, they have not created a world public opinion behind the competences which they legally create. The process by which international law has developed has not at the same time constituted a process whereby a public opinion has been created to give its institutions political power.

The consciousness that institutions must be directly related to the people if they are to have political power lay behind the American controversy as to the sources of the federal Constitution. Was the Constitution a compact of states or a constitution of the American people? Only with the latter interpretation could the United States have sufficient political power to overcome state nullification. While it is conceivable that international law might gradually and peacefully come to be considered the fundamental law of the human race, binding individuals as well as states, such a process is not likely. International unions do not as a rule grow gradually into federal unions, because in the transitional period states must rely for security either upon the balance of power or upon the jural authority of the federation. If they rely on the first, they prevent the federation from developing political power. If they prematurely rely on the second, they fall a victim to one of their number bent on domination. The federation must have adequate political power or its jural claims will prove a delusion and a snare.

International law is likely, therefore, to continue a primitive law, based on a balance of power, until historic events give the world community sufficient political power to maintain its law. It is to be hoped that this transition may be completed peacefully because of widespread comprehension of the needs of a shrinking world and the dangers of nuclear war. Federations, while in principle organizations of consent, distinguished from empires organizing violence, have actually involved a good deal of violence in their establishment and maintenance. Bismarck converted the German *Zollverein* into a federation by blood and iron, and the United States was formed only by a process which began with the Revolution and ended with the Civil War. War played a part in the creation of the Swiss confederation, the League of Nations, and the United Nations.

Effective government necessarily combines the principles of consent and coercion, but the proportion of each is not unimportant. The virtues of modern civilization—the spirit of liberty, humanity, toleration, and reason—can be better preserved if every stage of organization can be effected with a maximum of consent and a minimum of compulsion and if every institution can be sanctioned by a maximum of rational conviction and a minimum of threatened penalties. These conditions suggest that the world-community should accord a certain respect to individual, local, national, and regional autonomy.

The League reached people indirectly through the mediums of national governments. Consequently, popular insistence upon the observance of League procedures was at the mercy of government policies, and reciprocally, government policies were at the mercy of nationally minded publics. The United Nations and the specialized agencies have sought to remedy this defect; but like the League, they have accepted the prevailing doctrine of international law that sovereign states are its primary subjects and that, apart from special treaties, individuals are subjects of the sovereign state and not citizens of the world-community. Consequently, the United Nations procedures deal with sovereign states and are confronted by an almost insoluble problem whenever a powerful state resists the application of international sanctions, the enactment of international legislation, or the reconciliation of peace with justice.

2. INTERNATIONAL SANCTIONS

The word "sanction" has often been applied to measures of self-help taken by single states under circumstances which they deem render such action permissible under international law. It has also been applied to include all social, psychological, and physical conditions inducing respect for law, such as the pressure of public opinion, the inertia of custom, and the calculations of self-interest. "Sanctions" in the present connection is confined to organized sanctions, or positive action which a community has authorized in a particular situation for the purpose of inducing its members to observe the law to which they are bound as members of that community. Sanctions would thus be distinguished from war, which implies a struggle between equals. Sanctions can only be authorized by the community of which the state or other person against which the sanctions are directed is a member; they can only be utilized to enforce a rule which bound the delinquent state or person before its wrongful act; they should be utilized only after impartial procedures have demonstrated that the rule has been violated; and they must involve action taken with the purpose of such enforcement.

The League of Nations and the United Nations both required consideration of disputes endangering peace by their political organs or by the World Court if the parties accepted its jurisdiction. These procedures did not always provide adequate bases for the application of sanction.

Sanctions may be moral, involving appeal merely to the intelligence and good faith of the person, such as the judgment of a court, advice, or admonition by suitable authority, or they may be physical, involving promises to employ or actual employment of measures affecting the person's interests in order to control his conduct or to nullify the effects of his illegal acts. Execution against property, fine, imprisonment, and corporal or capital punishment are the best known types of physical sanctions in systems of municipal law.

International law has in the past rested upon unorganized sanctions or organized moral sanctions, and some writers have distinguished international law from municipal law on the assumption that the former was supported by no organized physical sanctions. The League of Nations Covenant, however, required member states to engage in economic sanctions, and it permitted them to engage in military sanctions in case of certain gross breaches of the Covenant. The Pact of Paris permitted its parties to engage in physical sanctions against violators of the Pact. The United Nations provides for physical sanctions to preserve peace and to enforce judgments of the World Court.

The difficulty of applying physical sanctions in international affairs has frequently been stressed. The analogy between the family of nations and the state is far from complete. As Madison and Hamilton pointed out in the Federal Constitutional Convention of 1787, sanctions against states are in danger of assuming all the characteristics of war, in practice and in result, however much they might differ in theory and in initiation. Both the League and the United Nations have applied physical sanctions but in only a few cases of alleged delinquency and with varied success.

3. INTERNATIONAL LEGISLATION

The problem of keeping law abreast of changing conditions becomes more difficult in proportion as societies become progressive and dynamic. The judicial development of law by fictions and ideas of equity and justice has been adequate in relatively static societies. Advanced societies, however, have needed a legislative procedure whereby an authoritative body can make general laws for the community to meet new conditions, can remedy injustices arising from the application of law in unusual circumstances, and can override existing rights when necessary in the interests of the community as a whole.

Although the League of Nations Covenant recognized that, in principle, territorial or treaty rights should be modified if necessary to preserve peace or to remedy injustices and authorized the League organs to propose general treaties on numerous economic, social, and technical subjects, the procedure for implementing these principles did not command the confidence of states dissatisfied with the status quo. The United Nations provides similar procedures but is assisted by recommendations of its International Law Commission and by the assertion of broad purposes guiding the proper direction of international legislation.

The doctrine of the political equality of states, attributing to each state equal weight in international conferences and the requirement of consent by a state before it is bound by a new rule of law have in the past proved serious obstacles to international legislation. While equality before the law or in the protection of rights is a necessary principle of any system of law, equality in political capacity is incompatible with effective international legislation. A state with a population of one hundred million will not recognize a state with one million as entitled to equal influence in developing the law. The requirement of universal consent, or *liberum veto,* mitigates this difficulty but presents another serious obstacle to general changes of law to adapt it to rapidly changing conditions. Although the League and, even more, the United Nations made recommendations possible without full observance of these limitations, general treaties binding only the parties continue the only mode of effective change in law or rights.

4. PEACE AND JUSTICE

The procedures established by the League system created the germ of a world public opinion and in many cases were able to sanction rights and to rectify abuses, but they failed to deal with major political demands backed by violence or threats of violence. Disputes between lesser states were dealt with satisfactorily, even, in some cases, after violence had been resorted to; but when great powers made demands for political change, the League faced the dilemma of peace or justice and failed to solve it. This dilemma was in fact invited by the Covenant articles which reserved the political procedures for disputes which threatened a rupture or threatened the peace. Until a state had manifested a disposition to break the peace, it was not permitted to invoke the League's procedures for modifying the status quo. The League was overreluctant to trouble the status quo in the interest of justice when there was no threat to peace, and it was overwilling to sacrifice justice when peace was seriously endangered. The issue of peace or justice was presented not only on the issue of the seizing of disputes by the League and of applying sanctions but also on the issue of recognition of the fruits of aggression and of the right of neutrality in case of aggression.

The British suggested on April 9, 1938, that the members of the League should be free to recognize the Italian conquest of Ethiopia, and the issue was debated in the Council on May 12, 1938. Lord Halifax said:

When, as in the present case, two ideals were in conflict—on the one hand, the ideal of devotion, unflinching but unpractical, to some high purpose; on the other, the ideal of a practical victory for peace—he could not doubt that the stronger claim was that of peace. In an imperfect world, the indefinite maintenance of a principle evolved to safeguard international order, without regard to the circumstances in which it had to be applied, might have the effect merely of increasing international discord.

On the other hand, the emperor of Ethiopia, Haile Selassie, said:

It was true that the League's fundamental object, as Lord Halifax had said, was the maintenance of peace, but there were two ways of achieving that object—through right or by peace at any price. The League of Nations was not free to choose. Set up to maintain peace through right, it could not abandon that principle. . . . He would ask the League to refuse to encourage the Italian aggressors by offering up their victims as a sacrifice.

Mr. Litvinov, the representative of the Soviet Union, insisted that the dilemma could be solved by adhering to the principle of collective action:

Of course, the League, at the request of individual members, could always correct its decisions, but it should do so collectively, and it was not the business of the individual Members to act unilaterally and anarchically. The Council should not only disapprove of activities of such a nature, but should severely condemn those of its Members who set the example. . . . Neglect of the considerations he had laid before the Council would endanger the very existence of the League.

The League offered superior facilities for ascertaining the requirements of justice in particular situations. Its procedures for dealing with political controversies offered more resistance than the procedures of other bodies to the toleration of injustice. It, however, lacked authority to make collective security work. Members of the League, therefore, on a number of occasions preferred to utilize other agencies to deal with certain grave emergencies. It was easier for these bodies to "appease" than it was for the League to do so. For the same reason aggressive states demanding change preferred to avoid League procedures.

Unlimited freedom to recognize the legality of titles arising from aggression is difficult to reconcile with the legal principle *jus ex injuria non oritur*—a principle which has been stated by a British court: "The law is that no person can obtain or enforce any right resulting to him from his own crime. . . . The human mind revolts at the very idea that any other doctrine could be possible in our system of jurisprudence." In so far as an aggression has the status of a crime, it would appear that an individual act according legal recognition to its consequences has the character of complicity. Yet undoubtedly, customary international law has permitted such recognition after the victim of aggression had ceased resistance. Peace and stability, it is said, require that facts be accepted.

A similar issue was raised by the declaration of the foreign ministers of Denmark, Finland, the Netherlands, Norway, Spain, Sweden, and Switzerland on July 1, 1936. They contemplated reversion to neutrality in spite of their obligations to participate in sanctions under the League of Nations Covenant because of the failure of the League to achieve disarmament and the aggravation of the international situation. Thereby the smaller states of Europe indicated that they would have to place their own peace ahead of the collective efforts which the Covenant called for to maintain justice.

The freedom of states to remain aloof from unjust demands upon their neighbors, to recognize fruits of aggression, and to be neutral in case of aggression, though tolerated by traditional international law, accords a legal position to war that was difficult to reconcile with principles of justice or with the law of the Covenant. The United Nations has made progress in solving this dilemma, but its inability to bring about a just settlement of major disputes has left the world with *de facto* boundaries considered unjust by one or both parties after fighting had been stopped in Kashmir, Palestine, Korea, Vietnam, the Formosa Strait, and Germany.

CHAPTER XII

NATIONALISM AND WAR

It is difficult to organize political power so that it can maintain order within a society which is not related to other societies external to itself. Order is a consequence of organization which, however, cannot easily exist without external opposition.

In every society the problem is difficult. In the world-society it is most difficult of all because that society is composed of nations of very diverse customs and values and because it is not under pressures exerted against each nation by its neighbors, inducing each to maintain a united front.

Certain interpretations of sovereignty have been a leading obstacle to the adequate development of international law. In the same way certain interpretations of nationalism have been a formidable obstacle to the strengthening of the community of nations.

The two ideas, sovereignty and nationality, have functioned at times to support each other and at other times to oppose each other. Both have at times tended to build up larger political structures and at other times to disintegrate existing political structures. Both, in their modern form, originated in the liberal and humanitarian tendencies of the Renaissance, in opposition to authoritarian Christian feudalism, and both have at times presented the main opposition to humanitarian and liberal tendencies. Both have been causes of peace and also causes of war.

1. WARS ARISING FROM NATIONALISM

Nationalism has contributed to peace by creating loyalties, throughout the population of a considerable area, above local community, feudal lord, or economic class, even, in some cases, above race, language, and religion. This larger loyalty has permitted political organization within the area capable of maintaining peace. With the rise of nationalism, private feuds, duels, banditry, and feudal, religious, and class hostilities have tended to decline. The feudal and religious hostilities of the type which harassed England, France, Spain, Germany, and Italy in the fifteenth, sixteenth, and seventeenth centuries have hardly existed in these countries since the Thirty Years' War. Similar types of hostilities in the Balkan and Arab countries and in India, Japan, and China and class conflicts such as were manifested in the American, French, Mexican, Russian, Chinese, and Spanish revolutions may be in process of subjection to nationalism. On the other hand, nationalism has been a cause of wars of a different type and of even more disastrous consequences. Several varieties of such wars may be distinguished.

a) Self-determination and Irredentism.—Wars have arisen from demands of "nationalities" to be organized in nation-states. Nationalities within a state have fought for independence, as did Switzerland, the Netherlands, and Portugal in the sixteenth and seventeenth centuries; the United States in the eighteenth century; the Latin-American countries, the Confederate States of America, the Balkans, and Belgium in the nineteenth century; Poland, Czechoslovakia, Finland, the Baltic and Arab states, Ireland, India, and many Asian and African colonies in the twentieth century.

Some of the existing states have fought to incorporate irredentas, or foreign areas deemed to have their nationality, as did France after the period of Joan of Arc; Aragon and Castile in hostilities against the Moors in the fifteenth century; Sardinia, Prussia, Serbia, and Hungary in the wars of Italian, German, Yugoslav, and Hungarian unification in the nineteenth and twentieth centuries.

b) *Solidarity and Prestige.*—Wars have arisen because of the utilization by governments of military preparedness, fear of invasion, pride in national prestige, and expansionism as instruments of national solidarity. Imperial wars of Portugal, Spain, the Netherlands, France, Great Britain, the United States, Germany, Italy, Japan, and Russia in the East and West Indies, the Americas, Africa, Asia, and the Pacific Islands since the fifteenth century may be attributed in part to this motive. Balance-of-power wars have often originated from an exaltation of national honor, prestige, and power above all values, as did the wars of Edward III and Henry V of England in the fourteenth and fifteenth centuries, the wars of Charles V and Philip II of Spain in the sixteenth century, the wars of Louis XIV of France and Charles X of Sweden in the seventeenth century, the wars of Charles XII of Sweden, Peter the Great of Russia, and Frederick the Great of Prussia in the eighteenth century, the wars of Napoleon and Louis Napoleon of France in the nineteenth century, and the wars of the Kaiser, Mussolini, Hitler, and Japan in the twentieth century.

c) *Self-sufficiency and Isolation.*—Wars have arisen because of the tendency of states seriously afflicted by nationalism to seek security from attack, stability of their economic life, and development of a distinctive character by economic isolation and self-sufficiency. Such policies stimulate each country to attempt to expand its territory in order to include essential raw materials and markets and a defensible frontier. This motive has contributed to the imperial wars of the period since 1870 and to the expansiveness of the totalitarian states since World War I. Policies toward self-sufficiency on the part of states whose territorial domains make such policies reasonable may contribute to wars among other states whose interests are adversely affected. Thus the extreme protectionism of the United States and the increasing protectionism of the British and French empires after World War I contributed indirectly to the aggressiveness of the axis powers, which, because of their lack of domestic sources of raw materials compared with other great powers, denominated themselves "have-not" or "proletarian" powers.

d) *Mission and Expansion.*—Wars have also arisen because of the tendency of a people affected by nationalism, especially when pursuing economic policies of the type just suggested, to acquire an attitude of superiority to some or all other peoples, to seek to extend its cultural characteristics throughout the world, and to ignore the claims of other states and of the world-community. In this characteristic, nationalism tends to resemble the missionary and crusading religions, such as Islam and Christianity. Such motivations played a part in the imperial wars of Portugal, Spain, and France, whose nationalisms, in the fifteenth, sixteenth, and seventeenth centuries, were linked with an intense Catholicism. The French "Mission Civilatrice," the American "Manifest Destiny," and the German "Place in the Sun" slogans contributed to the imperial wars of these countries in the nineteenth century. The aggressions of the fascist and communist states in the twentieth century have owed much to attitudes of this type.

Nationalism, affecting opinions and policies, has been an important factor in a considerable proportion of the wars of the last five centuries and in most of the wars of the last two centuries. In the modern period nationalism has progressively reduced the importance of feudal, religious, and dynastic demands and has become itself a major cause of war, although in most wars it has been linked with other factors.

2. CHARACTERISTICS OF NATIONALISM

Nationalism is a term which has a variety of meanings today and which has greatly varied in emphasis in different historic periods. Can any common significance be detected through all these varied usages?

"Nationalism in its broadest meaning refers to the attitude which ascribes to national individuality a high place in the hierarchy of values" (E. M. H. Boehm). How does a nation differ from a tribe, a city-state, an empire, a religion, a civilization? "The state," said Edmund Burke, "is not a partnership in things subservient only to the gross animal existence of a temporary and perishable nature" but "a partnership in all science, a partnership in all art, a partnership in every virtue and in all perfection." In this spirit the nation may be defined as a perfect community.

A community differs from other forms of association in including the entire population of an area. A perfect community is objectively one which manifests cultural uniformity, spiritual union, institutional unity, and material unification in the highest possible degree and subjectively one with which the members consciously identify themselves. Its members resemble one another closely in evaluations, purposes, understandings, appreciations, prejudices, appearances, and other characteristics which any of them consider important. They are all in continuous contact with group sentiment, contributing to group policy and accepting group decisions. The government of such a community is capable of preserving peace and justice within it and of assuring co-operation of the members in its constitutionally accepted policies. Such a community supplies all the needs of its members and is self-sufficient and isolated.

The concept "nation" implies that the identification of the member with the community shall be conscious. According to Renan, a nation is a "soul," a "moral consciousness" resulting from a "common heritage of memories" and "actual agreement, the desire to live together." A community is not a nation if different individuals within it identify themselves primarily with different groups, some with a church, some with a class, others with a family or a village. Furthermore, people may actually identify themselves with the community but lack consciousness of that identification. The nation has sometimes been differentiated from the state by the fact that it is "natural" rather than "artificial," in that respect resembling the tribe. This assumption, however, is not supported by history if naturalness is interpreted in the technical sense of unplanned creation. Nations have been made by continuous civic education and other devices. "Naturalness" in a psychological sense may imply that spontaneous feelings as well as calculated interests motivate the individual's attachment to the group. The idea of "nation" undoubtedly implies such a bond between the nation and the individual, but it seems also to require that the individual be conscious of his feelings. A tribesman who is loyal to his tribe because no alternative has ever entered his mind, or a Chinese scholar who feels the antiquity and perfection of his civilization because he has known no other, cannot be a nationalist until he has consciously compared his own tribe or civilization to a different one. When this is done widely and the territorial limits of the community have been defined, tribes and civilizations tend to become nations, a development often stimulated by the comparison compelled by hostile invasion. Contact with an out-group is no less necessary than cohesion of the in-group to create a nation.

The nation is therefore a consequence of technical conditions which make possible a community of a high degree of solidarity and self-sufficiency and of social conditions which bring about conscious identification of all or most of the members of that community with its symbols. It is a phenomenon of internal communication and economy stimulated by external contact and conflict.

The nation is distinguished from other communities in that it strives for perfection in all the characteristics of a community. A family may have more cultural uniformity, a state more institutional unity, a religion more spiritual union, a region more material unification. A nation, however, in striving for perfection in all, tends to dominate other communities and to fit them into its pattern. Once accepted, it becomes the social a priori by which cultural, political, spiritual, and economic activities and institutions are shaped.

This definition of a nation is clearly self-contradictory. No nation can precisely correspond to it, because efforts to achieve a correspondence in one characteristic will deprive it of correspondence in others. Efforts to make the nation self-sufficient militate against its uniformity, union, and unity. Efforts to make its members conscious of their identity with the nation may, in fact, emphasize local differences. Nationalization propaganda may develop self-consciousness among minorities, which militates against unity. The suppression or expulsion of such minorities or the cession of geographical sections will usually militate against self-sufficiency. For this reason Lord Acton, distinguishing the nation from the state, characterized the theory that they should be be coterminus as "criminal."

The self-contradictory characteristics of nationalism account for its dynamic influence in history, for its war-producing tendency, and also for the more limited character of most of the definitions which appear in analytical discussions. Writers have distinguished (*a*) legal nationality, (*b*) ethnic or cultural nationality, (*c*) nations or nation-states, and (*d*) nationalism. These are said to refer, respectively, to the legal relation between a state and its members or between states with respect to their members; to a group whose members have many cultural characteristics and sentiments in common; to a cultural nationality which is organized as a state; and to the sentiments or attitudes which give high value to membership in a nationality or nation-state and which give force to policies which aim to secure the nation's independence and to increase its powers.

a) *Legal Nationality* may be a concept of municipal law related to, but different from, citizenship, indicating the reciprocal relationship of protection and allegiance between the state and its member. It may also be a concept of international law related to, but different from, domicile, indicating a relation between states with respect to an individual whereby a state is entitled to protect and legislate for him wherever he may be. The tendency of states to claim as nationals all persons born in the territory (*jus soli*) in addition to all persons born of parents who are nationals (*jus sanguinis*) indicates the close relationship of nationalism at present both to the homeland (patriotism) and to the race (racialism). In the early Middle Ages political allegiance tended to be exclusively tribal or racial, but the monarchs began to consider themselves territorial rulers in the eleventh and twelfth centuries. John of England changed his title from *rex Anglorum* to *rex Angliae*. The tendency to drop the claim of perpetual allegiance, to acknowledge voluntary expatriation by naturalization in another country, and to recognize dual or multiple nationality in case of conflict between the *jus soli* and the *jus sanguinis* indicates the liberal characteristics of nationalism in the nineteenth century in contrast to the situation before and since. The frequent changes in nationality laws of most countries indicate that even in a legal sense the concept of nationality is very unstable.

b) *Cultural Nationality* has proved difficult to define. There has been much controversy as to whether race, culture, language, habitat, history, political sentiment, or some other characteristic is its most important index. The geographical boundaries of nationalities have proved to be very different according to the index selected. The results of plebiscites are influenced by the selection of the voting area, by the policing of the area, and by efficiency in propaganda. In some parts of the world any index used will produce enclaves of minorities surrounded by people of a different nationality.

c) *Nation-States* have been artificial constructions. Sometimes a state, in the sense of an area whose population is administered by an independent government and system of law, has made the population into a nation by developing civic loyalty and a consciousness of their difference from others, utilizing education, military service, historic heroes, fear of invasion, religious and patriotic symbols, social prestige, etc., to this end. At other times a cultural nationality within a state or including areas of several states has succeeded by propaganda and arms in achieving independent statehood. The first method was characteristic of nation-building in Britain, France, and Spain in the early modern period; the second, of nation-building among the Balkan, Baltic, and Slavic peoples of eastern Europe and in the Asian and African colonies in the nineteenth and twentieth centuries. Nation-builders in the Netherlands, Switzerland, Germany, and Italy have utilized both methods.

d) *Nationalism* suggests a condition of public opinion within a group which constitutes it a nation-state, which motivates its definition of legal nationality, and which accounts for its maintenance of cultural nationality. It is a sociopsychological force which varies in intensity and which may be measured.

In any group, whether it be a family or a tribe, a nationality or a state, a despotism or a democracy, a religious, business, social, or political association, there must be a condition of opinion which preserves the group from disruption. That opinion may be defined in terms of (1) the symbols toward which it is directed, (2) its intensity, (3) its homogeneity, and (4) its continuity. Nationalism differs from tribalism, patriotism, pietism, commercialism, localism, communism, socialism, and other opinions supporting the solidarity of groups only in respect to the symbols toward which it is directed. If all publics should acquire a very homogeneous, intense, and continuous opinion favorable to the symbols of religion, the age of nationalism would have passed into an age of religion. In fact, during the modern period, first in western Europe, then in America, eastern Europe, the Near East, and Asia, populations have become more intensely, homogeneously, and continuously favorable to the symbols of some nation-state than to other symbols. This is not to deny that other symbols, relating to religions, races, classes, and parties have been more important in certain times and places. It has often been the effort of nationalists to associate other symbols, commanding a certain following, with their own symbols. Thus Irish nationalism has utilized the symbols of Catholicism; American nationalism, those of democracy and liberty; German nationalism, those of the Nordic race; Japanese nationalism, those of the Shinto religion and the Yamata race; and recent Russian nationalism, those of proletarian communism.

3. MEASUREMENT AND BUILDING
OF NATIONALISM

Defining the intensity of nationalism within a given state as the degree of resistance which the population offers to disruption of that nation-state, James C. King attempted to measure and compare this intensity in a number of states in 1933. France and Japan were found at that time to have the most intense nationalism, and Yugoslavia and Spain the least intense nationalism, of the dozen states compared.

This method did not measure the homogeneity or continuity of national attitudes. Presumably the more intense the attitude, the greater the homogeneity, although in certain circumstances intense attitudes may tend to provoke dissident minorities. Important changes have probably taken place in these ratings since 1933 as a result of the intensive nationalizing efforts of most states, particularly the totalitarian states. In 1933 Italy and Germany appeared to be in the middle ranks with respect to intensity of national solidarity.

Evaluation of the influence of various factors upon the intensity of nationalism indicated that length of literary tradition, uniformity of language and religion, and centralizing influence of geography were most important. The length of historic tradition and intensity of internal communications were not important, but the degree of central nucleation in the systems of communication and travel seemed to have considerable influence upon the intensity of nationalism.

A historical survey suggests that the intensity of nationalism has had a relation to international tensions. In periods of war or danger of war, the individual has emphasized his identification with the dominant group, which, in the modern world, has been the nation, has sought its protection, and has yielded it willing obedience even at the expense of his individual liberty. In long periods of peace, on the other hand, demands for increases of individual liberty and insistence upon constitutional guaranties, assuring respect for private rights, have developed. In such periods men have been reluctant or unwilling to yield to the state on many matters. In a balance-of-power system either too much or too little nationalism in important states disturbs the equilibrium and causes international tensions.

Modern history has, therefore, alternated between spirals of rising nationalism and rising international tensions culminating in general war, and spirals of increasing internationalism, increasing liberalism in most states and international peace. Eventually, however, this provides the opportunity for some states to commit aggressions, thus reversing the spiral.

The warlikeness of a state is probably more influenced by the methods used to build nationalism and by the rate at which nationalism is intensifying than by the intensity or homogeneity of nationalism actually achieved. Though Italy and Germany probably had a less intense or homogeneous nationalism than France or England in the early 1930's, the governments of these countries employed methods calculated to intensify nationalism and to increase warlikeness. Factors such as common race, culture, language, geography, history, association, and the *Volksgeist* which develop apart from human design have had an influence upon the development of nationalism, but with the progress of social consciousness in modern civilization, the efforts of leaders, organizations, and governments have contributed more and more to supplement these natural conditions or even to create nationalism in opposition to the natural trend.

Governments have unified nations by advertising the national heroes and symbols, the national language and literature, and the national customs and institutions. Such methods may be contrasted with methods which emphasize the independence and power of the nation, its differentiation from and opposition to its neighbors, and its need of economic self-sufficiency and military preparedness against an enemy whose invasion is anticipated and feared. Nation-building proceeds more peacefully if it emphasizes internal solidarity than if it emphasizes external opposition.

The methods of nation-building actually used depend in large degree on the type of leadership at a given time. Despotisms have tended to utilize preparedness and fear of an enemy, pride in diplomatic triumphs, and centralized propagandas, while democracies have utilized numerous private associations, electoral procedures, public education, and the granting of political and economic privileges and rewards. Leadership in either case may come from different types of elite—politicians, businessmen, military men, lawyers, and literary men—each of which tends to employ characteristic methods.

It should be noted that the process of nation-building is not the only process of state-building. Instead of assuring the unity of the state by making it a nation, states may be held together by the opposite process of divide and rule. This was the method characteristic of the Hapsburg and Ottoman empires before World War I.

An intense and homogeneous nationalism is doubtless a stronger guaranty of unity within a state than is an equilibrium between hostile groups, and the latter has only been resorted to when the existing differences of language, culture, religion, and opinion and the inefficiency of administration were so great as to render attempts at nation-building of very doubtful success. Even when minorities have been small and the administration efficient, measures to incorporate them as an integral part of the nation have usually failed. Liberal measures, permitting the minorities full enjoyment of their cultural distinctiveness, have usually been more successful than oppressive measures attempting to coerce them into the acceptance of the majority culture. Practice as well as theory therefore indicates that conditions set limits to the effectiveness of the nation-building process. The United States has gradually molded fifty states and numerous migrant groups into a nation, and the Soviet Union has made progress toward creating a nation of 143 nationalities. It seems unlikely, however, that any of the continents other than Australia, much less the world, can be developed into a single nation.

4. EVOLUTION OF NATIONALISM

Nationalism has evolved since it began in the late Middle Ages, when kings sought to invoke the loyalty of the rising *bourgeoisie* in their struggle against feudal barons below and the pope and emperor above. Participation in parliaments by the third estate, summoned as an aid to the royal exchequer, broadened consciousness of the nation in the mind of the people. During the Renaissance the absolute monarchs utilized gunpowder to reduce baronial castles, the printing press to propagandize in the vernacular language, royal palaces to render the state as visible and dignified as the church, efficient administration and law enforcement to maintain order, and the concept of territorial sovereignty to maintain both the reality and the image of the nation, identified as the monarch supporting and supported by his loyal subjects.

Revolutionary nationalism emerged with the Puritan, American, and French revolutions, changing the subject to the citizen and identifying the nation as a corporate body of all citizens rather than the person of the king. This kind of nationalism merged into the liberal nationalism of the nineteenth century, anticipating peace, harmony, and individual freedom if state boundaries conformed to the boundaries of the nationality defined by language, customs, or opinion resulting from plebiscite. This idea led to the unification of Italy and Germany and the dissolution of the multinational Hapsburg, Ottoman, Russian, and overseas colonial empires.

In spite of the assumption by nationalists like Mazzini that the change of boundaries or movement of peoples so that every nationality shall constitute a state, would make for peace, the effort to achieve that result has, as noted, led to many wars.

In the meantime the sentiment of nationalism had been subverted by economic, ideological, military, and imperial interests seeking to protect industries from competition, to prepare for defense, to organize totalitarian governments, or to conquer territory. Under the banner of national prosperity, necessity, culture, ideals, or pride, nationalism became both a cause and an instrument of war.

This aggressive nationalism utilized the old verbiage. It interpreted "liberty" as the freedom not of the individual but of the state. If defined "humanity," not as the opportunity of all mankind to achieve self-determined ends compatible with a like opportunity for others, but as the opportunity of the "superior" nation to impose its standards upon all. Liberty thus became identified with sovereignty, and humanity with world-empire. New concepts of international law justified the repudiation of all obligations, exalted the role of war, and opposed international organization. The antagonisms aroused by aggressive nationalism made it self-defeating.

Nationalism in the period of its most intense development has seemed least able to function for the benefit of the people. The nations, with all their efforts, have not been able to create the "perfect community," harmonious, prosperous, self-sufficient, and isolated. With a few exceptions they have lacked the resources to realize the maximum prosperity possible with contemporary economic techniques. They have become as economically and militarily obsolete as the feudal baronies in the late Middle Ages. Thus regional or continental economic blocs were proposed, and efforts were made to achieve them by persuasion in the British Commonwealth, the Americas, and Western Europe and by force in central Europe and the Far East. World War II was concerned with the issue of whether the subordination of nations to larger groups should proceed by the method of voluntary organization or by that of conquest.

Nationalism, which for a century functioned with moderate success and supplemented the Renaissance ideas of liberalism, humanism, tolerance, and science, was so subverted by the Fascists and Nazis that it led to World War II and, unless reinterpreted, threatened to destroy civilization. Its continuance in this extreme form appeared to be incompatible with world economy and international security.

Nationalism, thus, is not necessarily linked to the idea of the perfect community. It may mean the opportunity for political self-determination of reasonably homogeneous groups, not as absolute sovereigns, but as claimants to legal autonomy in regional and universal organizations. Thus interpreted, all nationalities might develop their talents and supplement one another's contributions to the cause of human progress. Such a concept of liberal nationalism, professed by the democracies, though their practices have sometimes gone beyond it, is not incompatible with peace and human welfare and would assure the variety so essential for human progress. With such a concept, nationalism could be maintained by a system of civic education which creates pride in the national culture and achievements and appreciation of the national character and distinctiveness rather than by organized propaganda designed to develop fear of, contempt for, hostility to, and isolation from other nations. Attitudes of the latter type have, however, proved a stronger stimulant to nationalism than have those of the former. Lacking the sense of necessity, which opposition to other nations appears to present, people will not submit to the intense forms of nationalism which enable a government to suppress liberty, to control opinion, and to administer economic life.

If the latter methods were abandoned, the worst forms of nationalism might disappear; but there seems little prospect that national governments acting individually will be able to abandon them. The opportunity can arise only if international organization is so developed as to assure national security by law and to divert some of the individual's loyalty from the nation to humanity. Loyalties divided among many groups are essential if the world is to have both unity and diversity. These divisions of loyalty must, however, be reconciled by the consciences of many individuals who are citizens both of the nation and of the world. It is not to be expected that national governments will effect a just reconciliation of national and international claims. Though divided in their duties as subjects of international law and as trustees of the nation, they owe their power exclusively to the nation. The nation will insist that national power be placed ahead of international justice so long as national existence depends upon the power equilibrium.

CHAPTER XIII

THE FAMILY OF NATIONS

The integration or the disintegration of a political group may equally endanger the peace. Integration may arouse the anxiety both of neighbors and of minorities. Disintegration may encourage the aggression of neighbors and the revolt of minorities. Maintenance of the status quo may, however, be no less dangerous in a dynamic society with changing foreign contacts and domestic interests. Every society is continually on the brink of conflict. It must continually adapt its organization and its policy to changing conditions of internal opinion and external pressure. If changes intended to effect such adaptation are too great or too little, too rapid or too slow, to the right when they should be to the left, to the center when they should be to the periphery, trouble may be expected. The problem of adapting the family of nations, internally heterogeneous and externally alone, to rapidly changing conditions of technology and opinion has presented extraordinary difficulties.

1. TENDENCIES AND FORMS OF
FAMILIES OF NATIONS

The nations of today stem from a common ancestry in aboriginal man, and they are in actual contact with one another. They therefore constitute a family of nations. Human populations which are wholly isolated have tended to divide into subpopulations, so that each can have a potential enemy providing opposition against which it can integrate itself. Thus an isolated population tends to become a family of nations rather than a single nation. At the present time there is only one family of nations, comprising the entire human race with more than three billion members organized in more than one hundred and twenty sovereign nations. Formerly, when the natural barriers to human contact were far more significant than they are today, there were a number of civilizations, each one constituting during most of its life a family of nations. Families of nations in the Far East, in India, in the Near East, in Europe, and in America have, throughout most of recorded history, been distinct and relatively isolated.

Historic civilizations have tended to develop through typical stages. The nations which constitute a civilization in its emerging stage have tended to increase in population. As a result there has been an increasing wealth of contact and of communication among them. This has resulted in a diffusion of culture, goods, techniques, and migrants from one to the other, in an increase in the frequency and severity of wars, and in an increasing standardization of behavior patterns. Sometimes, under these conditions, each nation has come to recognize the moral equality of the others and further integration of the family of nations has proceeded through a stabilization of the political equilibrium and the development of international institutions. More often further integration has been effected by the processes of conquest and empire-building.

In the past, cycles of history and families of nations have been affected in their development by the fact that they were not entirely isolated. Although contacts with outside peoples might be slight, there were always some contacts on the peripheries of the historic civilizations. The Roman Empire at the time of its greatest extent was in contact with the Germanic cultures of the North, the oriental cultures of the East, and the African tribes of the South. The present family of nations, however, will remain isolated, unless, indeed, interplanetary communication is established.

History records the oscillation of periods of integration and disintegration, both effected by wars. There is, however, no exact precedent for a family of nations that is entirely isolated on the planet. It may be that mankind as a whole can control the development of his population and of his polity. Upon what he does in these two matters may depend the future of his culture, technology, economy, language, literature, art, religion, ethics, and law. This is not to deny the profound influence which advances in science and technology have upon his capability to exercise such controls. Population may be regarded as the substance of humanity, and polity as its form. From a combination of these two have developed other institutions and patterns since civilization was achieved and inventions and commerce began to emancipate societies from the immediate limitations of physical nature.

In the past, states have had to regulate their population and their polity in the light of political pressure from outside. The freedom of national statesmanship has thus been limited. The same is true of past civilizations and families of nations. The present family of nations has the freedom to make its population and its polity what it will, limited only by the inertia of its own past and the imagination of its own future.

Historic families of nations have assumed the following forms: empire, church, balance of power, and federation in the broad sense including confederation and general international organization.

World-empire is built by conquest and maintained by force. The Roman Empire, by an authoritative law and an efficient army, maintained the *Pax Romana* during the period of the Antonine Caesars. It acquired the idea of empire from the Orient, left the idea in its wake, and stimulated numerous conquerors to attempt its revival. The medieval Hohenstaufen, the Hapsburgs, Louis XIV, Napoleon, the Kaiser, Mussolini, and Hitler attempted with varying degrees of success to re-establish a universal empire.

The Christian church dominated Western civilization after the Cluniac revival in the tenth century, during the period of the Crusades. The idea of a world peace maintained by a unified church was given expression in the "Truce of God" and the "Peace of God, sanctioned by excommunication and the interdict. The church was a powerful influence unifying Europe during the twelfth and thirteenth centuries, but it was continually in opposition to another great organized religion, that of Islam. After crusading zeal lagged in the fourteenth century, the power of the church began to decline.

The balance of power may be observed in smaller families of nations, such as that of the Greek city-states of the Periclean period. In the Hellenistic civilization a balance-of-power system described by Polybius extended over the whole of the Mediterranean area. In the fourteenth century, while Dante was expounding the theory of world-empire, Boniface VIII the theory of a world-church, and Pierre Dubois the theory of world-federation, the balance of power was being exemplified in the wars and alliances of the rising monarchies of England, France, and Austria. Two centuries later the balance of power was not only practiced in the Renaissance civilization of Italy but was for the first time given detailed literary exposition by Bernardo Ruscellis and Machiavelli. Finally, in the European system which arose out of the ruins of the universal church in the seventeenth century, the balance of power was recognized as the basic principle of European organization. Balance-of-power politics has aimed less at preserving peace than at preserving the independence of states and preventing the development of world-empire, but in the century after Waterloo, Britain, with overwhelming sea power and an invulnerable naval base, was able to maintain comparative peace in Europe.

The characteristics of federation can be studied in numerous limited unions and in plans of European confederation. Pierre Dubois in the fourteenth century; King George of Podĕbrad in the fifteenth; Henry VIII of England in the sixteenth; Henry IV of France, his minister Sully, Emeric Crucé, and William Penn in the seventeenth; Saint-Pierre, Jeremy Bentham, and Immanuel Kant in the eighteenth; and numerous writers in the nineteenth produced plans for European or world federation. There were practical attempts to federate Europe after the Napoleonic Wars. Concerts and conferences of the nineteenth century kept the idea in practical politics, and the Hague Conferences looked toward its realization. Finally, the League of Nations and then the United Nations were established, the most successful of any of these attempts.

These four types—empire, church, balance of power, and federation—seem to have exhausted the imagination of men on forms of universal organization, though many combinations and variations have developed in practice. These four types differ in structure, in object, and in procedure.

The empire seeks to concentrate military and political power in a single authority with control over individuals enforcible by law. It emphasizes institutional unity.

The church claims divine authority and seeks to rule with moral sanctions alone. It emphasizes spiritual union. Although the papacy sometimes tried to deal with temporal as well as spiritual matters and to employ material sanctions, in theory it ruled by persuasion.

The balance of power, instead of concentrating authority, seeks to distribute it among independent sovereign states which remain in equilibrium because of their separations and oppositions. While permitting considerable material unification, such a system may hamper the development of other aspects of association in the world-community.

The confederation seeks to achieve the unity of the empire without sacrificing the autonomy of states which characterizes the balance of power. It does this by insisting on the supremacy of a constitution which limits the central authorities to matters of general concern. Confederations, leagues, and international organizations in which the central authority acts only on member-states as units have often developed into true federations or unions in which the central authority deals directly with individuals in respect to matters within its competence. Whereas empires are primarily organizations of violence, confederations are primarily organizations of consent, because all authority is derived from the constitution accepted by the people and the states. Although both state and central authorities can exercise coercive authority within the limits of their competence, the confederation resembles the church in that the constitution itself is maintained by persuasion rather than by compulsion.

2. CONCEPTS OF A WORLD-SOCIETY

For a concept of the modern world-society one would naturally look to international law, but the international lawyers are undecided whether the family of nations constitutes a society.

Vattel adhered to the atomistic theory which holds that international law is merely a series of contracts between wholly independent states. A contractual association, however, presupposes that all its members are also members of a society whose standards interpret and apply the terms of the contract. The majority of writers on international law have, therefore, indorsed the thesis of Wolff and his predecessor Grotius that the family of nations constitutes a society.

The family of nations, whatever may have been its fundamental character, was not, when they wrote, adequately organized to maintain its principles or to protect the interests of its members. The League of Nations and the United Nations were created to remedy this defect, but they have not been entirely successful in doing so.

The question whether now or at any point in time the family of nations constitutes a society is a question of point of view as much as of fact. A society may exist if people so recognize it, even if its organization is inadequate to achieve its purposes. The judgment that a group is a society is a judgment of attitudes as much as of structures; a judgment of the meaning of symbols as well as of the classification of conditions; a judgment of the direction and intensity of a movement as much as of the application of a definition.

It cannot be said that a world-society, including all nations, exists merely because some persons have conceived of such a society. It is clear, however, that no such society can exist unless some persons have conceived of it. A society implies consciousness by some persons of their participation in it. When the society is larger than a local group, in which all members are in continual personal contact with one another, such consciousness is hardly possible without a conception of the society.

How may the world as a whole or mankind be conceived as a social unity? Abstract conceptions are formed either through the association of a word or other symbol with concrete experiences or through the inference of one conception from others. The first is the method of suggestion; the second, of definition. Monotheistic religions have tried to conceive God by both methods. They have associated the word God with other signs, such as icons, images, and symbols; with subjective experiences, such as truth, goodness, beauty, sympathy, love, and religion; with impressive manifestations, such as miracles, rituals, and sacraments; with unique men, such as the king, the emperor, the pope, the prophet, or the saint; with unique groups, such as the state, the church, the nation, the human race, and the symplasm; and with the totality of observations, such as nature, the world, and the universe. They have also tried to deduce God from other abstract classes or ideas. There has been a tendency for such religions (1) to create a hierarchy of gods with one supreme god, (2) to create an opposing god or devil with whom he may be contrasted, (3) to develop a concept of evolution comparing God with himself at different periods, or (4) to analyze the attributes of God, comparing each with familiar entities or experiences in which the attributes are assumed to exist in lesser degree.

These methods of conceiving God do so only by making him like something else. They therefore defeat their aim, because they subtract from the uniqueness of God, which is supposed to be the outstanding characteristic of monotheism. They do, however, illustrate the inherent difficulty of conceiving something wholly unique.

Similar difficulties have been met with in attempts to conceive of a world-society. That society has been associated with such concrete things as the United Nations Charter, the World Court Statute, the Peace Palace at The Hague, the United Nations building in New York, and the United Nations flag. Such a process of association is probably the most effective method by which the average man may be led to a conception of the world-society, but knowledge of these things contributes little to an understanding of the nature of that society.

These things are themselves symbols of the world-community rather than indications or evidences of its condition. Even such entities as the corpus of international law, the system of international organization, the process of international commerce, or the practice of world communication, which the more sophisticated often associate with the world-society, should be regarded as stereotypes or in some respects even as fictions, knowledge of which falls far short of disclosing the actual condition of the world-community. Knowledge of these symbols and stereotypes scarcely justifies a decision whether, or in what degree, a world-society exists in any period of history.

The method of definition by classification and analysis has also been employed to create a conception of the world-society. It has been thought of as a world-state, greater than but comparable to national states, or as a universal civilization, greater than but resembling historic civilizations. It has been thought of in contrast to world anarchy or as the limit toward which the process of world history seems to tend.

It is clear that none of these processes can yield a wholly satisfactory conception. Obviously, a society in complete isolation, with no other society on its periphery, might differ radically from any of the lesser states or civilizations in history. The analysis of existing limited societies might result in emphasis upon the very characteristics which they share with one another but which they would not share with a world-society. Organization and anarchy are relative to each other; consequently, it is impossible to determine which term should apply to a given situation of the world. In view of the oscillations and transitions of world history, the process of extrapolation of past trends is hazardous. Unsatisfactory as they are, however, it is only through such devices of suggestion and definition that a rational conception of a world-society can be approached.

3. CONDITIONS OF A WORLD-SOCIETY

A society results not only from subjective conception but also from objective conditions, from the integration of a group, and from its differentiation from other groups. The more the members of a group act as a unit and the more they are differentiated from other groups in appearance, behavior, and values, the more the group becomes a society.

The world-society cannot be contrasted with any out-group; consequently, the degree of its integration can be studied only in the relations of its members with one another. Changes in group solidarity are difficult to study directly, but they may be indicated by certain observable phenomena.

Among such phenomena are (1) instruments of communication and transportation and statistics indicating the degree of interdependence among members and of self-sufficiency of the whole; (2) organizations, laws, and institutions subordinating the members of the group to the whole; (3) standardized behavior patterns, indicating the degree of uniformity among the members; and (4) acts and declarations of the members indicating attitudes toward one another and toward values imputed to the whole. The characteristics of a society indicated by these four kinds of evidence may be designated, respectively, material unification, institutional unity, cultural uniformity, and spiritual union. It appears that the world has developed toward the realization of all these characteristics during the last four centuries.

Communication and commerce have developed remarkably in modern history. Language, writing, printing, general literacy, statistics, the mails, the press, the telegraph, and the radio suggest aspects of this process. Today many persons in every part of the world are continuously aware of and materially and emotionally affected by what is going on in every other part.

The various sections of the world have also become materially interdependent in respect to economy and security with the development of more abundant and rapid means of transportation by sea, land, and air. International trade provides most people with essentials of diet, clothing, and work. With the increasing rapidity of travel, protection from devastating epidemics requires organization of health on a world scale. Depressions and wars in any area extend their effects rapidly to the most remote areas.

Some individuals may still be unaware of their relationship to the world as a whole, but the number has decreased dramatically in the atomic age. The fact that communication and transport have been in large measure nationally organized and directed has reduced the natural influence of modern inventions in integrating the world-community.

Institutional unity arises from habits of leadership and obedience, permitting the group to act as a unit on certain subjects. All societies have some sort of leadership, whether of old men, military heroes, hereditary chiefs, self-chosen tyrants, or democratically elected magistrates. There has been progress toward a world leadership through international councils, assemblies, and commissions with, however, some periods of recession. Three centuries ago an unorganized diplomatic and consular service was the only regular instrument of official international organization. Before World War II there were fifty public international unions, to some extent integrated through the League of Nations, the Permanent Court of International Justice, the International Labour Organization, and the Pact of Paris. Since the war, the United Nations and its specialized agencies have exerted a greater influence. Furthermore, there are so many unofficial and semiofficial international conferences and associations that several may be in session at any one time. These institutions are as yet imperfect. Since the official agencies have usually dealt with states rather than with individuals, they have created only the rudiments of a world public opinion. Consequently, they have lacked the efficiency of governmental processes within the state supported by powerful national public opinions. Nevertheless, the trend toward world unity, in spite of reversals in the Napoleonic and Hitlerian periods, can scarcely be denied.

Cultural uniformity in some degree must characterize the members of a society. The members need not be identical. Variety among its members is a characteristic of societies distinguishing them from organisms in which many cells may approach identity. The members must, however, be in some respects similar. They must have some sentiments in common, or there can be no spiritual union. They must have some standardized responses to language and other means of communication, or there can be no obedience or leadership. They must have some common aims, or there can be no co-operation. In the modern world-community there has been a movement toward greater uniformity both among individuals and among states.

The world-society is an international rather than a cosmopolitan society, but increasing similarity of the institutions and economies of states has tended toward increasing similarity of the national cultures and of individual behavior patterns. Furthermore, the intellectuals of all countries have at times formed the germs of a cosmopolitan society exerting a certain influence toward assimilation of the national cultures and the institutions of government. Penetrations of commerce and economic techniques, of religions and philosophies, and of art forms and literatures have not only broken down many local cultures but have resulted in a general borrowing from the common stock by all local communities, so that all peoples have tended to conform to a common type. This process has, it is true, been combated by national, cultural, and racial propagandas, including insistences on national languages, folkways, and racial purities, and the establishment of artificial barriers to trade, migration, and intermarriage. But these countertendencies have, on the whole, been less influential than the economic efficiencies and conveniences of the universal behavior patterns and, in recent times, the conscious efforts of such institutions as UNESCO in promoting educational, scientific, and cultural exchanges.

Spiritual union implies general recognition of the superiority of the values of the society over those of its members and, as a consequence, equality of the members in respect to their loyalty to these values. This characteristic is imperfectly developed in the world as a whole. It is more developed in the relations of states than in the relations of individuals of different states. Union of purpose and sentiment constitutes the basis of a general will and is doubtless the most important characteristic of a society. Its imperfect development in the world as a whole manifested in the frequent hostilities of peoples constitutes the most important reason for doubting whether the latter is a society. Yet there has been an increasing acceptance of common values such as international peace, human welfare, personal freedom, national self-determination, precision of thought, and tolerance of cultural differences manifested in the purposes and principles of the United Nations.

There has been a tendency toward greater equality in the mutual recognition which the states accord one another in spite of occasional retrogressions to practices of colonialism and hegemony. Sovereign states have increasingly accorded one another equality in diplomatic representation and in theoretic right to the benefits of international law and to the international procedures for protection of such rights, in spite of occasional policies of protracted nonrecognition of established states or governments.

With respect to individuals there has been less recognition of equality. Peoples of different races, languages, cultures, and religions have been discriminated against in respect to immigration, civil rights, rights of war, and other matters. Numerous conventions have, however, been made dealing with non-self-governing territories, slavery and the slave trade, minorities, and immigration, intended to eliminate such discriminations. The democracies have frequently incorporated constitutional provisions assuring equality of civil rights irrespective of race or nationality. The United Nations accepts the theory that individuals are subjects of international law and entitled to the equal protection of human rights by that law, although adequate procedures to implement the theory have not been achieved.

This analysis of the world-community indicates certain outstanding peculiarities: (1) The members of this community largely because of nationalistic propaganda have not risen to a full awareness of their actual interdependence. The people of many nations and regions still think, as did Vattel, that "they are able to provide for most of their needs" within their boundaries, although today this is seldom true. (2) The institutions of world-government have proved inadequate to regulate the conflicts and controversies arising from this interdependence because of the general acceptance of the absolute interpretation of sovereignty. (3) International law has overemphasized the equality of states and underemphasized the equality of individuals. This has resulted in important disparities between the requirements of international law and the requirements of natural justice as it appeals to individuals unincumbered by technical learning. (4) These circumstances have had a hampering effect upon the development of a common will to maintain order and justice throughout the world. Since every state is in a high degree dependent for its material and cultural needs upon an area which extends beyond its national boundaries, national interests have suffered no less than the interests of the world-society.

It appears that the world's population has become more integrated in spite of itself during the last four centuries and that the very rapidity of this progress has stimulated the growth of artificial barriers, such as the sentiment of nationality, the dogma of absolute sovereignty, the concept of the independence of states, and the policy of national self-sufficiency. The community of nations has not become an effective society, but the necessity for an adequately functioning world-society has been widely recognized as the inefficiencies and dangers of an anarchic world become more apparent in the nuclear age.

4. ORGANIZATION OF MODERN CIVILIZATION

Modern civilization has sought to solve this problem by developing the family of nations from a balance of power to some form of federal organization. National federations have had difficulties. They have tended to break up or to form unitary states. The Netherlands and Germany passed through the transitional stage of federation and became unitary states. Switzerland and the United States remain at the stage of federation, although each has steadily increased the power of the central government. The historian Freeman entitled his book written during the American Civil War *History of Federal Government from the Foundation of the Achaian League to the Disruption of the United States* and supported the thesis that federations are inherently unstable. Confederations have usually succumbed if unable to develop into true federations. Twice, in 1787 and again in 1865, the United States avoided disruption only by drastic steps toward centralization.

A world federal organization has difficulties which are not faced by smaller federations. It can have no external enemies to compel union. Many of the plans of general federation, as that of Dubois to rescue the Holy Land (1306) and that of Streit to rescue the democracies (1939), have sought to utilize the external enemy, but they have thereby renounced a genuinely world character. Federations have seldom been successful, unless their members have been forced together through fear of external states. Without fear of England, it is unlikely that the American federal convention of 1787 would have succeeded. The Netherlands, Switzerland, and Canada were induced to strengthen their unions only because they were afraid of their neighbors. Efforts at federation after World War II, such as those in the British West Indies and various areas of Africa and the Middle East, failed because a common need for defense was lacking. The movement for federation of western Europe fluctuated in intensity with variations in the common fear of Soviet attack.

Furthermore, in the family of nations as a whole, there is less uniformity among the parts, a greater diversity of economic and cultural interests, than has been true in the case of the lesser federations which have been formed. To create a world federal organization which attempts to unite oriental and occidental states, democracies and autocracies, industrialized and nonindustrialized states, large and small, stable and unstable states, is clearly a task of unparalleled difficulty.

Further difficulties are presented by the problem of representation of states of very different size and population, of sanctions to assure obedience of the parts to the whole, of the distribution of powers, benefits, and burdens, and, most importantly, of devising acceptable procedures for settling disputes peacefully and adapting standards and structures to changing conditions. In view of these difficulties it is not surprising that the League of Nations failed to organize the world sufficiently to prevent a major war. It is a tribute to human perseverance that after World War II another effort was made under conditions which superficially appeared less favorable but which indicated a greater necessity to succeed.

CHAPTER XIV

SOCIAL INTEGRATION AND WAR

The sentiment of nationalism has so increased in the modern world that the nations are usually considered more important than the family of nations. The parts claim to be, and sometimes prove to be, greater than the whole. The nations claim great power and acknowledge little responsibility. Yet with all their powers they have not been able to meet the economic, cultural, and political demands of their people within their own jurisdictions. Because of their irresponsibility, they have often attempted to exercise power in jural areas claimed by others. Jural conflicts may degenerate into war unless dealt with by a superior, regulative authority. The family of nations has lacked the power to exercise such a regulative authority.

This situation accounts for most modern wars, but it appears to be a consequence of historic contingencies rather than of inevitable conditions.

1. CONFLICT AND SOCIETY

Sociologists have attempted to understand social life by defining and analyzing such concepts as "society," "co-operation," "opposition," and "conflict." They have treated war as a species of conflict which is itself a species of opposition. Conflict may designate a duel, a household brawl, a strife between political factions, a fight between street urchins, a suppression of a rebellion, or a war between nations. Observation of any one of these forms of conflict may throw light on the others. The sociologist can understand why nations occasionally go to war by understanding why he himself occasionally feels like fighting. Each of these forms of conflict has, of course, its peculiarities, but the sociologist, by comparison and analysis, distinguishes the universal from the particular aspect of each conflict.

The word "organization" may be used in a general sense to describe the process by which a corporation, a club, a city, a state, an empire, or a league of nations is created, developed, and maintained. The sociologist can understand why it is difficult to organize the world for peace by observing the difficulties within such lesser organizations as families, associations, and nations.

By the application of this method sociologists have concluded that opposition is an essential element in the existence of any social entity, just as essential as is co-operation.

The conception of conflict has been applied to physical and biological entities, but it has been developed especially in relation to social entities, whose organization involves a general appreciation of certain values by the members. "Social conflict," writes Lasswell, "results from the conscious pursuit of exclusive values." Whenever two or more personalities or societies in direct or indirect contact with one another recognize exclusive goals and strive to attain them, opposition is to be expected. If they are in direct contact with and conscious of one another, opposition may become conflict. Even though such entities regard themselves as co-operating to achieve the same values and as acting within the same logical hierarchy of means and ends, yet, in so far as more than one freedom of initiative exists, differences in respect to interpretation, timing, or limits of competence are likely to arise. The only type of society in which internal conflict is unlikely is, therefore, one in which all initiatives emanate from one source, that is, a society in which integration and autonomy have reached a point at which not only freedom but also the desire for it has been eliminated from all members of the society except the leader. If such a society is in contact with outside societies, tendencies toward internal conflict will be transferred to the relations between the groups.

The word "community" refers to the organization of all the social entities, in direct or indirect contact with one another, within an area. As the progress of communication has established some contact among social entities throughout the world, there is in this sense a world-community. Some opposition, however, is inevitable among the many individuals, families, factions, parties, corporations, associations, classes, churches, states, and nations within that community. War, therefore, may be explained by examining the processes of the world-community to ascertain why international oppositions tend to assume the form of military conflict.

If the world's population is divided into many small groups, these oppositions are likely to be moderate, whereas if there are few large groups they are likely to be intense. In the latter case, while conflicts will be less frequent, they will be more violent.

Increases of population and improvements in means of communication tend to augment intergroup contacts within the world-community and to increase the probability of violent conflict, unless accompanied by improvements in means of adjustment and of education. More intense political organization of a nation, region, or other group will not therefore necessarily reduce the amount of conflict in which it will be involved. Such organization may merely divert opposition from its internal to its external relations. This may explain why the efforts to avoid the social dangers of conflict by more extensive and intensive political organization have failed to assure peace so long as that organization was less extensive than the whole family of nations. Such efforts, however, may account for the trend of rising civilizations toward a decrease in the number and increase in the size both of states and of wars. As the in-group becomes larger, less homogeneous, and more in need of a scapegoat to relieve internal stresses, opposition to the out-group serving as a scapegoat becomes more intense. Mutual fear induces more intensive preparations for defense, which in turn augment mutual fear. Philosophies valuing efficiency or struggle tend to prevail over those valuing reason or renunciation.

Sociologists have explained in detail the processes of accommodation and assimilation by which oppositions between individuals, classes, and groups are moderated. These processes have often involved identification of the opposed entities with an inclusive group and transfer of the opposition to an out-group. The more the oppositions within the state become concentrated into oppositions between great classes, parties, or regions, the more necessary it is to intensify opposition to an external scapegoat if the identity of the state is to be preserved. This tendency toward international war is combated, on the one hand, by the particularism of individuals, localities, and associations resistant to assimilation by the state and, on the other hand, by the cosmopolitanism of international conferences, associations, and institutions. Against the influences of liberalism within and humanism without, the state has been able to preserve its dominant position only by continuous preparation for war and occasional resort to war itself. In civilized as well as in primitive societies there has tended to be an oscillation in the relative importance of the opposing tendencies, on the one hand, toward state integration and a concentration of all oppositions in interstate war and, on the other hand, toward state disintegration and a diffusion of opposition among numerous associations.

2. COMMUNITY-BUILDING IN HISTORY

A study of the methods used in the past to develop small and large communities into organized societies cannot be expected to solve the problem of world organization in the future. A review of these methods may, however, be suggestive. Modern conditions permit and encourage social organization of many types, bringing together people belonging to different races and dwelling in widely separated countries. These possibilities, however, have not been fully realized, and the dominant societies have remained the geographically limited and consanguineously related communities. How has their solidarity been created and maintained? The conscious processes by which local communities have been integrated when social dynamism has shaken the power of unconscious custom and habit may be classified into four types, relying, respectively, on (1) opposition, (2) co-operation, (3) authority, and (4) opinion.

Integration has often been effected through the organization of *opposition*. By creating and perpetuating in the community both a fear of invasion and a hope of expansion, obedience to a leader may be assured. The method of opposition—competition, rivalry, or conflict with an outside community—has been used to consolidate every type of community, particularly those which have claimed to be independent, such as clans, tribes, city-states, nation-states, and federations. Even churches have united the faithful in a common cause against infidelity, heresy, and sin. A system of world politics resting upon a balance of power contributes to the integration of each power by maintaining among its people both fear of war and hope of dominance. Mutual fears and jealousies among factions within a state have sometimes split the state in two, but they have sometimes perpetuated the rule of an unpopular government by preventing united opposition to it. Fear and ambition have been the great integrating forces in the conscious building of political communities. Communities so integrated have tended to relapse into reliance upon custom and habit in normal times and to tolerate the expression by individuals of primitive behavior patterns in times of emergency.

Voluntary *co-operation,* because of rational appreciation of its advantages to each member of a group, has been more important in advanced than in primitive societies. This method of integration has been especially employed by associations with limited purposes. Pressure groups are held together by the common business, political, religious, humanitarian, or other interests of the members. Political parties are in part held together by the common interest in sharing the spoils of office. Industrial organizations are maintained by the expectation of the officers, capitalists, salesmen, laborers, landowners, and technicians that all will share in the prosperity of the enterprise. Local and national communities gain solidarity through the realization by the members that the group as a whole contributes to the security and welfare of each. The probability that the membership will envisage a community as a co-operative enterprise increases with the generality of participation in its policy-making. The notion of the social contract and the practice of democracy tend to augment the sense of participation and the co-operative character of communities.

In all independent communities *authority* has been organized through leadership of a hierarchy which can reach all members of the community. Habituation to authority develops a belief that the leader has customary or divine sanction to rule. This method of integration is based on the feeling of awe and reverence, the sentiment of loyalty, the disposition to follow leadership, and the reluctance to think originally. Fear and greed also play a part because usually the authority supports itself by threats of punishment for treason and sedition and by the giving of special advantages to potential dissenters who are influential. Custom and superstition have also fostered the prestige of the established authority. The method of authority has manifested itself most clearly in armies but is also important in the government of tribes and states. A ruler always insists that adjudications and legislative enactments in his name are authoritative and must be obeyed by his subjects. In democracies, authority is established and maintained by the consent of the governed and by their belief that its decisions reflect the general will of the group.

The organization of *opinion* has in reality been fundamental to all other methods of political integration. Opinion has been the source of fear, of authority, and of the spirit of co-operation. It has, however, been pursued less consciously in building political communities in the past than the other methods mentioned. In primitive communities opinion has been the product of custom and has not often been consciously manufactured. In the modern nation, however, common customs, languages, symbols, and sentiments have been consciously created both by governments and by minorities. Common attitudes have been developed by education, and common opinions have been propagandized by oratory and the press. Behavior patterns thus established will be repeated on the presentation of similar stimuli. Consequently, characteristic group responses to established symbols can be relied upon in most circumstances. Common opinion holds together social groups such as fraternities, clubs, lodges, and polite society. It has been relied on in political groups more consciously as the size of the group has increased and its means of communication have become more perfect. Propaganda and opinion control have become the most important methods for integrating social and political groups.

3. THE PROCESS OF COMMUNITY-BUILDING

The processes of organizing opposition, co-operation, authority, and opinion emphasize the methods, respectively, of politics, law, administration, and propaganda. Are these methods adequate, if applied, to build the world as a whole into a society?

a) The *Political Method* consists in a realistic analysis of the subgroups within a given community and continuous negotiation to minimize some of their oppositions by exaggerating others. The controversies within one political party can be for a time subordinated by emphasis upon the opposition of the party as a whole to the opposing party. Party conflicts can be kept within bounds by emphasis upon the opposition of the nation as a whole to foreign nations.

Skilful maneuvering of the in- and out-group sentiment and continuous redefinition of each, as circumstances change, are necessary elements in maintaining solidarity in a group large enough and free enough to have subgroups within it. This method alone, however, cannot build the world as a whole into a society, because the world-community lacks an out-group of its own kind. Politics applied in the world-community leads to a shifting balance of power in which the development of the sense of the whole is thwarted by the kaleidoscopic changes of groupings against the momentarily most powerful. This method cannot prevent occasional wars.

b) The *Juridical Method* consists in the continuous comparison of social relations established by law with social conditions discovered by observation, and the continuous adaptation of each to the other by legal procedures. It exalts procedural above substantive law in the sense that, on the one hand, the enforcement of law and social policy is subject to judicial procedure and, on the other hand, all social policies and principles can be changed by the appropriate legislative and constitution-amending procedures. Thus stability and change are reconciled.

This method has proved inadequate to unify the world as a whole because of the practical difficulties of developing an effective sanctioning and legislative system. Only within societies whose members generally agree on the meaning of justice have such systems been developed. A legal system cannot in itself create such a society. In a community lacking social consensus a legislative system can be only a system of voluntary co-operation in which law can be sanctioned only by good faith and changed only by unanimous consent. In such a system law is subordinate to the operation of the balance of power, and peace is precarious.

c) The *Administrative Method* consists in the analysis of means and ends within the community and in the subordination of the former to the latter, striving for efficiency in all activities. Unqualified acceptance of the ends of policy, planning to foresee the stages of achievement in time sequences and to prevent interferences among independent activities, and efficiency in minimizing the costs of achieving a given objective are the guides to good administration.

In times of great emergency such as war, when the objectives of a society's policy—defeat of the enemy—are clear and unquestioned, the integrating influence of administration reaches a maximum. An army in time of war is unified primarily by administration, but even in an army, politics and propaganda are also utilized. In times of peace, goals of material prosperity, individual freedom, religious idealism, and ideological unity compete with each other. If administration is to be efficient, the administrator must unify these goals and determine the means to realize them. He cannot do this unless his authority is accepted by, or enforced upon, the members of the community.

Administration as a method of community-building has, therefore, the serious handicap that it tends to rigidify the ends of policy. Concentration of attention upon the achievement of accepted ends and development of institutions which assume without question the desirability of those ends militate against easy adaptation of policy to changed conditions. On the other hand, acceptance of an authority to adapt ends and means to changing conditions militates against the efficiency of long-range planning.

It is a paradox, probably accounting in part for the oscillating character of human history, that whereas increase in the spatial and temporal scale of planning is a sign of higher civilization, planning on too large a scale destroys civilization. Such planning ossifies faith into dogma and thwarts utilization of the opportunities for human betterment which increasing knowledge makes possible. Tradition, which fixes the ends of a society, tends to come into conflict with science, which, with the expansion of human mastery of nature, suggests new ends. The essence of planning is the organization of a hierarchy of values for the longest possible time and the largest possible space. As knowledge augments the duration of this time and the size of this space, the upper ranges of the hierarchy become fixed; only ways and means remain flexible. The liberty of individuals and of lesser groups to experiment with new values, perhaps better adapted to changing conditions, is prohibited by the plan, because such liberty would interfere with efficiency in carrying out the set objective. Planning on too large a scale becomes despotism. Democracy in one aspect is a system of barriers to planning on too large a scale and an insistence that efficiency in achieving any policy shall be subordinated to certain individual liberties, to certain local, national, and regional autonomies, to certain functional oppositions, and to certain procedural and temporal requirements.

Application of the administrative method to unite the world would envisage universal acceptance of certain values of at least relative permanence. Believers in absolutistic philosophies assume that the axioms on which their systems are founded will eventually be so accepted and that social organization will be reduced to administering these truths. The administration of dogmas, however, has always led to disputes over interpretation and the problem of heresy. The system has either collapsed because it could not adapt itself to changing conditions or else procedures of interpretation of a legislative character have been developed. Furthermore, in comprehensive administrative systems, it has usually been found necessary to state basic ends in general terms such as social justice and economic welfare. This offers such a generous opportunity for interpretation that the method becomes less administrative than juridical. Administrative systems have usually found it necessary to utilize methods of propaganda as well as of logic in applying their interpretation of such general objectives.

Experience, therefore, supports the hypothesis that truth, in matters of social significance at least, is relative rather than absolute. This hypothesis precludes a unification of the world as a whole by the administrative method alone.

d) The Propaganda Method.—The political, legal, and administrative methods of community-building may all contribute to organizing the world as a whole, but none can be adequate alone. The effectiveness of each is dependent upon the existence of a world opinion that the welfare of each nation requires its subordination to the world-community in actions which affect all. With this opinion the world-community may become an organized world-society. How can such an opinion be propagandized? This has always been the basic problem of community-building.

4. THE ROLE OF SYMBOLS IN SOCIAL ORGANIZATION

An organization, especially a large one, rests in part upon the general acceptance of a symbolic construction or simplified picture of the organization as a whole. The simplest picture is the name, seal, flag, or other symbol to which are linked vague suggestions of the attributes of the organization. Somewhat more complicated is the declaration of independence, the constitution, or other document stating the general purposes and principles of the organization. Even more complicated are idealized histories and descriptions of the organization and of the characteristics of its leaders. Organizations have sometimes been symbolized by the personality of the leader, and the face of the titular sovereign still plays a part in political symbolization.

If a symbol is to contribute to a society's solidarity, the attributes which it suggests or asserts must be regarded as valuable by the members of the society. The symbol must therefore assume that the members have or can be brought to have values in common.

a) *Symbols and Conditions.*—Utopias, myths, ideologies, social analyses, histories, and other social expositions are of significance in community-building as propaganda symbols. They are to be distinguished from societies, groups, cultures, governments, businesses, associations, and other social conditions, which constitute history in the sense of what happens in a given time and place. Social conditions are the subject matter of social expositions. Although the latter may give knowledge about and attitudes toward social conditions, their descriptions always need to be verified by direct acquaintance with society in action. The exposition is a complex of symbols which may or may not correctly represent or suggest the conditions.

An organized human group is both a symbol and a condition. It has a name and is referred to by words which suggest sentiments, purposes, methods, achievements, advantages, and so on. It is also a grouping of persons according to their behavior patterns and a nucleation of stimuli activating these patterns and conditioning the lives of individuals. An organized group is different from the sum of its members in that it contributes to each member status and relationships and the power which comes from co-operation. Organization of a group also implies that opposition to out-groups will normally dominate over internal oppositions. The group is not organized unless it normally functions as a unit in external affairs pertaining to its purposes.

The symbolic character of an organized group ordinarily has some relation to its condition, especially in the minds of those in daily contact with the group's activities. But in the case of an organization functioning over a vast area, the average member may be quite ignorant of its institutions, activities, and personnel. In his mind the group may have a symbolic character which springs from sources unrelated to its actual structure and functions. Delaisi's study of the contradictions of the modern world emphasized the probability that the voluntary obedience necessary for social order should be sustained by general beliefs about the society more permanent and less complicated than the actual conditions of the society and the equal probability that periodically the divergence should become so great as to shatter either the beliefs or the conditions.

There can be no doubt that in the modern world a wide gap has developed between the dominant symbols of "sovereign nations" and the actual condition of states economically, technically, and politically dependent upon one another. This gap has created a revolutionary situation in which activities to realize the nationalistic myth and to destroy the world economy and general security are contending with activities, often by the same government, to destroy the myth and to perfect the conditions of general prosperity and security.

It is difficult to adjust symbols and conditions in social analysis, because neither can be taken as fixed, nor can either be changed at will. The natural scientist uses symbols as tools to designate concepts which fit his observations. He can abandon them at will and make new ones as the science develops. In science conditions dominate over symbols. The sociologists' symbols, however, have a meaning, life, and reality in the society quite apart from the conditions which they are supposed to designate. In fact, they often designate not conditions of the present but conditions hoped for in the future. The flag may suggest glory, honor, and protection to millions of citizens, even though at the moment it designates a defeated nation with a corrupt government administering unjust laws. The meaning of symbols, no less than the actual social conditions, must enter into all judgments concerning social groups and their activities. The problem is complicated because symbols mean different things to different people. The self-image of a nation may differ radically from the image others have of it, and all such images may be remote from actual conditions.

b) *Integration and Differentiation.*—The independence of social symbols and of social conditions makes it difficult to analyze the processes of social differentiation and social integration. A society may be becoming more integrated symbolically while it is disintegrating in fact. A group may be one symbolically and a dozen in fact. As a consequence, sociologists have come to think of groups not as entities but as continuous processes of becoming and disappearing. Portions of the world's population are continually differentiating from the rest because of such internal influences as proximity and communication, common descent and physical resemblance, subordination to common authority, and similar occupations, behavior patterns, or customs. Each of these differentiated bits of population becomes more closely knit or integrated by operation of many of the same influences which differentiate them from others. But since the differentiating groups fade into one another and overlap, one individual perhaps being the member of half a dozen, it is impossible to make rigid definitions and classifications of groups. Political judgment is necessary to determine what groups exist at a given moment and which are important. Such judgment involves consideration of the future as well as of the past. Symbols may be in process of realization, and realities may be in process of desymbolization. The group symbols may constitute a social a priori which the future may justify.

The sociologist's judgment of the existence of a society cannot, therefore, be easily divorced from his views of the good society. He may affirm the existence of a society which he likes and deny the existence of one that he despises because he realizes that the assertion of such a judgment may contribute to making it come true.

Social differentiation or group formation may therefore be defined as a process of change, both in the observer and in the conditions observed, whereby a group becomes more distinguished from its social surroundings. Group integration or society formation may similarly be defined as a process of change, both in the observer and in the group observed, whereby the position, the relations, and the activities of the parts become more efficiently adapted to group ends.

If the group itself is considered to be the observer, its attitude may be identified with the meaning of the symbols with which its members communicate, and group differentiation and integration will mean the realization of whatever distinctive symbolic structures or social ideologies prevail in the group—that is, the realization of group aspirations or self-determination.

c) *A World Myth.*—The larger the social group considered, the greater is the relative importance of the subjective or symbolic element in the process of differentiation and integration. Primitive societies resemble organisms. A man has a place in society almost as definite as the place of a cell in the organism. In the city-states of antiquity, this was true to a less degree. Even in modern societies sociologists find it most easy to study the family or the local community objectively. These have structures and processes which are relatively persistent and which are represented by symbols which closely conform to actual conditions. On the other hand, the state, the nation, the empire, the federation, and the international organization are in larger measure governed by opinion. With them the symbolic meaning differs from actual conditions. Policy derived from symbols tends, however, to shape conditions, institutions, and ideas. Policy, therefore, tends to determine the limits of such groups and the forms of their internal organization. But policy is founded on opinions which usually differ among members of the group. The larger the group, therefore, the more likely is internal conflict.

It is true that in the seventeenth and eighteenth centuries state policy in Europe welded different cultures into nations, and in the nineteenth century nationalities shattered states and established new states from the fragments. This, however, does not mean that policy is less powerful today than it was in the eighteenth century but that, with the rise of communications and the press, unofficial propagandas have sometimes been able to outstrip governments in the achievement of policy. Propaganda has risen, relative to coercion, as an efficient instrument of policy, and during the nineteenth century propaganda was still far from being a state monopoly.

The effort to integrate the human race as a whole can, therefore, expect relatively little assistance from study of the methods of integrating primitive peoples or even civilized states and nations. In the latter, customary structures and procedures have played a larger part than can be expected in the world as a whole. The great society is unique. The human race is a social unit which cannot be differentiated from external societies however much, at any moment, it may be differentiated from its physical environment, from its parts, from its past, from its potential future, and from its ideal representations.

Herein lies a paradox. The great society can be integrated only by general acceptance of common ideals, myths, or symbols. These symbols must represent the world-society as a whole, but conception is an analytic and comparative process which balks at uniqueness. Persons or groups attempting to achieve practical ideals have usually proceeded by analyzing persons into those favorable and those unfavorable to the achievement. The latter tend to become symbolized as an opponent, enemy, or devil to be struggled against. An enemy or antithesis thus appears to develop from the very nature of an ideal amid imperfect conditions. Such an opposition has usually been an essential factor in integrating those holding the ideal into a society, but at the same time it has made that community less than universal.

To avoid this paradox, if peace is to be achieved, the ideal should be conceived not as a grouping of favorable persons from which the unfavorable should be expelled but as a reorganization of all persons and groups. Unfavorable persons should be treated not as evil but as a consequence of an inadequate organization of all. Thus the community of nations must be built by a continuous development of the principles, institutions, and laws of the world as a whole, not by an organization of the angels, with the hope of ignoring, excluding, converting, or destroying the devils. Such a continuous development presupposes that the symbols of the world as a whole predominate over those of lesser groups in world public opinion.

Is this possible? A group is strong in proportion as the distinction between the in-group and the out-group is evident. Powerful social symbols usually manifest that distinction. The world as a whole cannot create a human out-group. Can it make out-groups of impersonal ideas or conditions such as war, disease, unemployment, and poverty? Can the preparation for and conduct of a campaign against such an out-group stimulate the discipline, cohesiveness, and enthusiasm which war has provided in the past?

5. THE ROLE OF VIOLENCE IN SOCIAL ORGANIZATION

The process of social integration has been an important cause of war and also an important cause of peace.

Human communities larger than the primary group have usually been organized by conquest, enlarged by more conquest, and integrated internally through the fear of foreign invasion. Within the communities thus organized, enlarged, and integrated, private war and civil war have become less frequent. Only within organized communities have peoples and groups been able to accept procedures assuring a peaceful settlement of all their disputes. War has thus tended to become less frequent but more severe as social organization has proceeded. The organization of greater communities has enlarged the areas and the periods of peace, but at the expense of bigger and worse wars when they have come.

Why has war been so important in the process of community formation? Although the political importance of war has varied under different conditions, it seems probable that war will continue to be of dominant political importance so long as the process of community formation and development remains a process of persuading people to accept symbols rather than a process of enlightening people on how to deal with unwanted conditions.

Statesmen feel obliged to take immediate advantage of favorable historical conjunctions to increase the acceptance of their own symbols and to diminish that of rival symbols. Effort rapidly to persuade a large group is likely to involve violence. This argument rests on three assumptions which concern, respectively, opinion, historic contingency, and violence.

a) Group Integration and Opinion.—The boundaries of membership and even the existence of human groups are not entirely determined by objective conditions, and the larger the group, the less are they so determined. Consequently, judgment as to whether a large group exists is more a matter of persuasion and faith than of fact and reason. Nature sets few limits to the scope of empire. A nationality might expand or disappear within a generation by the application of a proper system of civic education, provided the children were taken young enough and parental influence were eliminated. Such a process is to be observed in the creation of Soviet, Fascist, and Nazi nationalities. Furthermore, the internal organization of groups is not wholly determined by past conditions. Types of organization as different as socialism, communism, fascism, and liberalism have developed in a Europe all parts of which had experienced similar technological and economic conditions in the nineteenth century. Which type will prevail is a matter of opinion and depends in some degree on the relative intensity of the faith, belief, and loyalty of the adherents to the respective symbols. It is true that groups with similar technologies tend to converge in many other conditions, as Sorokin points out in comparing the development of Soviet and American cultures, but if the symbolic structure differs, the group will retain its distinctiveness in matters not closely related to technology.

The dependence of the conditions of human groups upon attitudes becomes progressively greater as civilization and the means of communication and invention advance. Primitive peoples are limited by material conditions, especially the state of communication, in their capacity to bring new people into the group or to change their organization. On the other hand, advanced peoples have extensive capacities of assimilation and change. Consequently, as science and law widen their capacities to control nature and the processes of civilization, the number of possible population groupings and organizational forms increases and the process of selecting from among these possibilities becomes more political, more a matter of opinion. A study of past or present conditions of populations is of exceptionally small importance in judging the future of the modern world. That future depends in considerable degree on present and future opinions and faiths.

b) *Persuasion and Historic Contingency.*—The process of molding group opinions is a historic process. Every step is dependent upon a particular conjuncture. Consequently, time is important. States rest upon opinion, and at moments, particularly after a devastating war, opinion may be malleable and revolution may be possible; but unless the opportunity is seized, the rigidities of vested interest will again develop. Furthermore, states are surrounded by other states, and the relations within the entire group at a moment in time may create the opportunity for radical change whether by conquest or by federation. But custom is always important in human relations. Consequently, an opportunity lost may be lost for an indefinite future. Statesmen believe they must strike while the iron is hot, that in politics time and tide wait for no man. Consequently, processes of persuasion, if they are to be effective, must be rapidly achieved when historic circumstances are favorable.

In world politics, particularly, the course taken by opinion—its symbolic fixations, direction, and intensity—during a historic moment of a few months may fix the structure of the world-society for decades or centuries.

c) *The Historic Moment and Violence.*—These two circumstances—that community formation tends to depend upon opinion and that the opinion which dominates at a historic moment may set the course of development for a long time—account for many wars, because war is the most effective instrument of rapid persuasion. Education is an instrument of persuasion. So also are propaganda, economic inducement, and invocation of traditions, laws, and beliefs. These methods, however, have frequently seemed inadequate to bring masses to an enthusiastic concurrence with political proposals while the historic moment is at hand. Consequently, people have been told that they must concur to defend themselves from invasion, to protect threatened interests, or to prevent the success of destructive ideas. The argument may often be true, but whether it is or not, its validation may require war.

Federations have sometimes been achieved by peaceful negotiation, but not often. Most of the great political blocs designated as sovereign states and most of the great changes in forms of organization have been effected through utilization of such processes of persuasion as war or insurrection at the critical historic moment.

d) Violence and World Organization.—Is there any solution to this problem? Is violence an inherent condition of large-scale political integration? If statesmen were certain that force would result in a stalemate or intolerable destruction of all belligerents, force might cease to be of value, and other methods would have to be used. It has also been suggested that force might be so regulated as to be relatively harmless. War might become a duel of champions, a competition in building military machines, or even a game played on a chessboard.

Integration of the human race as a single organization may for the first time be practicable in the age of universal, rapid mass communication. Such integration might make possible universal acceptance of pacific settlement for all differences. The problem of determining what is the supreme group would be eliminated because the human race, distinguishable by objective evidence, would be the supreme group. Furthermore, the element of time would be less significant because the danger of external invasion would not exist. A universal society would have all time before it to educate its people to accept solutions. The dangers would remain that a universal organization might breed conservatism and unadaptiveness to climatic, organic, and technological changes or that the impatient might seek to accelerate solutions beyond the potentialities of peaceful persuasion. Civil war might still occur unless the integrating symbols—the world myth—were accepted with a sufficiently vigorous faith. Such a faith might stifle social inventiveness and adaptiveness and might flag without the stimulus of external pressure. Communities without an enemy have tended to divide.

The question whether progress is compatible with the elimination of violence has divided different branches of communists, socialists, anarchists, nationalists, progressives, and other reformers. The answer to this question may lie in the details of world-organization, maintaining oppositions among whole and parts, among functional and regional groups, and among individuals, with a sufficient general solidarity to prevent the expression of these oppositions by violence.

CHAPTER XV

PUBLIC OPINION AND WAR

Among the causes of war is the difficulty of making peace a more important symbol in world public opinion than particular symbols which may locally, temporarily, or generally favor war. If only love of peace and hatred of war could be universalized, say the pacifists, war would disappear. The more practical minded hope that understanding of the increasing destructiveness of war may develop a world public opinion adequate to sustain an organization able to prevent war. Hatred of war has provided a rallying cry for popular "peace movements," particularly after general wars of great destructiveness.

A public opinion is a relatively homogeneous expression of preference by members of a group about issues which, though debatable, concern the group as a whole. A public opinion, therefore, implies the existence of a public or a group the members of which communicate among themselves on matters of common interest, of a leadership which formulates, publicizes, and concentrates attention upon the issues which are important at a given time, and of manifestations of opinion on the issues and an indication of a willingness to acquiesce in action conforming to the predominant opinion. The essence of public opinion is controversy coupled with acquiescence in eventual group action, diversity of attitudes coupled with unity of action. There is no public opinion about an issue on which there are intransigent minorities within the public. Nor is there a public opinion about issues on which there are no minorities at all. According to A. Lawrence Lowell, "If there is no disagreement on a proposition, there is no issue. The proposition has the status of fact or truth in that public. It is undebatable. In a 'crowd' or 'mass,' distinguished from a 'public,' all questions are undebatable."

1. SYMBOLS OF WAR AND PEACE

The theory that a suitable public opinion might eliminate war assumes that wars have been caused by opinions about symbols, that these symbols have usually had little relation to actual conditions, that a persistent and world-wide opinion favorable to the symbol "peace" might be developed, and that that symbol might acquire an intelligible and realizable meaning. Wars, according to the UNESCO constitution, are made in the minds of men, and it is in the minds of men that the defenses of peace must be constructed. This merely applies the sociological principle that action always proceeds not from conditions but from the interpretation of conditions by the decision-maker, from the image in his mind.

a) Opinion and Symbols.—In a town meeting public opinion concerns issues with which the entire public is acquainted through direct experience. The issues concern conditions rather than symbols. In larger groups, on the other hand, the greater part of the members can seldom have this direct acquaintance with issues. The issues necessarily concern symbols which carry to the average members of the public only vague suggestions of the conditions involved.

Many aspects of the policy of contemporary nations and international organizations are controversial, but if the group is to survive, general acquiescence of the members in group policy is even more necessary now than it was under less complex conditions. Controversial questions arise concerning the ends regarded as important to the group, the means to be employed for furthering group ends, the standards and rules which the group expects its members to observe, and the performances and rituals intended to manifest the existence and character of the group to its members and to outsiders. National and world politics concern in large measure the answering of such political, administrative, legal, and ceremonial questions. Such issues, especially in world politics, invite the use of vague symbols. Practical behavior, whether of a political or administrative character, aims at something potential but as yet unrealized. This "something" can be presented only by symbols. One can comprehend the unachieved ends or unapplied means for achieving ends only through the media of symbols. Formal behavior, whether legal or ceremonial, is guided by norms or rituals which can have only a symbolic existence. The larger the group and the less accessible all its members to direct sensory contact with all the others and their activities, the less available are instinct, custom, or universal acceptance as bases of group behavior, and the more symbols and opinions about them are the stimuli and guides for behavior.

In the large groups which make war in modern civilization, symbols have been responsible for initiating and guiding that particular behavior. Frontier guards may, it is true, shoot at one another from habit or caprice. Even large-scale hostilities may develop from accident, error, guerrilla activities, border hostilities, or civil strife. But war in the legal sense does not start without elaborate procedures of parliamentary or council discussions, declarations, orders, and proclamations dealing with its means, ends, modes, and justifications. War has therefore been intentional in the sense that symbolic acts which mean war and justify it have been indulged in by some government. Civilized war differs in this respect from animal war. The latter is stimulated by direct sensory experience by the intended victim and aggressor of each other, whereas the former is stimulated by an interpretation of events in terms of the rights, honor, interest, or policy of the group. The leaders who do this interpreting of symbols and the masses who accept the interpretations may have little or no acquaintance with the conditions meant by the symbols.

The subsequent experiences of the soldiers with the conditions of warfare are accidents of war not directly determining its origin or its termination. They are, it is true, elements which enter into the meaning of the determining symbols, but numerous other elements also enter into this meaning. War has meant the legal condition which equally permits two or more hostile groups to carry on a conflict by armed force. Sensory experiences of armed force are less important in this conception than political objectives, tactical and strategic movements, legal rules, and propagandistic characterizations of the enemy all expressed in an abstract vocabulary only remotely related to the sensory experience of actual fighting. War therefore has arisen immediately in the world of symbols, not in the world of conditions.

b) *Opinion and Conditions.*—The symbols behind wars are usually richer in affective than in informative meaning. They often refer to fictions, myths, and stereotypes with little relation to conditions. Opinions about such symbols are expressions of attitude. They manifest feelings which vary from individual to individual and are not necessarily related either to observation or to logic. They are, therefore, to be distinguished from truths which describe conditions verified by experience or which express the relation of such conditions through the logical ordering of symbols which have informative meaning. In questions concerning the most general objectives of group policy the experience of every member of the group is, according to democratic assumptions, equally important; consequently, the immediate test of the truth of a proposition on such questions lies in the universality of the acceptance of the proposition within the group by those who understand the meaning of its terms and who accept it on the basis of their own experience. Opinion may become so accepted as to constitute, for the time, truth, and truth may become so contradicted by new observations as to become opinion. But at a given time in a given group the distinction can usually be made. Truth is accepted as a fact; opinion only as a belief. Beliefs, it is true, in religions or propagandas are usually presented as historical facts; but in so far as both the facts and the deductions from them are controversial within any population, they lack the status of truth.

The military results to be expected from war are rarely certain, and the eventual economic, political, and cultural consequences can seldom be calculated with any approximation to accuracy. War is a gamble, and even if calculations are made, there is usually difference of opinion in high quarters and even more among the general population. There is almost never a universal acceptance of any proposition concerning the need or wisdom of a particular war. War is initiated or rejected because of the "weight of opinion" among those with authority to act for the group.

Even further, it can rarely be said that the particular arguments for war have any status as truths. Economic arguments, political arguments, and historical arguments are made by propagandists, but they seldom have the support of all the experts in these disciplines, even in the country utilizing them. Thus if such phrases as economic, political, and psychological causes of war are used, it is not because there is a direct relationship between the outbreaks of wars and the truths or facts accepted in these disciplines but only because propositions, good, bad, or indifferent, concerning economics, politics, or psychology have influenced an opinion favorable to war.

Arguments which influence opinion often have little support in social science, and truths affirmed by social scientists often have little influence upon the movements of opinion in contemporary societies. This suggests that little should be expected from studies of the statistics of population, commerce, finance, and armaments or the technicalities of law and procedure in explaining war. It is only as such matters affect opinion that they cause war, and opinion is moved by symbols of such vague meaning that no precise correlation with statistical series or refined analyses is to be expected. The causes of wars must be studied directly from indices of opinion, not indirectly from indices of conditions, even though the two have an overlapping vocabulary.

c) The Diversities of Opinion.—Opinion may be measured as to direction, intensity, homogeneity, and continuity with reference to symbols. It is clear that the opinions of groups vary greatly in all these dimensions. They are greatly affected by types of leadership, by methods of propaganda, and by economic, political, and other circumstances. But whatever the circumstances of particular groups, it is clear that the opinion of a group formed from sources wholly within itself will frequently differ from the opinion of any other group formed from sources wholly within itself. Such differences of opinion are likely to lead to opposition. If the groups are in close contact, conflict and war may result. Thus the only opinion which might assure peace is one held by a public which includes all the groups in contact with one another. Such public opinion must, under present conditions of interdependence, be a function of a world-group, and if peace is to continue, that opinion must be continuous. This does not mean that all the members of each nation must also be members of the world public, but it does mean that within each nation there must be enough persons whose horizons extend beyond the group to keep its policies consistent with the requirements of the world-community.

Even a world public opinion would be opposed by dissident minorities, unless, indeed, all political objectives had reached the status of truth, a condition which would end not only controversy but also human progress. With a vigorous public opinion, however, the opposition of minorities should not mean war. Public opinion may influence minorities to subordinate their opinions to the predominent opinion, to bide their time, and to keep the peace.

What if they do not? Those subscribing to the dominant public opinion will then be faced by the alternative of using force to suppress them or of acquiescing in the disintegration of the world public opinion. If peace is the symbol of world public opinion, which should they do? Does peace mean that coercion shall not be used, or does it mean that public opinion shall prevail?

d) The Meaning of Peace.—The dilemma just suggested indicates the importance of determining the meaning of peace. Advocates of peace have been divided into two camps—the pacifists and the internationalists. In times of peace they have tended to come together, but in times of war or threats of war the pacifists have urged nonintervention, whereas the internationalists have urged collaboration against aggression. States which pursue pacifist policies of avoiding war may encourage aggression and may become its victims. States pursuing internationalist policies, by seeking to prevent or to suppress aggression, may become involved in war. Support is thus lent to the hypothesis that peace is not an intelligible idea.

The unsophisticated interpretation of peace is that of pacifism. Peace is negative. It is the absence of war. The philosophers of this theory have pointed out that if everyone renounced intransigent opposition to existing conditions or opinions, no matter how oppressive or unjust they might be, there would be no war. Eventually, rational means of solution would be found. Peace, they say, can only be a negative symbol because, if any positive symbol were taken as the dominant ideal, war might seem necessary to achieve it. Wars, they point out, have been fought for the sanctity of treaties, for the preservation of law, for the achievement of justice, for the promotion of religion, even to end war and to secure peace. When peace assumes a positive form, therefore, it ceases to be peace. Peace requires that no end should justify violence as a means to its attainment; consequently, no person or group should believe in any end so firmly that compromise or at least postponement of realization is impossible.

The internationalists, however, reply that the desire for peace cannot be superior to itself. Although peace may require a renunciation of intransigent oppositions, it cannot require a renunciation of all oppositions or it becomes self-contradictory. Peace cannot dissipate actual war by wishful thinking. Peace which tolerates breaches of peace or encourages them by appeasing aggressors destroys itself. Peace which means merely the avoidance of war in any circumstances is self-defeating, because it encourages injustice which leads to war and it frustrates the co-operative handling of problems which alone can prevent war. To be either logically conceivable or practically effective, peace, they say, must have a positive meaning. It must mean international justice. International justice implies orderly procedures and a spirit of co-operation in dealing with international problems. These conditions can only be realized in a world-society. The symbols of peace are, therefore, the symbols of a world-society.

The internationalists concede that the achievement and maintenance of a world-society are certain to arouse opposition and to require the occasional use of force by the whole to control the parts. Consequently, the concept of peace, although it excludes war, cannot exclude all use of force. A peaceful society must anticipate occasional crimes and rebellions and must provide for defense and police to suppress them. The building of peace even involves risks of violence on such a scale as to resemble war in the material sense. The rejection of such risks, however, would stop work on the building. This concept of peace, presented by constitutionalism within the state and by internationalism in the family of nations, distinguishes crime, rebellion, aggression, and war from necessary defense, criminal justice, police action, and sanctions.

Although unanimity as to the meaning of peace has not been achieved, the weight of opinion has supported the internationalist point of view. Theologians, philosophers, psychologists, mathematicians, economists, jurists, and publicists who have considered the subject carefully have believed that, if peace is to attract public opinion and to fulfil its expectations, it must be a positive conception. It must mean justice and order, and it cannot mean those without organization. Experience has shown that in limited areas violence has been prevented only when peace was identified with an organized society which made justice and order its first concern.

The advent of nuclear weapons, however, has brought recruits to the pacifist position. Many, including "realists" who had thought "power politics" inevitable, have followed Pope John XXIII in believing that no situation could make nuclear war, or any hostilities which might escalate into such war, justifiable or rational. They have reconciled the two views by recognizing that peace requires a world organization to prevent such escalation even if human rationality and a sense of national interest were sufficient to prevent the deliberate initiation of nuclear war.

The conception of positive peace is not easy to grasp. It encroaches upon many established conceptions and interests. The world public is not likely to favor it sufficiently intensely, continuously, and homogeneously to achieve it unless the conception exists not merely in public opinion but also in private attitudes. If it is to be realized, peace must be accepted not merely in symbols and myths but also in personalities and cultures. To gain such acceptances presents a problem of propaganda and education.

2. PEACE AND WAR PROPAGANDA

Propaganda is the process of manipulating symbols so as to affect the opinion of a group. It may be contrasted with education, which is the process of manipulating symbols so as to affect the attitudes of an individual. The two are related, because opinion to some extent reflects attitudes and attitudes are to some extent influenced by opinion, but they are not necessarily identical. An individual's overt expression of his attitudes may not accurately indicate his actual attitudes. He may lie. He may be unconscious of his attitudes. He may be influenced by immediate associations, suggestions, or pressures without realizing it. His personality may be divided: his public or mass conscience may be disclosed by his opinion; his private or individual conscience—his "real" attitude—may be undisclosed.

Subtle conflicts between dispositions derived from heredity, from family training, from formal education, from the church, from business associations, and from introspection and reflection may remain unresolved in the personality, ready to manifest contradictory behaviors on different occasions. These conflicts roughly categorized by the distinction between impulse, conscience, and reason give warning that the distinction between opinion and attitude is over-simplified. It is, however, useful as marking the general distinction between propaganda and education.

In no field is the difference between attitude and opinion more marked than in relation to war. Private attitudes are likely to be affected by the personal aspects of war—death, destruction, killing, mutilation, glory, adventure, escape, economic advancement—and the evaluations of such events and possibilities from hereditary impulses of self-preservation and family affection, from social standards acquired through education, religion, and group experience, and from personal standards derived from past efforts to adjust impulses with social requirements. Public opinion, on the other hand, tends to emphasize the public aspects of war—national defense, national policy, national ideals, international law, world politics, human welfare, justice, and progress. Both pacifists and militarists, it is true, seek to utilize private attitudes in building public opinion about war and peace, but the wide divergence of their symbolisms indicates the extreme ambivalence of these attitudes.

The influence of attitudes and education on war and peace will be dealt with in the chapter on human nature and war. Attention will here be confined to opinion and propaganda.

Propaganda seeks to manipulate symbols so that opinions in a given population will maintain or change direction, become more or less intense, more or less homogeneous, more or less continuous. Propaganda is conducted through access to or control of instruments of communication—in modern societies, especially the press, moving picture, radio, and television.

a) *War Propaganda.*—Wars have always required propaganda for both their initiation and their conduct, and the methods have long been elucidated. They were exhibited in the histories of Thucydides and the orations of Demosthenes and have been analyzed in the studies of recent wars. The objects of war propaganda are the uni-fication of our side, the disunion of the enemy, and the good will of neutrals. Our unity is promoted by identifying the enemy as the source of all grievances of our people, by repeating and displaying symbols which represent the ideals which we share, by associating the enemy with hostility to those ideals, and by insisting on our own nobility and certainty of victory and on the enemies' diabolism and certainty of defeat. The enemy is disunited by accentuating the divergency of factions, by suggesting incompetence of the leaders, by demonstrating the certainty of eventual defeat, and by implying benefits to groups or individuals if resistance is ended. Neutrals are influenced by threats of invasion, by emphasis upon the loftiness of our war aims and the sordidness of the aims and methods of the enemy, and by emphasis upon special advantages to particular groups or to neutral nations by favoring our side.

The pressure of propaganda coupled with the pressure of events has frequently brought neutrals into war. In only three of the fifteen war periods of the last three centuries which involved one or more great powers on each side and lasted more than two years did a single great power avoid being drawn into war. If a war breaks out between great powers, it is to be expected that all the great powers will get in unless the war ends very rapidly. A belligerent disposition evolves from continuous whetting of the natural war interest in the news, from humiliating incidents, from political interest in the balance of power, and occasionally from the influence of special economic interests. Interest brings familiarity, and familiarity gradually brings acceptance. An American population with a tradition of neutrality rapidly became war-minded and eventually belligerent in the periods of the French Revolutionary and the Napoleonic Wars and of World Wars I and II. The development of this belligerency has been traced in detail through studies of the American press during neutrality periods. These studies indicate a gradual shift from objective war stories to stories relating the war to the United States, then, as the actual crisis involving American interests developed, to an emotional appeal.

b) Peace Propaganda.—Efforts have also been made among both primitive and civilized peoples to preserve peace by propaganda. The problem is more difficult than the problem of war propaganda because, to be effective, peace propaganda must gain attention simultaneously within all potential belligerents, and yet peace is intrinsically less interesting to human beings than war. On hearing of a conflict situation, people instinctively prick up their ears. Perhaps this is a biological inheritance. Perhaps those who were not alert and attentive in the presence of conflict situations were long ago eliminated in the process of natural selection. When actual conflict situations are not present, the same interest may attach to symbols suggesting them. The newspaper reporter and the historian know that they can claim the attention of their readers by accounts of conquest, war, and rumors of war. The artist, sculptor, or poet can produce a work of art which the untutored will at once label "war." It is difficult, on the other hand, to imagine a painting, statue, or poem that the average man would unequivocally label "peace." People will buy newspapers which explain the technical details or tactics of a battle or a ball game, but who, except the specialist, would read such a dissertation on the structure or procedure of orderly government? In spite of the efforts of peace propagandas to objectify peace as a particular religion, as international law, as a system of arbitration, as a treaty of disarmament, as the Kellogg Pact, or as the United Nations, the public thinks of peace as merely the absence of war and finds it uninteresting. They may even regard the positive idea of peace as dangerous. The League of Nations was opposed by some United States senators on the ground that it would lead to war.

The negative idea of peace has, however, often frustrated realization of such a positive peace. Peace propaganda has frequently in times of crisis urged particular groups to isolate themselves from areas of contention in order to avoid war and has thereby disintegrated the international community and assured the initiation and subsequent spread of war. In an interdependent world, propagandas of isolationism, neutrality, and absolute pacifism, however honestly pursued in the name of peace, have been causes of war.

Peace propaganda has also often defeated itself by denouncing the private rather than the public aspects of war. Emphasis upon the horrors of war has not, under all circumstances, created an attitude favorable to peace. It may instead stimulate an interest in war. It may stimulate intensive preparedness to avoid war and thus create conditions of military rivalry favorable to war. It may stimulate reluctance to accept the risks of war necessary for effective building of peace. Diversion of attention from war or threats of war to other interests may also endanger peace. Lysistrata's female strike against war might have contributed to the defeat of Athens rather than to the ending of war. The interest of the Athenians in business as usual in spite of Demosthenes' *Philippics* seems to have contributed both to war and to the end of Athenian liberties. Nothing is more promotive of war than diversion of the attention of the prospective victims from the aggressor's preparations.

Peace propaganda to be effective must present a suitable conception of peace simultaneously in all parts of the world. Peace must be pictured as rational progress toward a world-society maintaining justice and solving world problems co-operatively. War must be pictured as the irrational obstruction of that progress. On the other hand, force which forwards such a society must be pictured not as war but as a necessary instrument of peace.

Such a propaganda can proceed simultaneously in all nations only if managed by a world agency with access to all important populations. Reliance upon a just world-order by some of the states might induce them to neglect their defenses and so to increase opportunity of others for successful aggression unless opinion favorable to positive peace is sufficiently general to make the world-order actually effective. Propaganda, even for a positive peace, may therefore, if carried on only in a few nations, increase the probability of war in proportion to its success in the areas in which it operates. Obviously, the dominant control of communications and propaganda by the national governments seriously limits the possibilities of a sufficiently general and effective peace propaganda by a world agency. The nations acting individually cannot carry on a sufficiently general propaganda to be effective, but they can prevent a central agency from functioning.

Within a given area the success of propaganda for positive peace probably depends upon the position of conflict and violence in the personality types created by the culture. Propaganda, as a short-run activity distinct from education, cannot change personality or culture but only stimulate or suppress attitudes which exist. These attitudes may be classified according as they relate to impulse, reason, or conscience, that is, to the biological, the psychological, or the social man.

(1) *Appeals to the biological man,* that is, to the instincts of self-preservation and of family affection, are of little significance in preserving peace under conditions of high social tension. The response to the stimulus of such instincts may be pugnacious rather than cautious. Furthermore, biological instincts in the opposite direction, such as aggressiveness and sadism, may cancel them out. Men may be afraid of getting killed, but they may be lured by the love of aggression and dominance. In organized societies the biological instincts are sublimated by acquired dispositions and social ideals. The social man rules the biological man. Even though his fears are

not canceled by his aggressions, the soldier may go on from the greater fear of social disgrace. The propaganda of the military, based on social ideals, loyalty, and sacrifice, in time of crisis, has overridden the pacifist appeals based on the horrors of war. This may not be true when the destructiveness of weapons threatens not only the individual but the nation and mankind.

(2) *Appeals to the psychological man,* that is, to reasonable consideration of habitual interests, also have had rather slight influence in times of crisis. High tension levels exist because of widespread dissatisfaction with the normal. When there is much unrest, appeals to war override appeals to the humdrum of daily routine. Furthermore, appeals to custom may favor war as well as peace. The behavior of man in normal times is governed by social custom and by interests. Custom includes both non-institutionalized folkways and institutionalized mores such as systems of law and religion. The interests which guide the behavior of individuals or groups are those objectives which custom, culture, public opinion, and group procedures assume people are interested in. Why does a man in contemporary culture have an interest in the accumulation of property? It is not because of a biological drive, for many primitive men do not have it. It is because of the particular culture. In modern societies, interests include pecuniary gain and personal prestige through political, professional, or social recognition, but in times of group crisis interests tend to become social and symbolical. In most societies war is an institutionalized custom, and particular wars are associated with the preservation of group integrity, territory, and culture. Soldiers are drilled to obedience, and reserves are accustomed to the idea of mobilizing upon call. The entire population is propagandized into accepting the necessity of war and the justice of its cause. Thus in modern nations both custom and opinion support war more than they support peace in time of crisis, as illustrated by American opinion during the Berlin and Cuba crises of 1961 and 1962.

While in general men can make money or acquire prestige more rapidly in time of peace than in time of war, some may acquire money and prestige from war, and others may be persuaded that they can do so. Munitions manufacturers generally prosper in war, and military and naval officers advance more rapidly. Speculators may profit from war inflation, and many types of businessmen may for a time. Labor and the civilian population may profit from war industries and bases in their neighborhoods.

War itself may promise the satisfaction of normal interest to many, and the results of successful war may promise it to others, such as younger sons and experts looking for good jobs in colonial areas, traders looking for new markets, investors expecting concessions in undeveloped lands, manufacturers expecting access to cheaper raw materials, and entrepreneurs seeking privileged opportunities which may result from the conquest of foreign territories. Mention may also be made of farmers and laborers, who may anticipate opportunities to migrate to more favorable regions in case of victory and whose products and services command more return during the war itself.

Doubtless, the rationality of such expectations varies enormously according to the techniques of war and international intercourse at the time. The increasing destructiveness of war and the increasing complexity of international commercial and financial operations have decreased the probability of many people's making profits out of war. War could make money for Cortez in the sixteenth century and for the British East India Company in the seventeenth, but not for many Europeans in the wars of the twentieth century. War has tended, as economists have pointed out and bankers have agreed, to be a great illusion to the population as a whole and to stable economic enterprises.

The appeal to the psychological man—to normal interests—is gaining weight as an instrument of peace with the totalitarianization of war and the expansion of world intercourse. But such appeals are still overcome in times of high tension by appeals to ideals. Men will go to war for nation, for state, for humanity, or for permanent peace, even when they know it will give them personally nothing, either in cash or in prestige.

(3) *Appeals to the social man* are the strongest of appeals, especially in times of stress. War is propagandized by appeal to group symbols and social utopias. A peace sentiment may be propagandized if the prevailing ideal of human personality is pacifistic. Quakers and followers of Gandhi have resisted the appeal to war because of religious ideals rather than from fear or interest. The strength of ideal resistance to war is in fact indicated by the recog-

nition accorded to it in the military recruiting systems of many
states.

Newspaper studies indicate that as tension increases, as war approaches, appeals have tended to be on an idealistic level. Appeals
in the *New York Times* during the early days of American neutrality
in World War I were often legalistic or economic, but as interest in
the war increased, and the tension level in the United States became
higher and higher, the tone of editorial comment became more and
more idealistic. As the outcome of the war became doubtful, the
possible influence of the victory of one side or the other attracted
more attention, and the alternatives of peace or war shifted to the
alternatives of assistance to one side or to the other. This was soon
followed by entry into the war on the favored side.

3 . CONDITIONS FAVORABLE
TO WARLIKE OPINIONS

Among conditions that influence the peacefulness of warlikeness
of public opinion are those affecting the tension level.

a) *The General Tension Level* of a population, in its positive
phase, may be compared to the potential energy of a dynamic system and, in its negative phase, to the tensions of the materials in a
static system.

The tension level indicates the quantity of social energy available
to the leaders of a group, and it varies proportionately to the intensity and homogeneity of opinion. If each member of the population is
intensely loyal to the same symbol, the tension level is at a positive
maximum. If each member of the population is intensely loyal to a
different symbol, the tension level is at a negative maximum. Between the two is the condition of minimum tension level characterized by moderate loyalty to many symbols of overlapping meaning.

The positive maximum is approached in the totalitarian states,
where all other symbols are subordinated to those of the state and its
leader, and attitudes toward these symbols are intensely favorable.
On the supposition that attitudes vary in intensity in proportion to
opposition, such a condition requires opposition to an enemy external to the population. Intense and homogeneous attitudes cannot exist in a wholly isolated population. In such a population the
maximum tension level would be achieved by a comparatively equal

division of opinion between two factions or parties one of which is intensely and homogeneously favorable and the other opposed to the same symbols. If two such factions are equally favorable to different symbols, the tension level would vary proportionately to the degree of opposition between these symbols. Thus if two political parties are each in the middle of the road with only trifling differences of policy, the tensions will be lower than if there is one party to the extreme left and another to the extreme right. In the latter case the tension level may reach heights threatening a revolution and may also enable an adroit leader to externalize the high tension against an outside enemy.

As the number of symbolic formations within a group increases, the intensity and homogeneity of attitudes toward each tend to diminish. The tension level also tends to diminish and attains neutrality when every member of the population is moderately interested in all the diverse, overlapping, and sometimes conflicting symbols of importance within the group. Under such conditions leadership can only adjust conflicts within the group. Very little social energy is available for enterprises of the group as a whole. Energy is largely absorbed by the effort of each individual to adjust the conflicts among his own loyalties. Such is the ideal of democratic liberalism.

Below this condition of stability and peace, negative tensions may develop in proportion as the symbols attracting loyalty increase in number and diminish in number of adherents. Conditions of extreme negative tension place a strain on the stability of all social institutions, and thus a comparison may be made to the overloading of the materials in structural mechanics. Social institutions and myths subjected to heavy tensions because of the diversity of attitudes about them will crack. Negative tension levels reach a maximum under conditions of complete anarchy and panic, in which each individual is intensely interested only in his own self-preservation. This is the condition of *bellum omnium contra omnes* which Hobbes described as the state of nature, in which everyone is completely free and completely frustrated. In such a condition of high negative tensions the adroit leader may direct loyalties arising from self-interest to a single symbol offering security to all. All may regress and, in the Hobbesian social contract, sacrifice their liberties in exchange for the security which a dictator will give them. "Among embittered and reckless people the symbols and practices of the established order are imperiled, and the moment is propitious for the speedy diffusion of opposing myths in whose names power may be seized by challenging élites." According to Lasswell, the anarchy of revolution is succeeded by the oppression of tyranny.

Thus extreme tension levels, whether positive or negative, are closely related and favor violence, either external or internal. They may be contrasted to normal tension levels where opinions are moderate, social institutions are capable of regulating behavior, and society is stable. The latter, however, places a greater responsibility and a greater strain of individual adjustment upon each member of the community.

The advance of civilization tends to require more social energy and higher tension levels, but it also tends to increase the strength of institutions, the rationality of leadership, and the responsibility of individuals. Advanced civilizations may therefore be stable. Civilization makes possible the union of great social energy and stability to an extent impossible among primitive people guided mainly by custom. But this characteristic of civilizations may explain why their rise has tended to be accompanied by war fluctuations of increasing amplitude.

b) Extreme Tension Levels.—What are the conditions favorable to extreme tension levels and hence favorable to violence? It appears that extremes, either of general security or of general insecurity, may generate high tension levels. On the one hand, prolonged conditions of tranquillity and stability tend to decrease resistance to propagandas of violence, and on the other hand, conditions of insecurity, anxiety, and apprehension tend to create a receptivity to such propagandas.

Revolutions are concentrated manifestations of the conditions of opinion underlying all social violence, whether denominated rebellion, insurrection, or war. Starting as new symbols in local areas, revolutions spread ideas of violence by contagion and opposition. All revolutions start in principle as world revolutions. Their symbols and principles must, in the opinion of their initiators, become universal or nothing. Although the sobering experience of local success usually tends toward geographic limitation, before such limits have been established friends and foes of the new symbol will have come into conflict and will have heightened tension levels in remote areas.

Red-baiting, transmitted by the White opposition from distant Russia, caused excitement in an America trying to return to normalcy. The American Declaration of Independence, a century and a quarter earlier, had agitated autocratic Russia in much the same way.

Threats of, or resort to, violence, in any corner of the world, under modern conditions of communication, whether in support of established ideologies or of revolutionary utopias, induce a general rise in tension level.

High tension levels may arise not only because a minority is discontented as a result of too much stability but also because the majority is disgusted as a result of too much instability. General apprehension about the future of social, economic, cultural, and political institutions and practices creates a rising tension level. Such a condition may arise either from declining faith in the prevailing ideology or from explicit symptoms in the material processes themselves. The confidence of most people in the continuity of existing conditions results from beliefs in myths about them rather than from analysis of the conditions themselves, and loss of that belief may result from the propaganda of new myths as well as from changes in conditions. The extensive observance by an elite of an imported myth may develop a schizophrenic society of high tension if the public continues to believe in the old myths, a condition said to have characterized Japan after the revolution of 1868.

Economic decline may not cause great anxiety if slow, as has often been the case in India and China, because the energy of the population may be reduced more rapidly than their consciousness of deteriorating conditions. Rapid economic decline, however, may arouse the awareness of people while they still have energy and induce them to accept propagandas of violence and revolution. The French peasants and workingmen revolted in the late eighteenth century, although their condition was better absolutely than that of the similar classes in Germany at the time. The Russian peasants and workmen revolted after the rapid economic decline in the latter part of World War I, whereas their more gradual impoverishment during earlier periods had led only to sporadic incidents. The rapid development of unemployment after the world crisis of 1929 caused high tension levels, although absolute conditions were in many cases better than in much of the nineteenth century.

Rapidly changing technological and economic conditions which

seriously alter the position of economic classes cause exceptional unrest. The development of commerce and industry deteriorated the
relative position of landowners and peasants and caused much unrest in Renaissance Europe. World War I, inflation, and depression
deteriorated the relative position of the middle class in much of
Europe and caused serious unrest in the 1930's. The industrial revolution, though its ultimate effect was economically beneficial to the
working classes, produced extensive technological unemployment
and violence in Chartist England. Similar conditions led to violence
in China in the 1920's. An extreme lag between technological
changes and cultural adaptations causes high tensions.

International trade and dependence on distant areas for food and
raw materials through increasing living standards in time of peace
have caused great distress from blockade in time of war and extreme
anxiety lest such conditions be repeated by commercial barriers even
in time of peace. This anxiety has contributed to demands for national economic self-sufficiency. Efforts to meet these demands have
resulted in a disintegration of international trade, general deterioration of standards of living, and more intense anxieties. Such a vicious
circle precipitated by World War I contributed to the unrest before
World War II.

No less important than economic apprehensions have been apprehensions of a loss of cultural prestige. The future relative importance
of a type of culture is often considered dependent upon the population potentialities of the land which it occupies. If a people believes
that new lands are mainly destined to be overrun by alien cultures,
it may fear that its own "place in the sun" will be impaired. In the
1880's John Fiske wrote of the "manifest destiny" and "stupendous
future of the English race" when its far-flung areas, the United
States, Canada, Australia, and New Zealand, become fully populated by peoples of English-speaking culture, and "such nations as
France and Germany can only claim such a relative position in the
political world as Holland and Switzerland now occupy." At the
same time Heinrich von Treitschke was commenting: "What opportunities we have missed. . . . The whole position of Germany depends upon the number of German-speaking millions in the future.
. . . We must see to it that the outcome of our next successful war
must be the acquisition of colonies by any possible means." The
same thought was emphasized by Prince von Bülow in 1897: "We do

not want to put anyone in the shade, but we demand a place for ourselves in the sun." Land was wanted not to supply the economic wants of the German population but rather to assure that in the future German culture would have as large a role in human civilization as did British culture.

Apprehension of the loss of political prestige and relative power has been a major cause of popular anxiety under a balance-of-power system. This apprehension may arise because of differential rates of population growth, of economic development, of political unification, or of military development. While these changes are watched closely by statesmen, their influence upon war is mainly indirect, through their influence in creating popular apprehension and in inducing a rise in tension levels.

c) Intergroup Tensions can be estimated from changes in various indices of the friendliness or unfriendliness of each group toward the other. The sporadic outbursts of violence in China during the nineteenth and twentieth centuries was due to the frustration arising from the incapacity of what the Chinese considered their great civilization to defend itself from the encroachments of the Western "barbarians." Such tensions appear to increase if the material contacts between the two groups increase without integration of their institutions or if their institutions differentiate without a diminution of their contacts.

When two previously isolated peoples, whether primitive or civilized, with no common cultural symbols or institutions at all come suddenly into extensive contact because of trade, immigration, or invasion, material interdependence will develop and rising tensions between them may be expected. Tensions can diminish only as accommodations are made through a common acceptance of certain symbols, conventions, ideologies, and institutions. Contact first brings conflict and gradually develops accommodation.

Geographic separation tends to minimize material contact and dependence. Efforts to maintain intense institutional integration in overseas colonies have often led to tension and revolt. To avoid this, institutional autonomy may be accorded in proportion as material dependence diminishes, not only because of the increasing demands for it, but because of the difficulty of controlling colonies which have become materially independent.

Not only does this hypothesis throw light on the state of relations between pairs of states, but it also throws light on the state of the family of nations as a whole. The nineteenth and twentieth centuries have been remarkable for the increase of material contact between peoples in all parts of the world and for the development of material interdependence. At times this development has been paralleled by a tendency toward institutional and ideological acommodations. Whenever the latter process has lagged behind the first, high tensions have arisen. During the 1930's the world was divided as to whether tensions could best be reduced by diminishing material contacts through isolationist policies or by increasing institutional and cultural accommodations through co-operative institutions. Since World War II the latter solution has been more widely accepted.

4. OPINIONS, CONDITIONS, AND WAR

The conclusion may be drawn from the foregoing discussion that tangible, economic, and historic conditions, on the one hand, and symbolic, psychological, and ideological opinions, on the other, are interrelated in the causation of war. In an intelligent and reasonable world the conditions of and the opinions about a given situation would be parallel expositions derived from observation and analysis. In such a world the historic tendencies and the symbolic significance of a given situation would be consistent interpretations of that situation as a stage in a process viewed respectively from the past and from the future. In the actual world, opinions often differ from conditions: hopes and expectations often have little relation to historical trends. These inconsistencies vary with the degree of knowledge and of wisdom of the person or culture involved. John Dewey has suggested that peace will exist according to the degree in which cultural conditions are established that "will support the kinds of behavior in which emotions and ideas, desires and appraisals, are integrated."

In international relations the sources of opinion have been only remotely related to the conditions about which opinions are held. Apprehensions of the tendency of conditions have, therefore, had little relation to the actual tendency of those conditions; yet it is from the apprehensions that wars develop. The economist may analyze actual conditions of trade, prices, and technology and may make accurate predictions of their tendency, but through such activity he has been able to contribute little toward estimating the probability of war. The journalist, the politician, or the psychologist, ignoring such conditions and analyzing the apprehensions and opinions which are actually held, however irrational they may be, has been able to judge far better of the probability of war. In this sense it would seem that psychological rather than economic factors have been responsible for war. The economist, keeping within his field, cannot explain war. He may do much to prevent war in the future by enlightening opinion so that apprehensions, opinions, and ideologies will conform more closely to the actual tendency of events and conditions. If people only fought when they would actually better their conditions by doing so, war would disappear in the nuclear age. In modern civilization war springs from "emotions devoid of ideas and desires devoid of appraisals." To prevent war, the emotion-charged symbols which control opinion must everywhere be kept in closer contact with the conditions which people think they describe. Symbols must everywhere refer to conditions, not to myths, stereotypes, or fictions.

CHAPTER XVI

POPULATION CHANGES AND WAR

Population changes are measurable and are being measured to an increasing extent in all countries. They are also, given time, controllable by restrictive, expansive, or eugenic population policies. If the effect on international relations of such changes proved to be determinate, statesmen would have at their disposal a means which might be useful both for predicting and for controlling war.

Unfortunately, it appears that no such determinate relation exists. A general increase in the world's population may lead to closer cooperation among people or to more conflict. Extreme differentials in the density of population in different areas may lead to mutually advantageous exchanges and to the development of peaceful interdependence, as is customarily found in the relations of the city and the rural areas within a state or in the relations of motherland and young migration colony. Population differentials may, however, lead to tensions, mass migrations, aggressions, wars, and conquests, as did the relation of Europe to the American Indians in the sixteenth and seventeenth centuries. A country whose population is growing more rapidly than its neighbor's may start a war of conquest; and a country whose population is growing less rapidly than its neighbor's may start a preventive war. On the other hand, neighboring countries whose population rates are very different may live at peace.

Population changes, like climatic changes, geographical and geological discoveries, and technological and social inventions, greatly influence political behavior; but the more "civilized" peoples become, the less determinate is this relationship. Among primitive peoples, the possible alternatives, when confronted by such changes, are limited, definite, and predictable. Such people may be said to behave under "necessity," although ethnological investigation proves that the behavior is dictated not by physical or physiological laws but by tribal custom. These patterns have sometimes prescribed war or migration in case of population pressure. When desert Arabs increased in population beyond their pasturage, they raided their neighbors. When desiccation reduced the pasturage of nomads of the steppes, great hordes moved into the agricultural areas of Russia or China. When a Pacific island became overcrowded, certain of the Polynesian inhabitants took to their boats to find new islands. But "whether there shall be foeticide or infanticide, parricide, human sacrifice, blood feuds or war, is largely a matter of the mores," writes E. T. Hiller.

The essence of civilization is increased realization that there are alternative solutions to problems and increased opportunity to explore different alternatives. Civilized man is able to substitute "rational" for "necessary" solutions. What Great Britain, France, Germany, Italy, Japan, the Netherlands, Russia, or the United States will do in the presence of population changes is not predetermined.

In all these countries the mass of the population is normally so much above the starvation line that population pressure influences not the means of subsistence but rather the "standard of living." Remedies for incipient population pressure are explored before starvation or even a serious diminution of the standard of living is threatened. This is less true in China, India, Egypt, Indonesia, and other countries which under existing techniques for utilizing resources are overpopulated.

Japan, Italy, and Germany, with growing populations in the 1930's attempted conquest. Java and China, with even more serious population problems, attempted to intensify their agricultural methods and to develop rural industries. Russia, confronted by a similar situation in 1917, had a revolution, abandoned territory which it had possessed, suspended projects for further expansion, and changed the emphasis of its economy from agriculture to mining and industry. Belgium and Switzerland have met their population problems by continually expanding their industrial exports and their imports of foodstuffs and raw materials.

Few writers contend that international disturbances of a definite type will flow from the numerical population situation alone. Warren Thompson, who attempted to draw rather precise prescriptions for international policy from his study of population, realized that the tendency of certain states toward conquest does not flow from population pressure alone. It also depends upon whether the nation is at the "swarming stage of development," whether its people are literate and aware of superior conditions elsewhere, whether racially and culturally they are better adapted than the present possessors to develop available areas. In other words, he recognized that the international disturbance to be anticipated is a function of a number of variables of which population pressure is only one.

The indeterminateness of the situation is emphasized by the opposing influences upon population policy of economic pressure and the balance of power: "As soon as a population grows big, its leaders say: 'Our people are so numerous we must fight for more space.' As soon as war has taken place, the leaders invert this appeal, and say: 'We must breed more people in preparation for the next war,'" writes Harold Cox.

It is obviously difficult for the state to adopt a policy which both restricts population to the food supply and expands it to supply cannon fodder at the rate set by a growing neighbor. "The political doctrine exhorts man to propagate and prevail; the economic to be cautious and comfortable." War may result from the inability of statesmen to choose either horn of this dilemma. On the other hand, it may result whichever horn is chosen. The international consequences, however, will usually be different according as policy is directed toward economic welfare or toward military power.

It may then be concluded that population situations and changes are never necessary causes of war among civilized nations, nor are they rational causes of war, although theories about population changes and conditions have at times provided both reasons and rationalizations for war.

Even though no determinate international consequence can be predicted from given population conditions, an analysis may suggest certain tendencies to be anticipated from population changes on the assumption that other conditions remain constant.

1. PHILOSOPHICAL ANALYSIS

Philosophical analysis relies upon the logical deduction of consequences from a general proposition assumed to be true. Most writers on population accept the Malthusian theory that population tends to increase more rapidly than the supply of food and that population is kept down to the subsistence level by preventive and positive checks. They differ, however, as indeed did Malthus himself in succeeding editions of his work, as to whether the subsistence level means the maintenance merely of life or of the customary standards of living; as to whether rapid local or general technological advances may not, for considerable periods, augment the food supply more rapidly than the population increases, permitting a higher standard of living to become established; and as to whether the preventive checks such as postponed marriage, moral restraint, and birth control may not render unnecessary the positive checks such as vice, famine, pestilence, migration, and war. Malthus himself was skeptical of preventive checks and felt that social reform would be thwarted by the operation of positive checks.

Recent writers tend to insist that the desire to maintain a customary standard of living, not the threat of starvation, stimulates use of population checks, that the kind of checks utilized is determined by custom and cost, and that among primitive peoples these have been "preventive" (if infanticide and abortion are included in that category) as often as positive.

Opinions can, however, be cited suggesting that war is a necessary consequence of the Malthusian doctrine. General Bernhardi wrote:

The strong, healthy, and flourishing nations increase in numbers. From a given moment they require continual expansion of their frontiers, they require new territory for the accommodation of their surplus population. Since almost every part of the globe is inhabited, new territory must, as a rule, be obtained at the cost of its possessors—that is to say, by conquest, which thus becomes a law of necessity.

Even this quotation refers only to "strong, healthy, and flourishing nations," implying that there may be nations which need not engage in aggression, although they may be in danger of becoming victims of aggression.

Harold Cox is almost as positive as Bernhardi when he writes:

It is not conceivable that human beings would ever hesitate to kill one another when, as a result of the pressure of population, they find that war is the only alternative to starvation, yet that is the situation that must arise if different races of the world continue to use their inherent powers of multiplication without regard to the available resources of the earth.

In the contingency suggested it is hard to see how even war might prove a satisfactory alternative. Secerov tries to show how war may restore the balance between industrial and agricultural production, but he admits that it will make everyone worse off. Cox himself presents birth control as an alternative better than war. If one considers all the qualifications added to the original Malthusian doctrine, the idea of "necessity" to fight evaporates in all situations of the contemporary world. Even if the entire world should become overpopulated under the most efficient economic system, so that migration could not provide a remedy, the other positive checks—vice, famine, and pestilence—might operate within each state, and thus the overpopulation might have no effect on international relations.

However, in such a state of civilization, it is more likely that the preventive checks might eliminate the "necessity" for war. The birth controllers have emphasized this, although they view the alternative too narrowly, as does Cox when he writes: "The different races of the world either must agree to restrain their powers of increase or must prepare to fight one another."

There are still other alternatives. If the entire world is not filled up, co-operation to utilize the remaining land might be feasible, as indeed Sir Thomas More suggested, though the Utopians accounted it a most just cause of war if the inhabitants of such inadequately used land refused to co-operate. Furthermore, the limits of agricultural and technological advance have as yet not been reached, although doubtless the law of diminishing returns imposes such limits, given the limited resources and surface of the earth.

Technological improvements, such as transition from agriculture to industry, may for a long time permit both population and standards of living to increase, as in Britain in the nineteenth century and in Japan during the fifty years after the restoration of 1867. Increasing population differentials may tend to create tensions and lead to war between neighbors who are traditional rivals. However, beyond the most primitive human conditions, population changes affect war and migration only indirectly through the notions they engender in people's minds. Civilized men migrate or make wars because of their thoughts, whatever may have caused them to think that way, not because of "necessity."

Under what conditions is overpopulation likely to suggest such internationally disturbing policies as migration or war? In the first place, there must be another area which to the overpopulated area appears to be underpopulated. This does not mean that the area is underpopulated judged by the state of the arts or the standard of living of its population. California may, for example, have an optimum population for the Californian standard of living, and Massachusetts may, in 1620, have had an optimum population for the Indians' technology. But for the Japanese standard of living, California was underpopulated in 1928, and for the Pilgrims of England, Massachusetts was underpopulated in 1620.

Second, there must be knowledge of this area within the overpopulated area. Before Columbus, overpopulation in Europe caused no migration to America. Even today, knowledge of areas where people might better their conditions may be very limited among the people who are most depressed.

In the third place, there must be means of mobility. Horsemen and seamen tended to migrate and fight more than agriculturalists until the advent of the railroads and steamboats and organized armies with artificial means of mobility. Energy is also necessary. People who have long suffered from overpopulation may be so depressed and feeble that they lack the initiative either to migrate or to fight.

But even with knowledge, mobility, and energy, the physical obstacles to be overcome must not be too difficult. Geographical barriers to travel—seas, mountains, deserts—may be less deterrent than the difficulty of reducing the pioneer area to productivity. If the coveted area is inhabited, social and moral barriers may be even more formidable. Immigration laws and discriminations against aliens may augment the psychological desirability of the area, but in the presence of such obstacles war may have to be resorted to, and consequently the prospective migrants must have military instruments and habits which give promise of adequacy. Even with prospects of military success the practical problem of assimilating, governing, driving out, or exterminating the inhabitants may be a deterrent, to say nothing of ideas of humanity and respect for international law. Finally, there is a host of subjective conditions to be considered. Overpopulated and depressed as they may be, are the people prepared to sacrifice an accustomed way of life in order to endure vaguely perceived hardships in an unfamiliar environment?

Experience suggests that only rarely have all these conditions conspired actually to bring about large-scale migration, war, and conquest as a result of overpopulation. Apart from the gradual pushing-out from the center by primitive peoples, the adjective "necessary" hardly seems appropriate to apply to the behavior of those who migrate or fight for a new home.

Thus it appears that the Malthusian doctrine, properly qualified, leads only to the proposition that population pressures may or may not lead to international difficulties, depending upon a multitude of geographic, cultural, technological, physiological, political, military, psychological, and other factors in the particular situation.

2. HISTORICAL ANALYSIS

The most superficial historical consideration amply supports Cox's proposition that "a reduction of the world's population will not in fact necessarily prevent all wars." Some anthropologists believe that when the world was very sparsely populated by primitive hunters before the invention of agriculture or commerce, there was no war, but this conclusion is not generally accepted. Certainly, historic instances abound of falling population without peace, as in Europe from A.D. 252 to 700 and from A.D. 1346 to 1500. In both these periods political structures were disintegrating and smaller political units were engaging in wars. In the first instance the imperial wars of the Roman Empire gave way to smaller wars of barbarian groups, and in the latter instance the Crusades gave way to feudal wars and wars between the rising princes. Although in both cases depopulation was begun by epidemics, it was promoted by the political and economic disorganization which followed these disasters. Depopulation did not prevent but promoted war and international disorder.

On the other hand, the periods of most rapidly increasing population in Western history have been the first two centuries of the Roman Empire and the nineteenth century, the periods of the *Pax Romana* and the *Pax Britannica,* when international relations were on the whole most tranquil.

It is, of course, recognized that periods of declining population may be periods of increasing population pressure (in the sense of decreasing standards of living), because the production of food and other goods may be diminishing even more rapidly. Conversely, periods of rising population may be periods of decreasing population pressure, because, as was true in nineteenth-century Europe, the technology of production is increasing even more rapidly. However, consideration of the diverse foreign policies of neutralized Belgium, expansionist Japan, and commercial England in the latter part of the nineteenth century, during which they were all rapidly increasing in population and standards of living as a result of industrialization, suggests that many factors besides population changes contribute to foreign policy. The same suggestion would flow from a comparison of policies of dominantly agricultural countries with a rising population but a probably declining or stationary standard of living during the same period, such as disintegrating China, expansionist Russia, and colonial India.

It is very difficult to compare the degree of population pressure (or rate of change of standards of living) in different countries. It seems clear, however, that historic tradition, geographic position, stage of technological development, state of literacy and communication, and relative military power influenced the consequences of variations in such pressure upon foreign policy.

In fact, it would appear that population changes have more often influenced international relations because of their effect upon military potential than because of their effect on standards of living. A country growing in population more rapidly than its neighbor may be less belligerent than the latter because, with respect to relative military potential, time is with it and it feels increasingly secure. Whereas, on the other hand, a country increasing in population less rapidly than its neighbor may view with increasing alarm the shift of the balance of power against it. These conditions, which were obvious in the relations of France and Germany from 1870 to 1890, may, of course, be altered by the establishment of alliances, as when France, with a stationary population, allied itself in 1891 with Russia, whose population was growing more rapidly than that of Germany. Germany, which previously had viewed its relations with France with comparative equanimity, then became alarmed.

These two types of population influence have worked in opposite directions. In the period after 1871 it might have been supposed that France, with a declining population pressure, would be satisfied and non-expansionist, but actually, with its its declining military potential relative to Germany, it rapidly expanded in Africa to supplement its armies by black troops. Russia, on the other hand, with a rising military potential with relation to Germany—at least in respect to the supply of cannon fodder—was also continuously expansionist because of the need to find new lands for the extensive farming of a teeming, low-standard population. Germany, with a population growth between those of France and Russia, viewed its military position vis-à-vis France with equanimity and vis-à-vis Russia with alarm, although industrialization made it possible to provide a growing population with a rising standard of living if an expanding international trade could be maintained. The supposition that colonies and a navy would mutually help each other and both would help trade led Germany also to expansionism. The role of population change in each of these three cases was different, though the expansionist result in each case had a resemblance.

3. PSYCHOLOGICAL ANALYSIS

Psychologists examine the influence of conditions and their interpretation on attitudes, opinion, and policy. Population changes have frequently provided legislators, statesmen, and journalists with arguments in discussions of immigration, tariff, colonial, and military policy.

In the United States the assumption has commonly been made that population tends to flow from low- to high-standard-of-living countries and ultimately reduces the standard of the latter. Thus American immigration legislation was based on the theory that higher bars should be provided against Orientals than against Europeans because the economic level of the former was lower. In such discussions, however, cultural difference and the possibility of assimilation were also stressed. It is difficult to tell whether the dominant motivation was economic or cultural. On the other hand,

Italian publicists asserted (as did American politicians of an earlier period) that their low-paid industrious labor will cheerfully do work which American workers eschew.

In the tariff issue American protectionists commonly assume that the products from low-wage foreign populations would flood American markets and reduce the pay envelope of the American worker, whereas free traders stress the mutual advantage if each population produces what it is adapted to make the most efficiently and then trades.

Imperialist orators have suggested the need of colonies as an outlet for population as well as a source of raw materials and markets, whereas anti-imperialists have emphasized the insignificant migration from the motherland to most overseas colonies, the slight relief to home population pressure from such migration because the gap is rapidly filled by the workers left behind, the relative unimportance of colonial markets and raw materials, as compared with foreign markets and raw materials for most industrial states, and the generally unfavorable balance of the colonial account when the total advantages and costs are counted.

Most of the talk by politicians and publicists about the general economic value to a country of colonies was "rationalization." The "reason" for supporting such policies was to be found rather in the military advantage of having certain key raw materials, a source of cannon fodder, and perhaps a naval base or a strategic frontier under military control; in the hope for colonial jobs and concessions from which a very small minority of the home population could profit at the expense of the general taxpayer; in the realization that colonial jobs for younger sons and college graduates may be a preventive of revolution in a country where centralization of political and industrial responsibility steadily diminished the number of leadership jobs while higher education increased the number of those who thought themselves qualified to lead; in the expansiveness which the average citizen with a rather limited and humdrum experience felt in identifying himself with a growing area on the map, even if he had to pay for it by a diminished standard of living; in the need which the political and economic elite felt, in times of depression, for diverting the public mind to distant adventure as a protection against criticism or revolutionary impulse; and in the anxiety which both the leaders and the average citizens felt lest the national brand of culture might die out or diminish in relative importance unless it was growing in an ever larger section of the earth's surface.

The latter point does indeed frequently appear in political oratory on the subject. Thus Treitschke wrote:

All great nations in the fullness of their strength have desired to set their mark upon barbarian lands. All over the globe today we see the peoples of Europe creating a mighty aristocracy of the white races. Those who take no share in this great rivalry will play a pitiable part in time to come. The colonizing impulse has become a vital question for a great nation. . . . The consequences of the last half century have been appalling, for in them England has conquered the world. . . . It is the short-sightedness of the opponents of our colonial policy which prevents them from understanding that the whole position of Germany depends upon the number of German-speaking millions in the future.

Mussolini presented the same arguments in 1927 to the Chamber of Deputies:

I affirm that the fundamental, if not the absolutely essential datum for the political, and therefore the economic and moral power of nations is their ability to increase their population. Let us speak quite clearly. What are 40,000,000 Italians compared to 90,000,000 Germans and 200,000,000 Slavs? Let us turn toward the West. What are 40,000,000 Italians compared to 40,000,000 Frenchmen plus 90,000,000 inhabitants of France's colonies, or compared to 46,000,000 Englishmen plus 450,000,000 who live in England's colonies? Gentlemen, if Italy is to amount to anything, it must enter into the second half of this century with a population of at least 60,000,000 inhabitants. . . . If we decrease in numbers, gentlemen, we will never create an empire but become a colony.

This ambition for a growing place in the sun for a national culture explains the frequent union of demands for a growing population and colonies—a union which would be, to say the least, anomalous if the economic argument provided the sole motive.

Population projections in 1964 indicated that population was relatively stable in western Europe and Japan and was increasing at a moderate rate in the United States and the Soviet Union, but was increasing at such a rapid rate in Latin America, Asia, and Africa that the world's population would more than double by the year 2000. This situation, which, with the independence of nations in the latter areas, can no longer be dealt with by colonial expansion, may lead to political and racial problems both within and between nations.

A study of the legislative debates on military measures in the French and German parliaments since 1870 disclosed frequent allusion in the former to the growing population differential between France and Germany and in the latter, after the Franco-Russian alliance of 1892, to the huge Russian population.

A study of these discussions of immigration, tariff, colonial, and military policy creates the impression that population arguments, especially of an economic type, do not always express the real motives of the speaker. The economics is often so patently bad that one concludes that expansionist policies flow from the sentiment that national expansion and military power are ends in themselves. Economic arguments are advanced only because in an economic age it sounds more reasonable to act for greed than for glory. This is not to say that economic self-seeking by financial and commercial magnates, retention of political position by leaders and politicians, and military self-sufficiency for the army may not also be an undisclosed motivation behind some of this oratory, nor does it deny that many of the rank and file are persuaded that the nation and perhaps they, individually, will reap economic gains from the proposed policy.

Political proposals and discussions of the 1930's indicated wide acceptance of the theory that territorial redistribution was required by justice or expediency to relieve the population pressure of "have-not" countries. With the general anticolonial sentiment which has developed since World War II, such arguments have not been utilized.

4 . SOCIOLOGICAL ANALYSIS

Sociologists try to analyze a given population problem in its concrete setting with a view to prediction or control. Analysis and comparison of the composition of different populations disclose differences in respect to the proportion of each grouping (sex, age, race, occupation, income, health, education, social status, etc.) into which the population may be classified. The rates of change of these proportions usually vary in the history of a population and among different populations. As a result the character of every population and its relation to others is being continuously modified in time. These qualitative changes may be more significant in explaining the causes and estimating the probability of war than the changes in the total numbers of the populations. This method of analysis may permit the expression of qualitative changes in quantitative terms, facilitating the measurement of trends and the inference of causal relations. Its application may throw light on the conditions influencing the warlikeness of a given population and the development of maladjustments between different populations.

Applications of this method have suggested that the age composition of a population may have a significant effect upon the psychology of the nation. A rapidly growing population is a young population. According to Gini:

> A population in which young age-groups abound bears the imprint of their spirit of daring in all its social organization and in the trend of its collective policies; whereas cold, calculating prudence is the characteristic of populations in which the older age-groups prevail.

Other aspects of population composition may be isolated which influence aggressiveness. More comprehensive analyses may take into consideration the relation of many aspects of the population in order to establish the approximation of a given population to an optimum condition in relation to its physical environment, its international relations, and its social ideals. An approach by all populations to relative optima might minimize the danger of war. Ferenczi, who attempted to develop statistical indexes for determining the relative national population optima, emphasized the possibilities of this method.

> Historical evolution does not enable us to fix a uniform standard of the life of nations in the near future, not even for the nations belonging to the same civilization; nevertheless, an intimate and comparable knowledge of the respective situations of nations can further their social progress and the cause of peace and, at least, prevent false prophets from hampering a peaceful development. . . .

This approach is as yet a promise, not a fulfilment. Population studies have as yet yielded few secure generalizations concerning the relation of the character of populations and the probability of war.

The study of the population situation in particular areas of international tension may often assist in the practical solution of that problem. International commissions sent by the League of Nations and the United Nations to such areas have usually paid attention to the population situation in the area.

The number of factors which must be considered to estimate the international trends in such an area was well illustrated by the discussion regarding Manchuria in the Institute of Pacific Relations in 1929. The different character of the population movements from China, Korea, Russia, and Japan into this area, the differences in the stage of economic organization of the sources of these migrations, the political and economic interest in the area of states other than the three most interested, the problem of military defense, the nature of historic rivalries, and the character of international institutions for adjusting difficulties were discussed, with the following conclusion:

The problems of Manchuria are, therefore, complex. They present in a new area of striking and even dramatic development, all the problems of international intercourse which a modern world is groping to control. If economic necessities can be reconciled with national sovereignty, international co-operation with national security, population pressure with peaceful intercourse, a large part of the common problem confronting all nations will have been solved in one area at least.

Japan in 1933, with a population under twenty years of age 10,-000,000 greater than the population between twenty and forty years old, was confronted by the very real problem of finding 10,000,000 additional jobs in twenty years. Birth control could do nothing to relieve this situation. The possibilities of further intensification of agriculture in Japan were very limited. Emigration, conquest, industrialization, and trade expansion were suggested. If the general welfare of its people had been the object of Japanese policy, as most of the abstract economists assumed, the possibilities of various alternatives might have been explored.

With migration to Australia, New Zealand, Canada, and America barred by law or administrative practice under the law, and to the Philippines, Indonesia, China, and Korea barred by lower-grade and frequently denser indigenous populations, migration seemed to offer little relief without successful war against the overseas countries. This undoubtedly would be extremely expensive, even if the war were successful and if the problem of providing suffiicent tonnage to transport Japanese away more rapidly than new ones were born could be solved. With an annual increase of nearly a million, provision would have to be made for exporting about 3,000 Japanese a day, or assuming that birth control should at once prevent further increments, the problem of the potentially unemployed 10,000,000 in the next twenty years would require an export of 1,500 a day.

Conquest of territory seemed hardly practicable except in Asia, where it was undertaken. The prospect of large-scale migration to the occupied territory was slight, the raw materials were more expensive to exploit than those which were available to Japan by trade in other places, and the indigenous population, although large, did not provide a market for high-grade manufactured goods. The ultimate defeat of Japan in World War II left it with less territory than it had before the war. Aggression had not paid.

The alternative of further industrialization and expansion of trade throughout the whole world, importing more and more foodstuffs and raw materials and exporting an increasing proportion of manufactured goods, seemed to offer the best economic solution—one which had the economically desirable feature that industrialization tends to urbanization and reduction of the birth rate, so that the problem might be permanently ended if the 10,000,000 additional workers already born could be cared for during the next twenty years.

Perhaps the policy embarked upon by Japan in 1922, which seems to have been along this line, would have been persisted in if the United States and others had not seen fit to slap Japan in the face with discriminatory immigration policies which were interpreted as implying racial inferiority and to hamper the enlargement of Japan's industrial exports by ever higher tariffs. It is not surprising that the military faction in Japan, which had always scoffed at the liberal policy, gained more and more popular support until it was able to embark upon a policy which had little to offer economically but might induce the Japanese to lower their standards of living in exchange for glory and might even slaughter some of the 10,000,000 in war. In spite of the anti-industrial tone of the military party which came into control, Japan inflated its currency and expanded its export trade to a large extent. Since World War II, Japan, with much economic assistance from the United States and success in a policy of population control has become prosperous through industrialization and trade expansion. Aggression had not paid as expected, but its consequences—military defeat, large-scale aid from its ex-enemy, and reorientation of its policy—had paid economically.

After discussing four policies open to Germany to meet the problem of "an annual increase in population of almost 900,000 souls" (birth control, domestic colonization, territorial conquest, and further industrialization), Adolf Hitler concluded that "taking with the fist," of new soil "at Russia's expense" was a "healthier" course than the policy of industrialization and trade followed by the German Republic, because it would preserve a healthy peasant class, promote economic self-sufficiency, destroy "pacifistic nonsense," and enlarge the homeland. This policy was attempted with no success, but after World War II, with a loss of a third of its territory and a 25 per cent increase of its population by refugees from the east, West Germany, with extensive economic assistance from its recent enemy, increased even more in prosperity than did Japan.

A rational study of the alternatives in any population situation of the modern interdependent world from a purely economic point of view seldom suggests a military or colonial policy—a fact which confirms the conclusion that the objectives of foreign policy are generally only in small degree economic among the leaders who understand and who make the policy. The rank and file who do not understand may frequently be influenced by bad economic arguments.

If, instead of assuming general welfare as the end of policy, some other end is assumed, such as national self-sufficiency, augmentation of relative military power, maintenance or increase of relative position of national culture in the world, or retention of the present relative position of rulers and classes within the state, similar exploration of the best alternatives for attaining this end in a given population situation could be made. The actual policy by which most states meet their population problems is likely to be a compromise between the results of these different analyses.

5. INFLUENCE OF POPULATION ON WAR

The conclusions to be drawn with respect to the relation of population changes to war in the contemporary world are in the main negative, but seven points may be noted.

First, the rapid growth of the world's population during the past century has stimulated international communication, interpenetrated cultures, increased international co-operation, and tended to bring the entire human race together into a single community. But it has also—by increasing population pressures, especially in the under-developed two-thirds of the world, by increasing the economic gap between the poor and rich countries, and by augmenting contacts between people of different cultural and political allegiances—increased opportunities for friction between nations, each of which often places retention of its cultural individuality, its ideology, its political unity, and its relative power position above its economic prosperity. Thus, while becoming more united, the world has become less stable and tensions have increased. Both of these trends may be accentuated by the "population explosion" after World War II, resulting from a rapid decrease of death rates by the application of health measures, especially in the underdeveloped areas of the world, unaccompanied by a similar decrease in birth rates. If present birth and death rates continue, the world's population will more than double by the year 2000, but the experiences of Japan and western Europe indicate the possibility of controlling the rate of population growth.

Second, policies of war and expansion have been less influenced by population changes than by the willingness of people to accept unsound economic theories on the subject. A more general knowledge of the economic value of the various alternatives for meeting particular population problems would under present conditions make for international peace and co-operation rather than for war, provided people really wished to make general welfare the object of policy.

Third, differentials of population pressure in neighboring areas, if generally known to the inhabitants of the overpopulated area and if maintained by artificial barriers to trade and migration, tend to international violence, provided the people of the overpopulated area have energy and mobility, are accustomed to the use of violence as an instrument of policy, and are dominated, as people in the mass usually are, by political rather than by economic objectives.

Fourth, population has been one factor in military potential, and differential rates of population growth in neighboring states have tended to disturb the balance of power if such neighbors are in positions of traditional rivalry and depend for their defense upon their own resources rather than upon the mutual jealousies of others. The changes in military technology since World War II have greatly diminished the military importance of population differentials.

Fifth, in accord with the two preceding propositions, imperial wars have been initiated by countries with the most rapidly rising populations, whereas balance-of-power wars have been initiated by the alliances with the less rapidly rising populations, provided other factors of the military potential are being equally affected by time.

Sixth, differential rates of population growth, of migration, and of imperial expansion have influenced the relative importance of national cultures, languages, and institutions in the world and have created anxieties inducing colonial expansion. If these differential rates of growth, which in the twentieth century have tended to diminish the proportion of the world's population in Europe and North America and to increase the proportion in Latin America, Asia, and Africa, continue, anxieties about the position of races and cultures in the world may precipitate serious political problems.

Seventh, although population conditions in the broad sense are a major factor in international politics and establish limits to the possibilities of international relations during any historical epoch, the possible variations of policy within these limits steadily increase as civilization develops, and today such variations are very great. Consequently, today the character of the influence of a particular population change is so dependent on other factors that it is impossible to predict from a study of population phenomena alone what international policies or occurrences to expect. E. F. Penrose, writing on the population problem in 1934, supported the thesis later incorporated in the constitution of UNESCO: "It is not in the circumstances of the external world but in the minds of men that the mainsprings of violent social conflict lie."

Without denying the seriousness of the population problem which faces mankind, statements without qualification of the international consequences of, or the remedies for, population conditions do not assist in solving that problem. Alarming statements regarding the relations of population conditions to international conflict have often been made as propaganda for policies of value to the few rather than to the many; consequently, it is in the general interest that the indeterminateness of the actual relationship should be understood.

CHAPTER XVII

THE UTILIZATION OF
RESOURCES AND WAR

1. COMPETITION FOR THE
MEANS OF LIVING

A people cannot live if it cannot get the means of life. Nature does not provide all the means of life everywhere in unlimited abundance. From these two propositions it has been inferred that the *struggle* among *peoples* for the *limited resources* provided by *nature* inevitably leads to war. This theory of the cause of war has often been called economic because it argues from rational motives and natural conditions. Economists have, however, usually rejected this theory. The position of different economic schools differs, but in general the argument may be analyzed by considering the ambiguities lurking in the key words in this proposition: (*a*) "struggle," (*b*) "peoples," (*c*) "limited resources," and (*d*) "nature."

a) *Struggle* is a word which may apply to either competition or conflict. The effort of a number of individuals or peoples to gain the lion's share of limited resources is competition, but war is a form of conflict. Competition may occasionally lead to conflict. If two lions are each trying to get the same antelope, the situation is one of competition. A may get it, and B may depart either to find another antelope or to starve. A may get it, eat his fill, and leave the rest for B. A, seeing that the antelope is escaping, may enlist the co-opera-tion of B, and the two may capture the antelope and share it. A having captured the antelope, B may attack him, drive him off, and eat the antelope, leaving A to starve or find another antelope. Only in the last pattern has competition led to conflict.

The struggle for existence among members of the same species, which Darwin regarded as a factor in evolution, was competition usually resulting in the starving-out of the less fit according to the first pattern. It seldom resulted in conflict. The relation of the lion to the antelope is one of conflict, but the relation of the two lions usually is not. The economic struggle among business firms and individuals in a civilized community is also competition. Some firms fail because they cannot capture the market. Some individuals be-come unemployed because they cannot capture a job. The discon-tent and misery may lead to conflict, but usually they do not.

The struggle among human beings between classes, ideologies, cultures, races, and economic systems, and between the powerful and the weak, is usually competition resulting in the unsuccessful retiring to a narrower area of activity, as in the first pattern of com-petition, or in getting something, as in the second. This may lead to conflict and violence by the underprivileged or unsuccessful, but it has not usually done so. "Experience hath shown," according to the Declaration of Independence, "that mankind are more disposed to suffer, while evils are sufferable, than to right themselves by abol-ishing forms to which they are accustomed." Many unsuccessful sys-tems of thought, life, and economy have been confined to limited areas or have disappeared in human history without conflict.

Among many animals, as Kropotkin, Allee, and others have pointed out, the struggle for existence has sometimes led to co-operation within the family, the community, or the aggregation, in accordance with the third pattern of competition. Among human beings there has always been such co-operation, resulting in permanent villages, tribes, and nations within which individuals and families, in varying degrees, collaborate in production and share the proceeds. Business competition has tended to a similar result in the formation of mergers, trusts, holding companies, and trade associations. International political competition has often resulted in federations, unions, and leagues. The slogan of political party competition has been to "fight 'em or join 'em," and the latter has been as frequent as the former.

The struggle between similar individuals or groups for limited resources, through competition, has not necessarily resulted in conflict, though it has more often done so among sovereign nations competing for power than among other groups. Competition for resources should not, therefore, be identified with war.

b) *People* is a word which may apply to a population of individuals or to an organized group of individuals. War is a conflict between organized human groups. The proposition that struggle among peoples for resources leads to war, therefore, assumes that the word "peoples" refers to organized groups, but in this sense peoples are dependent upon particular social and political institutions, not upon the distribution of resources. The latter dependence can only be attributed to "peoples" in the sense of populations of individuals. Food is eaten by individuals, not by organizations. Competition for resources on which to live therefore takes place ultimately among individuals. An economic treatment of human competition for resources should, therefore, examine either what behavior is most beneficial to an individual in competition with other individuals under given social institutions or what social institutions are most promotive of a distribution and utilization of the resources beneficial to the people in a particular group or to the human race as a whole.

The competition of "peoples" in the sense of organized groups is a sociological and political rather than an economic problem. Nation-states, the dominant groups in modern history, exist and compete with other nation-states for power rather than for the welfare of peoples.

c) *Limited Resources* is a phrase which may refer to useful goods and services available at a given time and place or to the total material and human resources of the world convertible with a given technology in a given period of time and at a given place into useful goods or services. The difference in the meaning of the word "limited" in these two senses is enormous. An Indian village of fifty people on the site of Chicago was in far more danger of starving to death with the technology available to it in the eighteenth century than are three million people on the same site with the technology available to them in the twentieth century. To the Indians, resources were limited to the game and fish which they could take within a few square miles. To the modern city, except in time of war, the resources of the world are available, capable of modification in form and transportation to Chicago in a few days or weeks. Under modern conditions of transportation and communication, limits of resources can be defined not by what can be obtained from the monopolistic utilization of a fixed area of land and subsurface deposits but by what can be obtained from the world system of production, transportation, and distribution, utilizing mineral, vegetable, and animal products from remote areas.

In such a dynamic order, resources cannot be thought of as limited in the sense of a loaf of bread from which only so many slices can be cut but as limited mainly by the social obstruction to human ingenuity, foresight, and co-operation. The usefulness of the raw materials of the earth is not a quality of the material per se but of human inventiveness and co-operativeness. Economic activity has thus acquired the peculiarity that one man's gain is not another man's loss. Exchange of things and ideas is mutually beneficial.

Competition for a livelihood tends, therefore, to be of general advantage in proportion as trade is conducted as a form of co-operation and of general disadvantage in proportion as it is pursued as a form of conflict. Conflict, instead of being one of the possible ways of winning in the competition for existence, tends to become a way of losing—inevitably, if it takes the form of nuclear war.

In this respect competition for a living differs radically from competition for political power. The latter is relative. One man's superiority of power is another man's inferiority. Political competition therefore tends toward conflict, whereas economic competition tends toward co-operation.

d) *Nature* is a word with a multiplicity of meanings. In the present context it may refer to the earth, its deposits of minerals, its vegetable and animal life existing without human intervention, or it may refer to those resources capable of utilization by a given system of production. In the first sense nature really provides nothing useful at all. Some technology, even though no more developed than finding and collecting, must be employed to convert minerals, animals, and plants to use. Nature is economically meaningless apart from its relation to a productive technology. What things are resources depends upon man's knowledge of resource utilization.

Nature in the physical sense, it is true, may largely influence types of utilization among primitive peoples, as it does among animals. Among them, deserts produce one sort of economy, forests another, seashores another; the tropics one, the arctic another. With civilization, however, technology dominates over resources, over topography, and even over climate.

A study of the influence of resources upon war becomes, therefore, a study of the influence of particular productive systems upon war. What is the influence of agrarianism, of feudalism, of capitalism, of socialism, upon war? What are the specific causes of war within these systems? What is the influence upon war of the contact of different economic systems?

2. TYPES OF ECONOMY

Economic systems are continually changing. Such words as "agrarianism," "feudalism," "capitalism," and "socialism" each designates a type of economy which at some time and place has been dominant. Agricultural economies have tended to be either agrarian or feudal. Commercial and industrial economies have tended to be capitalistic. Socialistic economies have developed in dominantly agricultural as well as in dominantly industrial regions. Hunting, fishing, and pastoral economies have distinctive characteristics, but they have been in the main confined to primitive peoples.

The technological and economic system does not determine all aspects of a culture. The economy, religion, and politics of a people may spring from different origins and appear in strange combinations. Thus, in considering the tendency of economic systems in respect to war, it must be recognized that these tendencies may be arrested or diverted because of the combination of the economic system in a particular instance with a religion or a government of different tendency. The symbols of a society are no less important than its conditions.

Agrarianism began among primitive peoples and has existed in most civilizations after nomadic tribes settled down to agriculture, as the barbarian invaders did in Europe in the early Middle Ages. Its organizing principle has been the voluntary co-operation of landowners of equal legal and not too unequal economic status. It cannot easily extend beyond the village of personal acquaintances, so that larger organization has depended not upon economy but upon political or religious principles, usually combined in the kingship.

Agrarian village economies have been relatively peaceful but agricultural economies controlled by feudalism have tended to be belligerent and those controlled by kingdoms, empires, or nation-states have tended to be more belligerent than commercial or industrial economies so controlled.

Feudalism has arisen, on the one hand, from the need of the small landholder for military and economic protection from the great after there have developed a considerable differentiation in the wealth and power of individual landholders and a considerable dependence of agriculture upon urban markets and, on the other hand, from the incapacity of the central government because of lack of communication, transport, and efficient administrative services either to give this protection or to collect taxes directly. Thus the most powerful landholder in every locality acquires a combination of economic and political power from the acceptance of his protection by the lesser landholders, the landless, or the village as a whole and from the farming-out to him of taxing and military power by the central government. Feudalism has emphasized the spirit of personal loyalty, developing a system of unequal obligations which tend to become hereditary and to establish relationships of status rather than of contract. Freedom is thus sacrificed for security.

Feudal societies have therefore tended to be highly militaristic, in the sense that military activity has carried high social prestige, that the ruling class has been dominantly engaged in military activity, and that wars have been frequent, whether private wars between barons or public wars between kings.

Capitalism has usually developed from agrarian or feudal economies when commerce and industry have created accumulations of wealth in forms more mobile than land. The owners of this wealth have acquired a political influence comparable to or greater than that of the landowners. Capitalism has sometimes dominated in towns within states which were dominated as a whole by agrarian or feudal economies. Considerable state socialism has often been a transitional stage before the development of capitalism, and capitalism has tended to proceed from commercial to industrial and financial stages.

Capitalistic enterprises have been organized to compete in a market regulated by a price system that is capable of equating the values of capital, labor, management, and commodities. This competition has, however, been regulated by a common law forbidding fraud and violence and protecting private property and contracts. This law has been enforced not by self-help but by a powerful state which, apart from enforcing the common law, ought under capitalist theory to disinterest itself from economic activity. The freedom of the capitalistic enterprise has not extended, as has the feudal manor, to military activity, except in the great trading companies organized by imperial states to operate in colonial areas.

Capitalistic societies have been the most peaceful forms of societies yet developed. The bourgeois, who have been their organizers, have usually held military affairs in contempt and have considered war as the great destroyer of wealth and the great obstacle to the expansion of economic enterprise. They have generally sought to expand their enterprises not by the forcible seizure of land but by successful competition in markets, foreign and domestic. They have regarded the role of political and military power as the maintenance of domestic order, the enforcement of law, and the prevention of invasion. Capitalistic entrepreneurs have on occasion sought to influence laws and the use of military power in favor of their enterprises, and in colonial areas their enterprises have themselves sometimes used force. Governments in a regime of capitalism have on occasion sought to direct economic activity, particularly in times of active military preparation or of war. The central idea of capitalism, however, has been the separation of government and business. The theorists of capitalism—the classical economists—considered good government that which secured justice and order with the least interference with individual freedom, and good economy that which utilized resources to provide what individuals wanted with the least waste. Although the necessity of state defense was not denied, the initiation of war was considered both politically and economically irrational, and it was anticipated that war would disappear as civilization advanced.

Wars have occurred during the periods of capitalistic dominance, but they have been least frequent in the areas most completely organized under that system. The increased control of nature and the vulnerability of the economy to commercial stoppages, incident upon the evolution within capitalism of an advanced industrial technology, have, however, been utilized by the state for war purposes. Consequently, when wars have occurred among capitalistic states they have usually been more destructive than among agricultural states.

In the modern period, in which alone capitalism has been fully developed, war has more frequently been initiated by states dominated by agrarianism or by socialism than by those dominated by capitalism. Nationalism in agrarian Serbia was an important cause of World War I. The spearhead of German militarism was, in 1914, the Prussian *Junker*, not the Rhineland industrialists. In 1939 war was begun by the National Socialists, not by the capitalists. Japanese militarism sprang from the peasantry and the army, not from the bankers, merchants, and industrialists. In the United States the Revolution, the War of 1812, and the Civil War were pressed by the agrarian West and South more than by the commercial East and North. British imperialism was supported in the eighteenth and nineteenth centuries by the conservative landed aristocracy rather than by the liberal merchants and industrialists. Businessmen, bankers, and investors have generally urged peaceful policies in time of crisis. Dominantly agricultural countries like Russia and the Balkans were more ready to spring to arms during the nineteenth century than were the more industrialized states.

Socialism has been commonly used to describe a utopia or a program of reform. These programs have been various but have all emphasized the control of economic life by the organized community and the elimination of private property in production goods. Historical economies manifesting these characteristics may, therefore, be appropriately described as socialism, even though they lack other characteristics which socialist propagandas have attached to their utopias.

Historical socialisms have arisen when communities have encountered practical exigencies which seemed to require government to engage in public works, production, trade, and social welfare so extensively as to dominate the economy. Socialism has more often developed from necessity than from theory, though the proposals for co-operatives by Robert Owen and of communism by Karl Marx have exerted increasing influence in the last century.

Pioneers with a common faith and a meager and hostile environment have sometimes been able to survive by communistically pooling their resources. Governments have felt obliged to construct public works and engage in warehousing and large-scale relief in time of famine and depression. States have felt it necessary to engage in arms-making, regulation of external trade, and control of consumption in order to prepare for or to wage war. They have also found it necessary to intervene extensively in business, banking, and finance to preserve capitalism itself from monopolies and depression.

Empires pressed by rivals and native lethargy have controlled the economy in their colonies in order to hasten the introduction of advanced techniques of transportation and industry. Similar conditions have driven national governments of economically backward countries to hasten the tempo of industrialization by government initiative. In times of exceptionally rapid technological change all governments have extended their initiative into branches of economy. Humanitarian sentiment has induced governments, even of capitalistic states, to undertake extensive programs of social security, medical services, unemployment relief, assistance to the underprivileged, and development of depressed areas—called by Walter Lippmann "the agenda of liberalism" and by others "the welfare state"— in addition to general services, such as education, postal and telecommunication, and rail and air transport, which are commonly conducted by governments.

The spirit of socialism is the dominance of group welfare over individual interests—a spirit which thrives in the presence of obvious threats to the group as a whole. The spirit, therefore, resembles the fealty of feudalism more than the freedom of agrarianism or the acquisitiveness of capitalism. Theoretically, it substitutes the impersonal state, society, or community for the personal lord as the object of loyalty. Actually, systems of socialism have tended to develop around a personal leader who embodies the community. The unifying spirit of national socialism and fascism has resembled that of feudalism. Even Russian communism emphasized loyalty to the leaders, Lenin and Stalin.

Socialist technology, whether dominantly agricultural or dominantly industrial, has been characterized by the centralized planning of the economy of the community. In state socialism the area of planning has been larger than that of the agricultural village or the feudal manor, and the functions planned have been more comprehensive than those of the capitalistic enterprise. As under feudalism, political and economic authorities have been combined.

A socialistic economy cannot long survive unless its planning and administration are both done efficiently. Socialism has been the most self-conscious and highly integrated of all forms of economic organization. It has tended to subject all activities, not only economic but also religious and cultural, to the dominant control of the state. Natural rights of men and of communities have been denied. Rights have been said to exist only by grant of the state, whose interest and welfare are the supreme good of the society.

Political movements have often propagandized for ideal representations of all these systems. The period since World War I has been characterized by the struggle of propagandas of agrarianism and socialism against capitalism in Europe and North America and against semifeudalism in most other parts of the world. These propagandas have sometimes accompanied or induced actual changes of economy. They have also stimulated or supported conflict, especially in the "cold war" after World War II, between a miscellaneous group led by the United States under the banner of capitalism, or free enterprise, and a group led by the Soviet Union under the banner of communism, an extreme form of socialism.

Historic periods of transition from one economy to another have been warlike. Agricultural classes accustomed to a dominant position have usually resisted violently the rise to dominance of commercial or industrial classes. The latter have resisted the rise of labor and a socialist bureaucracy. Geographic frontiers marking the transition from one economy to another have also often been the scene of war. An industrial state in close contact with an agricultural state tends to expand its commerce and industry into, and to draw its food and raw materials from, the latter. Regarding this process as subversive of its culture and dangerous to its independence, the agricultural state is likely to resist by arms.

3. CAUSES OF WAR UNDER SOCIALISM

State socialism has been the economy of the most warlike of all societies. The socialistic empires of Assyria and Peru were the most militant of ancient civilizations. Socialistic Sparta was the most warlike of the Greek states. Italy, Germany, Russia, and Japan increased in militarism as they adopted forms of socialism in recent times. The autocratic states of post-Renaissance Europe with semisocialistic mercantile economies were engaged in continuous wars.

Military policy and socialistic economy appear to have influenced each other reciprocally. Preparation for war has required governmentalization of the economy, but a centrally administered socialistic economy has usually required warlike preparation. Administrative, economic, political, and psychological conditions combine to account for this. It is possible that the spirit of socialism might be realized without central economic planning through adjustment of the relations of autonomous local or industrial co-operatives by a price system. It is possible that a free economy might be maintained without the profit motive. Whether such economies should be characterized as socialism or capitalism is a question of definition.

Since the rise of communism in Russia after World War I, the capitalistic countries have adopted many features of socialism, creating "mixed economies" and "welfare states." Since World War II the more mature communist states of Europe have adopted features of capitalism such as privately operated market gardens in collective farms, economic incentives in industry, increased autonomy of and competition between enterprises, and better legal protection of individual freedoms. The sociologist Pitirim Sorokin has indicated the convergence in most aspects of life of Soviet and American societies as they have approximated each other in advanced technology.

Socialism as a distinctive system, however, exists only if the competitive market as the regulator of production, consumption, and distribution is replaced by the effective administration of a comprehensive economic plan, possible only if there is a government with sufficient political and military power to administer the plan. As an operative alternative to capitalism, socialism has meant state socialism, and if conducted as its theory requires, it tends to belligerency, because administration of a large-scale plan requires (a) precise goals, (b) self-sufficiency of the economy, (c) coercion of the people, and (d) control of opinion.

a) Precise Goals.—Administration of an economic plan for a large area has required a more precise formulation of objectives, a more efficient subordination of individual, group, and local freedom to those objectives, a more thoroughgoing command of economic resources, and a more complete exclusion of incalculable external influences from the area in which the plan operates than has the maintenance of a common law and the prevention of violence. Consequently, socialist states have tended to be more dictatorial, regimented, self-sufficient, and isolated than liberal states.

b) *Self-sufficiency.*—The economic objectives of a large population cannot be precisely formulated. The aims of the individuals, groups, and local communities composing that population are certain to differ considerably. The only economic objective which has in practice proved sufficiently precise to permit of long-time general planning has been that of national defense. Military boards have been able to state the economic requirements of defense in advance, to plan a national economy to supply those needs, and to command general support for the plan in a way which civil authorities or legislative bodies interested only in welfare have not. National economic plans, therefore, have tended to become national defense plans.

An economic plan cannot be achieved unless the resources which it requires at every stage are assured under the conditions foreseen in the plan. The planner must, on the one hand, guard against the interference of external circumstances in the area of his plan and, on the other, assure his control of an area which contains all the resources which he will need. Since all developed economies must draw resources from some areas outside the national domain, national economic planning has led to the dual policies of national economic self-sufficiency and territorial expansion, both of which develop high tensions and continuous danger of war.

c) *Coercion.*—Successful economic planning requires that the activities of the population accord with the plan and not with the spontaneous desires of individuals or groups. In order to carry out economic plans of wide scope, governments have found it necessary either to increase their coercive authority over individuals or to create a situation in which individual loyalty to the government may be expected. Usually they have done both. New crimes such as economic espionage and economic sabotage have been introduced, and new stimuli to loyalty such as nationalistic propaganda and an aggressive foreign policy have been disseminated. Even liberal states have recognized that in crisis situations the powers of government must be extended and civil liberties curtailed. In order to sustain the degree of solidarity necessary to administer a completely planned economy, governments have found it convenient to perpetuate crisis conditions, often by pursuing foreign policy which continuously maintains an external enemy, latent or active, and creates the conviction that the life of the nation is always in jeopardy.

d) *Control of Opinion.*—Centralized economic planning has required controls of opinion not only to assure loyalty to the government but also to control consumers' demand and to prevent the interference of external influences. A planned economy must dispose of the goods produced in accord with the plan; consequently, the population must be persuaded or compelled to want those goods. In capitalist economies national advertising performs this service for producers with results sufficiently disastrous to a free economy when the producing units are very large. In a planned economy the police power of the state is also available for this purpose, and the coercion of the consumer becomes much greater. The control of internal opinion is much easier if free external communication is prevented or allowed to enter only through the filter of the national censorship. Governments controlling the national economy have therefore been especially active in efforts to promote economic self-sufficiency and the psychological isolation of the population in the planned area. Freedom of speech, of press, and of ideas has proved incompatible with large-scale economic planning.

A government which needs a precise objective for the economic activity of its population, which needs firm political control of an area containing all the economic resources which its population requires, which needs the intense loyalty of the population of that area, and which needs general acceptance by that population of the plan and the goods which it is to produce can hardly avoid becoming warlike. It almost inevitably adopts a bellicose foreign policy, eliminates all internal opposition to that policy, subordinates economic welfare to economic preparedness, and accentuates the economic significance of political boundaries. Such policies create the distinction between "have" and "have-not" states, the demand by the latter for territorial expansion, and a preparation of learning and opinion to achieve that demand by violence.

States at war have tended to become socialistic, and socialistic states have tended to be at war. Modern socialism has in fact been the war organization of capitalism, in the same sense that feudalism has been the war organization of agrarianism. The modern socialistic state resembles the feudal state in its spirit and its organization. It resembles the successful capitalistic enterprise in its efficiency and its technology. Its rise has been accompanied by an increase in the frequency and the destructiveness of war. The warlikeness of the twentieth century may be attributed in part to the corruption of capitalism by large-scale controls of production and consumption. Monopolistic combinations, price-stabilizing policies, mass advertising, legal barriers to trade, and extensive governmentalization of industry and opinion have tended away from capitalism toward warlike national socialism.

After the defeat of the Nazis and the rise of the Soviets the situation gradually changed. The expansionist activities of the Soviets in eastern Europe, China, and Korea and the threatening reactions of the West induced the "cold war." But the belligerent tendencies of communism internationally, its oppressive tendencies internally, and its difficulties in organizing efficient agricultural production became alarming to Soviet leaders as the destructiveness of nuclear war and the demands of their people for more consumer goods became apparent in the late 1950's. Russia and the European satellites began to liberalize communism and to seek peaceful coexistence, disarmament, and an end to the "cold war." China, less alarmed economically and politically and desiring to maintain a theoretically more complete system of communism, manifested the belligerency, tyranny, and agricultural inefficiency which Russia had faced earlier.

4 . CAUSES OF WAR UNDER CAPITALISM

In spite of the relative peacefulness of capitalistic societies, popular theories have frequently cited capitalism as the major cause of war in modern times. These theories have sprung primarily from socialist writers who have wished to supersede capitalistic systems by socialistic ones and so are to be received with caution. Yet there are tendencies within capitalism which make for war.

Theories have related capitalism to war in general, to imperial wars between capitalistic and agrarian economies, to civil wars between classes within capitalistic economies, to international wars between dominantly capitalistic states, and to general social disintegration within capitalistic economies, providing conditions favorable for war. These theories emphasize, respectively, the problems of (*a*) war profiteering, (*b*) expansionism, (*c*) depression, (*d*) protectionism, and (*e*) materialism.

a) *War Profiteering.*—The theory which attributes wars to the greed of special capitalistic interests, able to profit by war preparations or war itself, may be distinguished from the remaining theories which emphasize the war-provoking tendencies of capitalism as a system. This theory does not distinguish between classes of war. The war profiteer can gain from war preparations or activities whether in a colonial area, a "Balkan area," or among great powers; whether civil or international; and whether involving his own or other countries. His liability to disadvantages from the war or war scare may, however, vary in these different situations. This type of influence seems to have been important mainly in backward areas and in the relations of small states, though on a few occasions it may have affected the relations of great powers.

The charge of exercising such influence has been leveled especially against arms- and munitions-makers and traders, against international bankers, and against international investors. It is obvious that arms-makers or traders can increase their markets by war scares and wars, and there is evidence that they have on occasion evaded embargoes and international controls, bribed officials to get orders, sold arms simultaneously to both sides in wars and insurrections, stimulated armament races, and maintained lobbies to increase military appropriations and to prevent national or international restrictions on arms or arms trade.

Mixed firms which manufacture steel, vessels, airplanes, explosives, and chemicals for peace as well as war purposes are clearly under a temptation to expand the military side of the business in times of depression, when the demand for their peace products falls and high tensions facilitate warmongering. There is evidence that occasionally firms have yielded to the temptation.

The dependence of much industry and labor on war manufacturers when defense budgets reach 10 per cent of the gross national product, as they did in the United States in the 1960's, may induce these interests to oppose disarmament. When the United States Defense Department ordered the discontinuance of certain unnecessary bases and installations in 1963, there were vigorous protests from the interests affected.

Bankers can make profits from loans to actual or prospective belligerents which may be distributed to the public before defaults occur. Loans by neutral bankers and sales of war materials by neutral manufacturers and traders may eventually create an interest in the victory of the side with the greatest debt and the greatest trade. This interest may extend to farmers, miners, the general investing public, and manufacturers of numerous non-military articles purchased by the belligerent. The evidence indicates that this type of interest has been of relatively slight importance in drawing neutrals into war.

Investors in foreign bonds or enterprises suffering from defaults, from adverse laws, or from inefficient police in the investment area may seek the aid of their government to collect debts or to protect their interests. The practice of diplomatic protection has been fully recognized in international law, as has the danger that it may lead to hostilities. Numerous interpositions by powerful states in the territory of lesser states have occurred, but they have seldom led to major wars, unless associated with political objectives.

The voluminous evidence adduced by the League of Nations, by national commissions, and by private investigators indicates that all these abuses have occurred. Their relative importance in the causation of modern war has probably been greatly exaggerated and it is probable that some of the remedies proposed, especially those in a socialistic direction, would aggravate the abuse.

Only ten states of the world have important arms manufactures, and only four make nuclear weapons. Regulation of the arms trade by these states might increase their imperial dominance in certain areas by the control of internal policy of the governments dependent upon imported arms for police and defense. Government monopolies of arms production would move governments toward state socialism, because modern arms, munitions, and war materials constitute an important part of the national economy. Such monopolies would extend the control of the present arms-producing states even more than would international regulation of the private industry. Control of the arms trade might stimulate all states to establish an arms industry and to increase the total quantity of the world's productive capacity devoted to this essentially uneconomic activity.

Transfer of the arms industry from private to government hands would accentuate the national character of the industry and might make the balance of power less stable. When great international arms firms peddled their inventions among governments, each government knew what was available to the others. With national monopolies and secrecy of inventions, each state would continually be alarmed by rumors of new and devastating inventions by its rival. This situation became apparent with the development of nuclear weapons. International monopoly of the manufacture of these weapons was attempted in 1946 but failed. As the danger of spread of these weapons to irresponsible hands and of nuclear fall-out became known, further efforts at control were initiated. The limited nuclear test ban treaty of 1963 resulted and was immediately ratified by most states except Communist China and France.

b) Expansionism.—Socialist writers have charged capitalism with the vice of expansionism or imperialism, which, they say, leads not only to exploitative wars by advanced against backward peoples but also to wars between capitalistic nations struggling to exploit the same backward area. The tendency of capitalism to expand in backward areas is said by some to be due to the progressive attrition of the domestic market as the capitalists deprive labor of labor's fair share of the products of industry and decrease its purchasing power. Foreign markets, it is said, must be found to absorb the product of the ever increasing capitalistic plants.

Economists have denied the theoretical reasons adduced for such a development of underconsumption, and some socialists repudiate this theory. Although purchasing power has been inadequate to provide a market for existing productive capacity in periods of depression, it is not clear that serious and protracted depressions are an inherent characteristic of capitalism or that depressions have been the major factor in promoting imperialistic expansion.

The more orthodox socialist theory attributes the alleged expansive tendency of capitalism not to the necessities but to the greed of the entrepreneurs. Opportunities, they say, exist in undeveloped areas to utilize richer resources of raw materials, to exploit more helpless labor, to develop larger markets, and to make more profits out of investment than is possible at home. Consequently, when communication and transportation make it possible, the profit motive urges capitalists and entrepreneurs to exploit such areas and to seek protection through the diplomatic and military power of governments which, according to socialistic theory, the dominant capitalistic class will control.

This theory generalizes from too few facts. A general historical survey indicates that most capitalists and entrepreneurs have preferred domestic to foreign or colonial investment. Bankers and investors have, it is true, sometimes urged governments to assist them in imperial enterprises, but more frequently imperial-minded politicians have utilized bankers and investors as unwilling tools to justify or assist in expansions desired for strategic or political reasons. Although such imperial ventures have required military activity against natives, and although, in the early stages of capitalism, the division of newly discovered lands in the Americas and East Indies led to many international wars between European rivals, yet in the nineteenth century, when capitalism was more developed, rival imperialisms in Africa and in the Pacific were usually settled peacefully. It cannot be said that imperialistic rivalries contributed much to the causation of the Napoleonic Wars, the nationalistic wars of the mid-nineteenth century, or the world wars of the twentieth century.

Expansion of business enterprise to new lands can take place, and has in the main taken place, by peaceful trade, investment, and development. Agricultural expansion can occur only by migration or invasion, supplanting the existing population, and so is likely to involve violence. In practice and in theory the expansion of capitalism has been less productive of war than has been the expansion of other types of economy. Capitalism has figured in the imperial process, but the impetus of that process has more often been nationalism, agrarianism, or a missionary spirit.

After World War I the process of colonial emancipation, begun with the wars of American independence a century earlier, was continued peacefully by the imperial countries themselves. Britain recognized the independence of its dominions, and the League of Nations mandates system prepared certain colonies for independence. After World War II the United Nations Charter accepted the principle of colonial self-determination, and empires came to an end, in most cases peacefully.

c) *Depression.*—It has also been charged that capitalism tends inevitably toward periodic depressions of increasing amplitude, which, because of the miseries of the unemployed, tend toward civil war or, as a preventive, toward international war.

Depressions have been variously attributed to the extreme commodity price advances and burdens of debt caused by wars themselves, to the tendency of industrialism to decrease the internal market by exploitation of labor, and to fluctuations in the expectation of returns from capital. Explanations such as these in terms of political, industrial, or financial practices do not reach the heart of capitalist economy. If war is the cause of depressions, the difficulty lies in international relations rather than in capitalism.

The economic explanations, which relate depressions to progressive limitations of competition and to progressive lengthenings of the productive process, both of which may be inspired by the effort toward economic efficiency, suggest inherent weaknesses in capitalism. They assert that capitalism in larger enterprises eventually defeats itself by pursuing its economic end of eliminating inefficiency and increasing division of labor. There can be no doubt that protracted depressions have been a danger to peace. Unless capitalism can succeed in giving steady employment and rising standards of living, it will be in danger. Forms of government intervention have proved to be effective remedies.

d) *Protectionism.*—Capitalism has led to technologies giving greater control of natural forces, has conquered distance by new means of transportation and communication, and has stimulated trade between all parts of the world. These developments have built up an interdependence of national economies far beyond anything achieved by other economic systems and have also created military techniques greatly augmenting the social and economic costs of war.

The monopolistic tendency inherent in capitalism has urged domestic producers to demand protection through tariff or other economic barriers. National defense demands have added to these barriers. A high degree of economic interdependence of states, when associated with rising national barriers, has produced the problem of "have" and "have-not" states. The latter, unable to trade manufactures for necessary raw materials and foodstuffs, have felt oppressed in an inadequate living-space and have fought for more land. Agricultural countries faced by declining world prices for their products and rising prices for the manufactured goods they need have sometimes resorted to revolution.

Capitalism has contributed to these situations, as has nationalism. Neither is responsible in itself. The incompatibility of the two has proved disastrous. International and national agencies as well as programs to reduce trade restriction and to assist underdeveloped territories have been developed since World War II to meet these problems.

e) *Materialism.*—Perhaps the most serious charge against capitalism has been that it destroys the sense of social values by its emphasis upon individualism and its depersonalization of economic activity. Peace requires effective political organization, and that requires not only respect for and protection of individual rights but also constant loyalty to the symbols of the group. In so far as capitalism has tended to disintegrate all political loyalties, it has tended toward disorder and war.

Capitalism certainly has not built up community loyalties capable of sustaining a political organization operating effectively over the area which it has integrated economically. Instead, by its tendency to concentrate human interest on the business enterprise, on individual profits, and on impersonal productive processes, it has tended to minimize community values and to disintegrate political organizations dependent upon those values. A good economic man tends to be a bad citizen.

As a consequence, political organization during the period of modern capitalism has been sustained by sentiments unrelated to capitalism—sentiments of tribal and cultural solidarity, geographic unity, and historic tradition. The good citizen has tended to be a nationalist and a bad economist.

The natural ethic of capitalism is liberalism and humanism, as was realized by the classical economists who elaborated this ethic in their creed of utilitarianism. In spite of Richard Cobden and Cordell Hull, active capitalism with a laissez faire tradition was lukewarm in its support of those ideals. By accepting protectionist loaves and fishes from national states, it paved the way for its own destruction.

Marxian socialism took up what capitalism had abandoned. It preached internationalism and tried to put the individual and humanity (interpreted as the laboring class) above the nation. Thus the ethic of liberalism continued in the British labor party and in German social democracy. But the natural ethic of socialism is nationalism, since its program can be achieved only by a strong government supported by a powerful sense of group solidarity. Socialism in practice became "national socialism," destructive of both liberalism and humanism. Support for the universal ethical consciousness, essential for the preservation of peace, must be sought outside of either capitalism or socialism.

It may be concluded that, although capitalism and socialism have each claimed to be the most peaceful form of civilized economy, the difficulties of capitalism with depressions and profiteers and of socialism with oppressions and inefficiencies, the subordination of both to nationalism, and the incapacity of either to sustain a universal ethical consciousness in a world of economic and political interdependence, have made each productive of war. Mixed economies have tended to develop in technologically advanced states, converging the two systems and recognizing international objectives.

CHAPTER XVIII

HUMAN NATURE AND WAR

To the question, "Do you as a psychologist hold that there are present in human nature ineradicable, instinctive factors that make war between nations inevitable?" 346 of the 528 members of the American Psychological Association in 1932 replied "No," 10 replied "Yes," 22 replied ambiguously, and 150 did not reply at all.

The posing of such a question implies a picture of the world as a population of human individuals, each of which behaves according to a pattern derived from the interaction of heredity and experience. A great majority of professional psychologists assume that there is nothing in the heredity, and it is not necessary that there should be anything in the experience, of the members of this population which compels them to organize warfare.

The human population has spread over most of the world, but this spread has been quite uneven, and the inhabitable area exhibits great variations in the quantity and quality of its human blanket. Viewed from a distant planet, this spreading of *Homo sapiens* through most of its history would seem little different from that of the spread of other organic forms.

The behavior of other organic populations of the world is determined mainly by heredity and changes, very slowly, in the process of organic evolution. *Homo sapiens,* however, has learned to communicate general ideas by speech, writing, printing, and electricity. Each human individual has come to live in an infinitely vaster environment, both spatial and temporal, than does the individual of any other species. Consequently, human behavior is extraordinarily variable and changeable and extraordinarily difficult either to predict or to control.

Hostilities among animals occur between single individuals, between flocks, or even between societies, but only a limited area and an infinitesimal part of the species are involved in any such combat. Modern wars occur between alliances of nations and tend to involve the whole world and a large proportion of the human species. For any other organic species war appears like frequent but small eruptions on the skin, but for modern man it resembles a general fever involving the whole body.

In spite of this difference, the drives of animal war can be observed in human war. The defense of the home territory from invasion is a common situation in which insects, birds, fish, and mammals fight others of the same species. War for territorial defense is especially characteristic of human groups. Defense of territory, however, cannot start a war. Someone must have committed an aggression or be about to commit one before there is any need for defense. Among individual animals the drive for such aggression is usually the search for food or a nesting site, but the invasion of a defended area is usually accidental, and the intruder usually flees before the hostilities become serious. If individual men are found trespassing upon the property of others by inadvertence or with criminal intent, the behavior is usually similar. Only among certain social insects and among politically organized men has aggression been intentionally and habitually undertaken for predation upon the territory of the same species. In the entire organic world such aggression seems to be characteristic of societies rather than of individuals. War is in the main a sociological rather than a psychological phenomenon. It is primarily a product not of the organic structure but of the customs and traditions of societies.

The individual and the society are, however, closely related. Group-inspired propagandas and educational procedures continually influence the individual. Biologically rooted needs and wants of the individual continually influence the culture of the group. Human nature is the general aspect of human behavior which emerges from the interaction of any individual and any group, neglecting the peculiarities of a particular individual and of a particular group. It is, on the one hand, a generalization of all personality types and, on the other, a manifestation of the most general aspects of culture. Personality may be analyzed into motives and classified in types. Culture may be analyzed into attitudes, and these attitudes may be generalized into patterns, values, and ideals. The influence of human nature on war may therefore be studied by considering the relation to war (1) of personal motives and personality types, (2) of cultural attitudes, and (3) of personality of ideals.

I. PERSONAL MOTIVES AND PERSONALITY TYPES

There is no specific war instinct, but numerous motives and interests have led to aggression by human populations. Leaders have sought wealth, revenge, adventure, prestige, glory, the deflation of internal revolt, the stimulation of external revolt, and the expansion of religion, nationality, state, or dynasty. The masses have usually supported them under the influence of slogans and of social and legal compulsions. Individual followers have been influenced by expectations of adventure, plunder, better lands, higher wages, feminine approval, or sadistic orgies; by the hope to escape financial, matrimonial, or legal difficulties or simple boredom; by loyalty to leader, fatherland, religion, or ideals; by anxiety to test courage, capacity, or character; by habituation or pride in the military craft or profession. The motives are to be explained by the history of the particular individual and by all aspects of the particular situation and are difficult to generalize. Their complexities can be understood by an examination of the letters of recruits at the front, particularly of volunteers in foreign legions who have gone to war without any of the usual patriotic or social pressures. Literary men and psychologists have often explained the subtleties of such motivations.

Harold Lasswell's notion of a continuous struggle of each individual to remain or to become of the elite in the safety, income, or deference pyramid of a given community is suggestive but probably oversimplifies the complexities of human motivation. Even more oversimplified are statements by a minority of psychologists relating war to a primitive fighting instinct. G. W. Crile writes:

Soldiers say that they find relief in any muscular action; but the supreme bliss of forgetfulness is in an orgy of lustful satisfying killing in a hand-to-hand bayonet action, when the grunted breath of the enemy is heard, and his blood flows warm on the hand. . . . In the hand-to-hand fight the soldier sees neither to the right nor to the left. His eyes are fastened on one man—*his man.* In this lust-satisfying encounter injuries are not felt, all is exhilaration; injury and death alike are painless.

As I reflected upon the intensive application of man to war in cold, rain, and mud; in rivers, canals, and lakes; under ground, in the air, and under the sea; infected with vermin, covered with scabs, adding the stench of his own filthy body to that of his decomposing comrades; hairy, begrimed, bedraggled, yet with unflagging zeal striving eagerly to kill his fellows; and as I felt within myself the mystical urge of the sound of great cannon I realized that war is a normal state of man. . . . The impulse to war . . . is stronger than the fear of death.

Many observers emphasize the influence upon the soldier's motivation of the close proximity of his fellows. According to Fritz Kreisler:

The very massing together of so many individuals, with every will merged into one that strives with gigantic effort toward a common end, and the consequent simplicity and directness of all purpose, seem to release and unhinge all the primitive, aboriginal forces stored in the human soul, and tend to create the indescribable atmosphere of exultation which envelops everything and everybody as with a magic cloak.

Controlled studies designed to define, describe, or measure the situations, drives, or motives of war have utilized several types of material.

Studies of monkeys and children have disclosed the typical situations in which fighting occurs—rivalry for possession of a prized object, jealousy for the attention of an individual, frustration of an activity, and intrusion of a stranger in the group.

Comparative studies of animal, primitive, and civilized warfare have suggested that primitive drives of self-preservation and territory, of food and activity, of sex and society, and of dominance and independence have an influence on war and that they are related to the political, economic, cultural, and religious motives.

Psychoanalytic and anthropological studies have indicated the influence of such psychological mechanisms as identification, rationalization, repression, displacement, projection, and the scapegoat in transforming natural human affections, annoyances, ambivalences, and frustrations into group hostilities.

Psychometric studies have been made utilizing carefully devised questionnaires and interviews. These have attempted to ascertain the relation of warlike attitudes to other characteristics of the individual. Although the samplings on which these studies have been based have not been adequate, they suggest that men who fought frequently in childhood are more favorable to war than those who did not, that people with education beyond the high-school level are less favorable to war, that people are favorable to war in proportion to the amount of military education and military service they have had, that people are more favorable to war between the ages of thirty-five and forty-four than in any other period, and that men are more favorable to war than women.

Motives are combined in innumerable ways to form distinctive personalities. Efforts have been made to classify the latter in personality types identified by physical characteristics, past behavior, dominant attitudes, or developmental history. Some of these types of personality in positions of leadership are more likely than others to seek military solutions of problems. The device actually utilized by a leader is, however, usually the consequence of a total situation in which his personality is only one element.

The political type which seeks power by discovering general advantages for a group has been distinguished from the bargaining type which tries for special advantages in a transaction. The reactionary, the conservative, the liberal, and the radical types have been distinguished, as have the agitators, the theorists, and the administrators. The politician seeking to unify his group is more likely than the bargainer to focus hostilities upon an out-group. The reactionary and the radical are more likely to disturb the balance of power than the moderate conservative or liberal. The agitator is more likely to value military policy or to augment conflict than the theorist or administrator.

Particular personalities may manifest one type only, or they may present a mixture of several. Understanding of personality cannot be complete without knowledge of its developmental history. Such histories may be classified, thus providing another basis for personality typologies. Compensating and canalizing types are distinguishable. Leaders whose energy derives from the continual push of a feeling of physical or psychic inferiority frequently overcompensate by aggressiveness. They appear more likely to accept violence as a solution of problems than those whose energy derives from the pull exerted by acquired skills, inducing them to canalize drives of dominance or ambition into effective effort.

From the point of view of long-run prediction or control, it is less important to understand the behavior to be expected from personality types than the cultural and institutional conditions which tend to bring one or the other type into leadership. Societies dominated by industrialism, by liberalism, by constitutionalism, and by federalism have tended to give leadership to administrative and canalizing types, whereas societies dominated by feudalism, totalitarianism, absolutism, and nationalism have tended to accept and support agitators and compensating types. Democracy has usually been associated with the former group and has frequently selected rulers of the conciliatory type, but the election process often gives the agitator an advantage. There is little correlation between capacities useful in getting elected and those useful in administering.

Periods of crisis and high tension tend to perpetuate themselves by the favorable opportunity they present to the rise of agitators, whereas times of tranquillity similarly tend to perpetuate themselves by enhancing the influence of the administrative and conciliatory types.

2. CULTURAL ATTITUDES

The point of view which considers the individual personality as the center of social action and study may be supplemented by that which emphasizes attitudes as culture traits which may be studied irrespective of the personalities in which they appear. The biologists have given a parallel emphasis in supplementing the study of organism by the study of genes or bearers of biological traits.

From this point of view a personality is a complex of attitudes, each with a certain intensity and direction and each inducing the individual to behave in a certain manner when his attention is drawn to a given psychological object. The personality is not the unit of investigation. The attitudes themselves are regarded as entities, which together constitute the culture of the population, and in so far as they are publicly manifested with considerable homogeneity on controversial subjects, they constitute its public opinion.

Opinions have been measured through analyses of responses to questionnaires or interviews with a fair sample of the public; through analyses of responses to questions or interviews with experts deemed to have a sound judgment as to the attitudes within the public in question; and through the analysis of "attitude statements" copied from newspapers and indicative of favor or opposition to a given symbol, such as another country.

Four dimensions of opinion have been exhibited by graphs constructed by the latter method: direction (whether the opinion is for or against a symbol), intensity (degree for or against), homogeneity (distribution of attitudes at a given time about the average), and continuity (invariability of the attitude over a period of time). Such graphs have indicated changes in the prevalent opinion in the United States toward France, Germany, China, and Japan in different periods before World War II. The results of these studies have shown a high degree of reliability and of validity in the sense of conforming to expectations derived from a study of the historical facts. They indicate that the opinion prevalent in one country with respect to another tends to fluctuate in time, tends to be manifested by active hostility when it passes below a certain threshold, tends to be friendly toward other nations when it is hostile to one, tends to respond to hostility by hostility, tends to be interested in proportion to the intensity of the opinion and tends to be homogeneous when intense but divided when moderate. Such studies, if carried out with respect to a number of symbols, might supply evidence for charting the changes in the general tension level of a population.

A chart comparing the opinion of numerous experts in regard to the attitudes dominant in a large number of states with reference to other states during the period from 1937 to 1941 displays a fanning-out tendency of the opinions toward greater intensities of friendship or hostility, which suggests that the general tension level was rising during this period.

The interest of such studies is not only in their theoretical results but also in the assistance they might offer to practical action in propaganda and education. A continuous charting of changes in public opinion in the principal populations upon political questions and particularly upon questions concerning other states would be of value in the art of statesmanship and in the work of any world organization devoted to the regulation of international relations. Such charts would not often show anything qualitatively novel. Statesmen and journalists know roughly how opinion is moving in the important areas toward the important symbols. But, as in predicting weather, it is worthwhile to know the temperature, pressure, or wind velocity precisely, so in political prediction and control it would be extremely helpful to have opinion movements precisely charted from week to week and month to month.

Opinion, or public expression of attitude, normally reflects the actual attitude of the subject, because people tend to say what they believe and to believe what they have been saying for a long time. Efforts have, however, been made to measure attitudes directly. Questionnaires have been so worded that the subject will not be aware of the psychological object toward which his attitude is being tested. Such studies indicate that "real" attitudes may differ from consciously expressed opinions and that rigorous government controls of opinion tend to accentuate this condition. Although government decisions give consideration to public opinion regardless of its conformity to actual attitudes, the behavior of people in a crisis may more nearly follow their attitudes.

Such studies might throw light on the attitudes behind warlikeness and the influence of educational methods upon them. Do states go to war because leaders are hostile to a particular enemy or because they want war? There is evidence that attitudes of the latter type are sometimes of great importance. Such attitudes may spring from discontent with an existing situation inducing irrational violence, or they may spring from habitual preference for dictatorial rather than conciliatory modes of dealing with problems. Internal circumstances, such as depression or party feud, and general conditions, such as the existence of foreign war or the long passage of time since the last national war, may predispose a population to war. The particular state selected to fight may be largely fortuitous.

Conflict may, however, arise from attitudes toward a particular enemy, perhaps in reaction to an injury, perhaps to remove his obstruction to achieving a desired goal, perhaps because his aggressive intention is assumed, and perhaps because he is conceived as an evil to be destroyed. The role of false images, assumptions, stereotypes, and disparities between attitudes and opinions is significant in international conflict.

3. PERSONALITY IDEALS

A culture may give preference to particular modes of dealing with conflict situations. These modes may be classified from the point of view of the individual as renunciatory, conciliatory, dictatorial, or adjudicatory, according as the individual is disposed to yield to those who oppose him, to observe an understood rule or to compromise, to dominate over opponents, or to submit to group decision or demands. From the point of view of the group, they imply, respectively, its aloofness from, its passive regulation of, its incapacity to deal with, or its active intervention in the controversies of its members.

The last method puts a severe strain upon the society, which is continually obliged to resolve conflicts within itself. The first method puts a severe strain on the non-resisting individual, who is continually obliged to check the natural expressions of his personality. The one passes the problem of ambivalence to the group; the other leaves it with the individual. The one may lead to group revolution; the other, to individual insanity. Psychoanalysts stress the danger of oversuppressing aggressive dispositions. Groups made up of non-resisters are rare; and groups which attempt to resolve internal conflicts by active intervention frequently find it necessary to wage external war in order to prevent internal revolution. Group intervention based on the concept of group planning and administration differs from that of common law and adjudication in degree. It assumes the disposition of man to affirm his will unless compelled to desist, whereas common law assumes the disposition of men to observe rules or to conciliate in most cases.

A particular culture may emphasize attitudes favorable to one or the other of these procedures. Although public opinion within a group may be rapidly modified by propaganda designed to inculcate group ideals and utopias, personal attitudes appear to be influenced primarily by the personality ideals (superego) which the individual has acquired from early family, religious, and educational contacts. Cultures differ in warlikeness according as they idealize non-resistance, rationality, aggressiveness, or authority.

a) *Non-resistance*, illustrated in Jesus' Sermon on the Mount and in certain Buddhistic and Hindu writings, was accepted by the early Christians, the Quakers, the Mennonites, and the followers of Tolstoi and Gandhi. Its creed of renunciation has not, however, been indorsed by the bulk of mankind. A complete following of the ideal of non-resistance implies a renunciation of all the material ambitions of life, as recommended by Buddhism, the exact reverse of the Nietzschean creed of the superman.

b) *Rationality.*—A much larger proportion of the human race has recognized the necessity of common law and has accepted the ideal of the rational man, who voluntarily restrains his oppositions to the requirements of necessary laws, illustrated in the rationalist philosophies of Locke, Hume, Kant, and Bentham, or the ideal of the economic man, assumed by Ricardo and other classical economists. Self-interest, it is thought, will lead men to abide by contracts and laws, if those laws do not go beyond the constitution of the liberal state. Self-interest, it was optimistically anticipated, would lead governments to abide by treaties and international law.

c) *Aggressiveness.*—In times of rapid social change it is difficult for common law to maintain sufficient control over the aggressive activities of individuals or groups who first perceive the opportunities offered by changing conditions. Thus conditions of anarchy have sometimes resulted. Struggle, without limitation of means, has been actively advocated by radical champions of the oppressed masses and has been practiced by reactionary entrepreneurs who wish no limitations set to their opportunity to increase profits. The masses of men, however, have not for any length of time indorsed philosophies of individual, class, or international violence, such as those expounded by Clausewitz, Proudhon, Nietzsche, Bakunin, and Sorel. Although militarists, nationalists, and neo-Darwinian social philosophers have sometimes perceived a continuing need for violent struggle, most radical philosophers of violence, including Marx, Lenin, and Trotsky, have regarded violence and unbridled conflict as temporary expedients, justified only as a necessary means for ushering in a new order of peace.

d) *Authority*, planning and controlling the behavior of loyal subjects, is the ideal set up by autocracy, nationalism, fascism, and communism. In this philosophy the bee is nothing; the hive is everything. This philosophy, by concentrating all authority in the state as the sole object of individual loyalty, has been of little value in the propaganda for international peace, except by those who believe world conquest and empire are the roads to peace. National loyalty is hard to transmute in times of crisis to a higher loyalty to the world-community. The task is especially difficult because of the tendency for the Nietzschean ideal of the superman above good and evil to be accepted by the ruler of the highly organized state.

Numerous factors—geographic, economic, and historic—have to be considered in accounting for the future of a particular group, but in all, the ideal of human personality occupies a large place. Probably the ideal of the rational man in the liberal state is that which has best adjusted human nature to continuous peace. This ideal has been illustrated among a few primitive peoples such as the Yurock Indians of California, whose culture combines economic individualism with remarkable peacefulness. It was characteristic of the Chinese in much of their history and of the periods of the *Pax Romana* and the *Pax Britannica,* which witnessed the flowering of the two most widely accepted systems of law in the contemporary world. Education directed toward that ideal might create attitudes which could be invoked in times of crisis to prevent war. Non-resistance in practice puts too severe a strain on human nature, and complete loyalty to the group puts too severe a strain on the group and is no cure for intergroup war.

The rational ideal, however, is not attractive to a human race that is only partially rational. The ideal of the economic man or the reasonable man looks pallid beside a fasting Gandhi, a Light Brigade loyal to the death, or a Faustian hero in titanic struggles against the world. A tranquil world "sicklied o'er with the pale cast of thought" is not generally appealing. The maintenance of continuous peace may, however, depend on the acceptance by the masses of mankind of the ideal of the reasonable man—the man guided neither by an all-consuming ambition, an all-consuming loyalty, nor even an all-consuming asceticism, but ready to exercise his reason to maintain world conditions in which his type, preferring reason to violence, can prevail.

PART THREE

THE PREDICTION
OF WAR

CHAPTER XIX

ANALYSIS OF INTERNATIONAL RELATIONS

Each point of view with respect to war, to some extent, falsifies reality. Efforts to predict or to control war must estimate the relative weight to be given to each point of view and to numerous causal factors.

Practical prediction of the time and place of the next war is a process involving interpretation of the existing situation in terms of a given analysis, criticisms of the assumptions of that analysis by comparison of the results of that interpretation with the developing facts, modification of the analysis and reinterpretation of the situation in view of this criticism, criticism of the new analysis, and so on. Interpretation of facts by analysis and modification of analysis by facts may proceed ad infinitum, gradually approximating the truth as the outbreak of war approaches. Statesmen make history by this process—the initiation of war may be a self-fulfilling prophecy—but in view of the continuous development of new factors, correct analysis has never been able to get much ahead of events.

Is it possible to develop an analysis more adequate than those of the past for dealing with war in our time? Such an analysis should, in a single formula, relate the factors emphasized in each of the points of view about war. Factors inherent in a given period of history may be called "distances" between states, and factors dependent on the decisions of actors may be called "policies." It may be assumed as a first approximation that the probability of war is a function of the distances between states and of the policies which they pursue. Afghanistan is not likely to get into war with Bolivia because their contacts with each other are so slight. The United States is not likely to get into war with Canada because the policies of these two states with reference to each other are so friendly.

Distances between states in their various aspects, objective and subjective, vary with changes in the condition of the world, especially in respect to technology, law, opinion, and the power position of states. The policies of a state, if rational, seek to utilize available procedures and techniques to achieve values and goals with the least cost in view of the situation it faces.

1. ASPECTS OF DISTANCES

Eight aspects of distances between states may be distinguished: *Technological* (T) distance diminishes with increasing communication and transportation, tending in the long run to increased trade and understanding. *Strategic* (St) distance diminishes with increasing vulnerability to, or capability of, attack by one or the other, leading to fear or ambition. *Intellectual* (I) and *legal* (L) distances are of less material character. Intellectual distance diminishes with increasing similarity of rational processes, facilitating comprehension by each nation of the other and negotiation of controversies. Legal distance diminishes with increasing similarity of standards of justice in the law of the two states and increasing recognition by each of the equal status of the other in international law. *Social* (S) and *political* (P) distances involve appraisal of attitudes and opinions. They diminish respectively with increasing acceptance by the people and governments of the two states of the same or similar social and political institutions and values. *Psychic* (Ps) and *expectancy* (E) distances also refer to attitudes and opinions and diminish respectively with increasing friendliness and decreasing expectation of war by the people and governments of the two states.

Although none of these aspects of distance can be measured precisely, rough measurements can be made by the use of statistics of communication, travel, and trade; analyses of military logistics and preparations; measurements of opinions and attitudes; comparisons of institutions and legal systems; and appraisals by experts.

Distances between great powers tend to be the same viewed from either state, but this is not necessarily true between states differing in power and culture. A may be more vulnerable to attack from B than B is from A. A may recognize B as legally equal, but B may consider A of lower status. A may expect B to attack it, but B may have no such anxiety about A.

2. POLICIES AND DISTANCES

If the family of nations is considered as a whole, it is clear that, on the average, technological and intellectual distances between states have been decreasing during the modern period, whereas this is not so clearly true of psychic and social distances. There are, however, great variations among pairs of states with respect to each of these distances. The relationship of two states to each other may be described, on the one hand, as a function of their distances from each other, of the rate of change of these distances, and of the degree of reciprocity in the contacts and opinions accountable for the distances, and, on the other hand, as a function of the policies (*a*) of each of the states toward the other and (*b*) of outside states toward both of them.

a) *Policies of Disputing States.*—Policies and distances are clearly interrelated. States widely separated technologically have little interest in one another and are not likely to have any policies at all toward one another. States that are friendly will have very different policies toward one another from those that are hostile. On the other hand, policy may influence distance. A policy of co-operation tends to produce commerce and friendliness, and a policy of aggression, the reverse. The relationship of policies to distances may be considered both as cause and as effect. Friendly acts are likely to be reciprocated and hostile acts retaliated against.

The policy of a state when in controversy with another may be to seek solution (1) by delay, in the hope that conditions may be more favorable in the future; (2) by negotiation, in the hope that a satisfactory compromise or bargain may be made; (3) by adjudication, in the expectation that by an impartial application of law and accepted standards it may gain its ends; or (4) by dictation, in the confidence that by a strong stand, perhaps using threats or even violence itself, it can permanently settle the controversy in accordance with its wishes.

Dilatory tactics or negotiations are likely to be utilized if states are widely separated technologically, as have been Oriental and Occidental states until recent times, but if strategic relations are non-reciprocal, dictation is likely to be used. Negotiation promises success in proportion to the intellectual, legal, and psychic closeness of the parties. Adjudication or dictation is likely to be utilized only if technological and strategical distances are short, and adjudication only if psychic and legal distances are also short.

In general, therefore, as technological distance decreases, there is likely to be a movement from relative predominance of methods of delay to relative predominance of methods of negotiation, dictation, and finally adjudication. As psychic distances decrease, there will tend to be a movement from predominance of methods of dictation to predominance of methods of delay, negotiation, and finally adjudication. These relationships will, however, be affected by rates of change and conditions of reciprocity. Dictation is likely to be used if technological distance is decreasing more rapidly than psychic distance, and arbitration will be employed if the reverse is true.

b) Policies of Third States.—The policies of third states confronted by violent controversy may be classified as those of isolationist neutrality, prudent preparedness, balance of power, and collective security. The isolationist neutrals scatter from the conflict like a flock of chickens attacked by a hawk. The prudent preparers appease the powerful aggressor in order to divert his attention or to profit by his conquest, like the jackal following the lion. The balancers of power spontaneously help the weaker like a band of apes assisting one of their number in danger. The adherents to collective security collaborate in a prearranged plan against aggression as in human societies enforcing law. The effectiveness of any of these policies for a particular state depends on many circumstances of which its distance from the belligerents and the policies pursued by other states are important.

The policy likely to be followed in the group of states as a whole seems to depend mainly upon the average distances among states in the group. The policies followed tend to change those distances. As technological distances decrease, there will tend to be a movement from reliance upon isolationism, to balance-of-power policies, and finally, as states become technologically very interdependent, to appeasement or collective-security policies. If psychic distances are great, appeasement will be preferred. If they are small, mutual confidence may be sufficient to permit collective-security policies.

In a pioneer community composed of self-seekers with little psychic solidarity, as the California forty-niners, everyone seeks security through his own arms, and the spectators scatter whenever gunplay is in prospect. As economic interrelationships increase, this gives way to an era of vigilantism in which all combine *ad hoc* against dangerous characters. This may be followed by a feudal regime in which the weak attach themselves to the great, sacrificing liberty for protection. If, however, a general sense of community develops, a regime of law and order may be established in which the community as a whole suppresses banditry. As technological interdependences increase, if the sense of social solidarity does not increase proportionately, agitation for radical social change may gain support. If belief in the success of the agitators develops, increasing numbers may flock to the revolutionary bandwagon and a new order may be set up. Revolution may, however, so shatter social order that technological distances increase and a condition of anarchy is reverted to, starting a new cycle.

Latin America, according to Victor A. Belaunde, has gone through similar stages. After the colonial period, characterized by isolation and nationalistic differentiation, the independent states sought to develop relationships, first, of *convivencia,* or coexistence, and, then, of "economic and cultural co-operation." Leading spirits looked forward to the eventual formation of "a permanent international organism." This was partially achieved in the Organization of American States but has been threatened in the 1960's by movements of revolt led by Cuba.

The development of the community of nations in the modern period has been similar, from the isolated and warring princes of the Renaissance, through the balance of power of the seventeenth and eighteenth centuries, to the movement of concert and collective security culminating in the League of Nations. With increasing psychic and decreasing technological distances, movements of agitation and revolt were initiated by certain dissatisfied powers, and there was a tendency for others to jump on their bandwagon as the revolt of the axis gained success before World War II. The tendency toward the organization of collective security continued, however, in the United Nations in spite of the renewal of power balancing in the "cold war."

Anarchy and isolationism, *ad hoc* collaboration to maintain power equilibrium, political organization on a despotic or democratic model, and revolutionary agitation urging a new leadership—these stages appear to mark the normal political trend as technological distances decline within groups of individuals or of nations. The trend may become cyclical because revolution may widen technological distances and reintroduce anarchy.

If psychic distances decline more rapidly than technological distances, the cycle may be indefinitely stopped, and political organization may indefinitely maintain order. It is the lagging of psychic behind technological distance that causes aggression to be generalized into revolution and anarchy.

3. ANALYSIS AND PREDICTION

Analysis of the factors relevant to war and of their interrelationship is possible, but such an analysis does not permit precise prediction. Distances and their rates of change are imperfectly measurable, and the actions of states are primarily dependent on their policies, which may change rapidly, rather than on their distances. Long-range prediction of a state's policy may be assisted by study of the trends of its changing distances from other states, of the changing conditions of the community of nations, of the changing procedures and techniques of peaceful settlement and military strategy available to it, and of its changing ideology and objectives, but the interpretation of a particular situation by the policy-maker and his co-ordination of means and ends is so affected by personality traits, imperfect information, stereotyped images, and mercurial domestic opinions, often introducing irrational factors, that a prediction would have an extraordinarily large margin of error. Nevertheless, attempts have been made to determine the probability of wars in the future as indicated in the next chapter.

CHAPTER XX

THE PROBABILITY OF WAR

The phrase "probability of war" may refer to the probability that within a given time (1) a particular state will become involved in war, (2) a particular pair of states will get into war with each other, (3) any state or a state in a particular area will become involved in war, or (4) a general war involving all or most of the states will occur. This discussion will, in general, deal with only the first two of these probabilities, although some reference will be made to the last. The third probability is so vague that it cannot be discussed intelligently unless the terms "state" and "war" are defined very precisely. With broad definitions it could be said that at least one war occurred in the world every year from 1920 to 1963. On the other hand, with a very narrow definition, only one war occurred during that period. A common-sense judgment suggests that twenty wars occurred during the period. The most precise question, if attention is confined to the great powers, concerns the second probability. It is usually clear what is meant by a war between a particular pair of great powers.

The probability that a possible event will occur in the future increases in proportion as the time considered increases. To have meaning, predictive probability must be confined to a limited period of time marked by two future dates or by the present and one future date.

The relations of friendliness and unfriendliness between states appear to be closely related to probabilities of peace or war, but these relationships fluctuate widely in short periods of time and exhibit little relationship to the more stable factors in international relations, such as geography, trade, state of the arts, and population, included in the concept "technological distance."

Since war results from opinions and policies unrelated to processes, patterns, or time series which can be reliably projected far into the future, long-time prediction of war is not possible. Bismarck doubted the usefulness of attempting to predict international politics beyond three years.

Predictions may be based upon a projection of the present as a whole, with all its complications, for a few months or years into the future. Predictions may also be based upon an abstraction of the elements of history deemed to be persistent through centuries or millenniums. Between these two types of prediction are those based upon analyses distinguishing the degrees of stability of the factors constituting international relations over a decade or a generation. The latter type of prediction, which reflects the usual perspective of the social sciences, can be based only upon a synthesis of the data and analyses appropriate to the other two, and its reliability can rise little above that of its sources.

If we bear in mind the wide margin of error in short-, long-, and medium-run predictions respecting war, four methods may be considered for estimating its probability: (1) analysis of the opinions of experts, (2) extrapolation of the trends of certain indices, (3) ascertainment of the periodicity of crises, and (4) analysis of the relations of distances between states.

1. OPINIONS OF EXPERTS

In January, 1937, a schedule form was circulated in connection with this study of war to two hundred and twenty persons selected because of their knowledge of world affairs. They were asked to rate from 0 to 10 the probability of war (within the next ten years) for eighty-eight pairs of states. Eighty-two judges filled out the schedule, and the scale values for each pair of states were calculated with a range of 0 to 1. The results were borne out by subsequent history. Within six months of January, 1937, war (defined for the judges as "military operations on a large scale designed to compel submission of the opposing government") broke out between Japan and China (scored as the highest pair, .94). Russia and Japan (.89) carried on rather large-scale border hostilities, particularly in August, 1938 (Changkufeng incident), and May to August, 1939. Germany-Russia (.87), Germany-Czechoslovakia (.81), and Germany-France (.78) had serious crises in September, 1938; Germany occupied parts of Czechoslovakia in October, 1938, and March, 1939; and war with France began in September, 1939. Next on the scale was Germany-Great Britain (.66). This relatively low score might be interpreted as predicting England's efforts at "appeasement" in 1938, which, however, failed, and war began in September, 1939. Germany-Poland, Germany-Belgium, and Hungary-Czechoslovakia were next (.64). During the two years following the expressions of opinion, hostilities or a major crisis occurred between the states in the fourteen highest pairs (all above .60), with the exception of Italy-Yugoslavia (.65), Hungary-Yugoslavia (.63), Hungary-Rumania (.62), and Soviet Union–Poland (.60), and in two years more the states in all of these pairs had been engaged in hostilities with one another. Of the wars predicted with a probability above .60, 100 per cent occurred during the five years following the prediction, if important border hostilities are counted as wars; of those with a probability of from .50 to .60, 58 per cent occurred during this period; of those with a probability of from .40 to .50, 50 per cent occurred; and of those with a probability of from .30 to .40, only 18 per cent occurred. If the Soviet-Japanese border hostilities of 1938 and 1939 are considered war, all of the great-power wars which occurred during this five-year period (January, 1937–January, 1942) began in approximately the order of the predicted probabilities, with the exception of Germany-U.S.S.R., which was postponed about two years beyond expectation, presumably by the German-Soviet pact of August, 1939. The internal consistency of these judgments was relatively high.

This study dealt with the probability of wars between designated pairs of states, including all combinations of the great powers. From these data the probability of each one of the great powers getting into war during the period was estimated. The order of this prob-

ability for the fourteen highest states in January, 1937, was Germany
(.999), U.S.S.R. (.994), Japan (.993), Hungary (.95), China (.94),
Czechoslovakia (.93), Yugoslavia (.87), Poland (.86), France (.78),
Great Britain (.66), Italy (.65), Belgium (.65), Rumania (.62), and
Lithuania (.60). The most probable wars for the United States were
with Japan (.56), with Germany (.46), and with Italy (.38). These
results correspond closely with the actual order in which the states
entered war in the next four years, though Yugoslavia was too high
in the list and Poland too low. The Netherlands, Denmark, and Norway failed to appear in the list at all.

The instructions requested a rating of the probability of a pair
being drawn into war against each other through any circumstances.
The judges were therefore asked to consider not only the relation of
the members of each pair to each other but also the relation of each
member of the pair to other states which might participate in a general war. The influence of the general orientation of the policy of a
state is undoubtedly an important factor in estimating the probability that it will become involved in war. This influence is especially
important in the modern period because of the tendency of wars to
spread and of all powers to polarize around one or the other side in a
war between two great powers. Nothing in the relations of Germany
and Cuba in 1915 would have been likely to suggest war between
them; yet in two years they were at war, largely because of the relations of Cuba to the United States and of the United States to Germany.

This study by Frank Klingberg was based on an appraisal of the
probability of war for each pair of states, but a study could be made,
using psychometric procedures, by asking the judges to rate the
relative probability of war among the various pairs. In later studies
judges were asked to rate pairs of states according to their relative
friendliness or unfriendliness. This would provide an index of the
psychic distance between states, which is closely related to their
war expectancy and war probability but is a less complicated concept and easier to rate. A comparison of the results of these ratings
made at five intervals from 1937 to 1941 among all pairs of the great
powers indicated considerable fluctuations and a general tendency
for both enmities and friendships to increase in intensity as the crisis
deepened.

These studies suggest that predictive results of some value for a few years ahead can be obtained from an analysis of expert opinions upon questions related to the probability of war. A moderate number of qualified judges would seem to be adequate to give useful results.

2. TRENDS OF INDICES

Several types of observable facts have been thought to indicate a trend toward war. These facts include incidents initiated by private individuals or officials involving violence against or contempt for the nationals, agents, or symbols of another state; diplomatic correspondence and official public utterances of unfriendly or hostile tone; declarations of policy, conclusion of treaties, and enactment of legislation adverse to the political interests or prestige of another state; mobilizations of forces and movements of warships into strategically significant positions; legislative or other action reducing trade with another state; increase in military appropriations and development of preparedness programs; and violent expressions in the press or other mediums of public opinion in regard to other states. These types of action have been dealt with descriptively by historians and journalists and analytically by jurists. They doubtless provide a most important basis upon which statesmen estimate the probabilities of war. The last three types of activity relating to trade, armament, and opinion are more susceptible of quantitative treatment than the others and have provided the basis for numerous discussions of economic, military, and moral armament and disarmament.

The commercial statistics and armament budgets of modern states are usually ascertainable. Commercial retaliations and armament races have often preceded war. L. F. Richardson developed an elaborate theory of international politics by an analysis of the influence of rising military budgets (positive preparedness for war) and rising trade (negative preparedness for war) on the relations of states. Although some of his assumptions might be questioned, his conclusions support the frequent observation that the eventual consequences of a foreign policy, because of the tendency of other nations to retaliate or to reciprocate, may be the opposite of that intended.

Before a situation can be controlled, it must be understood. If you steer a boat on the theory that it ought to go towards the side to which you move the tiller, the boat will seem uncontrollable. "If we threaten," says the militarist, "they will become docile." Actually they become angry and threaten reprisals. He has put the tiller to the wrong side. Or, to express it mathematically, he has mistaken the sign of the defense-coefficient.

This is overstated. There have been circumstances in which preparedness increased security, and there have also been circumstances in which increased trade decreased security. In other words, the defense coefficient may be positive or it may be negative, and statesmen will have to take numerous circumstances into mind in judging which it is at a given time. Preparedness of Britain would probably have contributed more to peace than did appeasement of Hitler in the 1930's. Statesmen should, however, consider the danger of armament races and the ameliorating influence of reciprocal trade increases emphasized by Richardson, as it was a century earlier by Richard Cobden and more recently by Secretary of State Cordell Hull. They should also avoid superficial comparisons, such as have been made by Western statesmen between the Hitler and Khrushchev periods without considering the great differences in respect to military technology, power positions, ideologies, opinions, and personalities between these periods. Richardson's application of his simultaneous equations to armament and trade statistics in the period before World War I did, however, yield significant results.

A classification and analysis of "attitude statements" copied from newspapers may give an indication of the changes during a period of time in the direction, intensity, and homogeneity of opinion in one country toward another. The accuracy of this index depends upon the degree to which newspapers are selected which either reflect or mold public opinion. Such studies were made of opinions in the United States toward France, Germany, Japan, and China and of China toward Japan in the 1930's. These studies suggested that opinions, like armament-building programs, tend to be reciprocated and that, when they pass below a certain threshold, active hostilities are likely to occur. They also suggest that public opinion may fluctuate widely within a short time. Thus it is risky to extrapolate trends of public opinion for any length of time. Nevertheless, a continuous charting of the changing characteristics of the opinions manifested by the press of each of the great powers toward the others, paralleled by a chronology of events, would give valuable evidence concerning the political importance of events and incidents.

Such indices might provide a basis for short-range forecasting of political crises and hostilities better than that provided by any indices now available. Opinions undoubtedly provide a more delicate index of international relations than do armament budgets or commercial statistics. The more complete preparation, analysis, and use of such indices by foreign offices and international organizations might be of importance for purposes of control even more than of prediction. Such indices, if up-to-date and comprehensive, should have a value for statesmen similar to that of weather maps for farmers or of business indices for businessmen.

Such indices could be used not only for studying the probability of war between particular pairs of states but also for ascertaining the changes in the general tension level within a state or throughout the world. They could quantify such assertions, often made by statesmen and journalists in times of crisis, as "tensions are increasing in Europe" or "during the past few days the crisis has substantially abated." More precise measurement of such changes would be of value in predicting war.

3 . PERIODICITY OF CRISES

Attempts to discover a precise periodicity of economic fluctuations have not been attended by complete success, and the determination of sufficiently precise political cycles to serve for prediction is an even less hopeful task. That important political fluctuations take place, no one can doubt, though many would say that they are completely irregular and unpredictable.

A certain periodicity in the frequency and intensity of war in particular states and in particular state systems has been observed, but such fluctuations have not been sufficiently regular to permit prediction with any exactness. Data are lacking on the periodicity of strained relations between states. With such data and with data indicating the gravity of successive crises, a persistent pattern might emerge. The probability of war between two states during a period of time is a function of the number of crises and the probability of avoiding war in each crisis.

A short political cycle of four or five years is suggested by the usual life of a political administration in most countries and the average duration of a war between great powers. A longer political cycle of from forty to sixty years has also been suggested by the average dominance of a political party in democratic countries and by the periodicity of general wars during epochs dominated by an expanding economy and a balance-of-power system. The tendency to postpone a new war until there has been time to recover economically from the last, coupled with the waning resistance to a new war as social memory of the last one fades with the passage of a generation, may influence this tendency toward periodicity.

Even longer periods in from two to three centuries have been detected, marking the phases in the development of a civilization, and periods of a thousand to fifteen hundred years, marking the life of a civilization. The historical evidence for such periodicity is far from adequate.

The factors responsible for political fluctuations have not been sufficiently analyzed to permit prediction, but an understanding of their normal course and of the conditions likely to increase their amplitude may assist in developing political controls. The types of study applied to business cycles might be applied to political cycles, utilizing as primary materials the fluctuations of opinion as disclosed by chronologies of political events as well as by the statistical treatment of attitude statements in the press, of responses to questionnaires or interviews, or of votes in elections or legislative bodies.

Such studies might disclose correlations between economic and political fluctuations. In fact, such correlations have been suggested on the theory that a major war is the fundamental cause of economic crises, which follow each other in waves of decreasing severity until a new war occurs, and in the theory that long economic fluctuations are the main cause of wars and revolutions. Materials, however, are as yet inadequate to demonstrate either of these theories.

As economic fluctuations are contingent upon the particular form of economic organization, so the character of political fluctuations is dependent upon the particular form of political organization. The political fluctuations characteristic of the medieval hierarchical organization would be expected to differ greatly from those characteristic of the balance-of-power structure of modern history or those characteristic of the collective-security structure attempted in the periods after World Wars I and II.

Similarly, the diminution of the number of independent political organizations in a political system, resulting in concentrations of political power, may tend toward accentuating political oscillations. In a balance-of-power system, if leadership becomes concentrated in two or relatively few hands, tensions, instead of being manifested by diplomatic exchanges or international conferences, tend to be manifested by war. On the other hand, a more far-reaching centralization of authority able to prevent violence, as in an effective federation, may maintain political stability even when there are only two opposing parties, as in some parliamentary democracies.

The specific conditions under which diplomatic tensions develop into war may also be analogous to those under which business recessions take the severe form of panic. The latter occurs when evidence of recession in key industries, coupled with extensive speculation, induces all traders on the stock market suddenly to sell in the same direction, thus causing a collapse in values beyond the capacity of the credit system to endure and a cumulative series of bankruptcies. Similarly, general war occurs when serious diplomatic tension involving great powers induces many governments simultaneously to try either to isolate themselves from world politics in spite of previous commitments or to intervene in the controversy, thus causing a collapse in the sense of security beyond the capacity of international law and tradition to endure and a cumulative series of aggressions. In either case regulative effort may hamper these simultaneous and cumulative movements by establishing moratoriums or cooling-off periods.

4. RELATIONS AMONG DISTANCES

In chapter xix it was suggested that the distance between states with respect to intercourse, defense, understanding, legal recognition, social symbols, political union, attitudes, and expectations of war could be measured by analyzing expert opinions upon these subjects. The relative distances between pairs of the great powers from each of these points of view were obtained by the rank-order method used in measuring psychic distances, so far as that method could be used by one person. The results lacked the objectivity which might have been obtained by employing the psychometric method of averaging many judgments. However, they may be used to illustrate a method of estimating the probability of war. Considering first the probability of war between a pair of states, attention will be given to the influence upon war (*a*) of changes in distances and (*b*) of non-reciprocity of relationships. Combining these considerations, an estimate will be presented of (*c*) the probability of war between pairs of the great powers in July, 1939. Attention will then be given to (*d*) the probability of war for a single state and (*e*) the probability of general war.

a) *Changes in Distances.*—A tabulation made in July, 1939, suggested certain relations between the various aspects of "distance" between pairs of states. A correlation was suggested between expectancy (*E*), psychological (*Ps*), political (*P*), and social (*S*) distances, all of them dependent upon subjective factors. There appeared, on the other hand, to be little direct correlation between these distances and technological (*T*) and strategic (*St*) distances, both of which depend on objective factors. Intellectual (*I*) and legal (*L*) distances did not appear to be closely correlated with either group, though intellectual distance was closer to the objective group and legal distance to the subjective.

Expectancy of war, though closely correlated with psychic distance, tended to be greater when psychic distance was greater than technological distance or when social distance was greater than intellectual distance. There will, apparently, be a trend toward war if the interest of one state in another, promoted by technological inventions, such as improvements in communications and transport, is proceeding more rapidly than the development of friendly opinions and attitudes and if the development of common intellectual understanding is proceeding more rapidly than the acceptance of common social symbols.

Peace would be probable if the order of change were reversed, if friendliness and mutual acceptance of common social symbols preceded the development of material interdependence, the reduction of strategic barriers, and the equalization of intelligence and understanding. Peace is obstructed if policy is shaped in accordance with existing material conditions rather than in accordance with future social needs. If statesmen neglect to use foresight, they may find themselves confronted by a condition, not a theory, and, under pressures of necessity, improvise policies whose long-run effect is to augment the probability of war.

International law before World War I was shaped mainly by traditions of the past and was little influenced by requirements of the present and future. It was assumed by the dominant school of thought on the subject that international law had as its prime object the maintenance of the legal independence of nations (sovereignty) and their freedom from responsibility for the world-order (neutrality) rather than the maintenance of a world-order promoting international peace and justice. International law thus interpreted tended to increase the social distance between nations and to thwart the development of policies toward world solidarity. It appears, however, that peace is promoted if psychic and social relations are decreasing more rapidly than technological and intellectual relations, that is, if the subjective relations of states lead the objective. The new international law of the Covenant and the Charter was designed to have this tendency.

b) *Non-reciprocity of Relationships.*—The analysis up to this point has proceeded on the assumption that the relations of states can be measured by distances which may be represented by points in a linear continuum. Though tending to be reciprocal, these relations are not necessarily so.

Some of the consequences of a possible lack of reciprocity may be considered. If in the relations of A and B, A is becoming less expectant of war than B, B's growing expectation of war will induce it to arm, but A's diminishing expectation of war will induce it to defer defense expenditure. The strategic situation will, therefore, tend progressively to favor B, and in time its conviction that war is inevitable will induce it to initiate war or to make demands likely to precipitate war. Such a situation appears to have led to the Munich crisis of September, 1938. Germany, during the preceding period, had been more expectant of war than had England and France and had prepared more rapidly, with the result that Germany made demands which nearly precipitated a war. This augmented the expectation of war of the Western powers and also that of Germany. The Danzig crisis arose, eventuating in general war in September, 1939. In such a case as this Richardson's "defense coefficient" would be negative. It appears to be a type of situation which is more likely to arise with the advance of the cost of and the moral objection to war, provided the influence of these factors on foreign policy is excluded in some states. Democratic governments have tended to ignore military defense and balance-of-power considerations more than autocracies when faced by growing budgets and peace propaganda. This want of reciprocity in respect to war expectancy may be anticipated when these different types of governments face each other.

c) *Probability of War between Pairs of States.*—The aspects of their relationships affecting the probability of war between two states have been combined in a complicated formula.* Application of this formula should indicate the relative probability of war between pairs of states during a given period of time, so far as that probability is determined by the relationships of the members of each pair with one another. The formula ignores the influence of third states and of the general structure of the family of nations. Its accuracy decreases in proportion as international relations become multilateral relations.

Application of this formula to estimates made of the distances between the great powers in July, 1939, indicated that the relative probability of war at that date was highest for Japan-U.S.S.R. (.96), Germany-U.S.S.R. (.86), and Germany-France (.82). This order is the same as that obtained by a different method in January, 1937, although, except for Germany-U.S.S.R., the probabilities were greater at the later date. Minor hostilities were actually in progress between Japan and the U.S.S.R. in July, 1939.

The remaining pairs also followed a similar order in the two estimates, though the probability of war between a party to the anticommunist agreement (Germany, Italy, Japan) and a democracy (Great Britain, France, the United States) had in every case increased, while the probability of war between two parties to the anticommunist agreement had in every case decreased. The prob-

$$* \quad \frac{dx}{dt} = k \left\{ \frac{dE}{dt} + \left[2 \frac{dPs}{dt} - \frac{dT}{dt} \right] + \left[\frac{dS}{dt} - \frac{dI}{dt} \right] \right.$$
$$\left. + \left[\frac{d(E_{ab} - E_{ba})}{dt} + \frac{d(St_{ba} - St_{ab})}{dt} \right] + \left[\frac{d(P_{ab} - P_{ba})}{dt} + \frac{d(L_{ba} - L_{ab})}{dt} \right] \right\}.$$

x is the probability of war between a given pair of states at a given moment. If the derivative dx/dt is positive, the probability of war is increasing; if negative, it is decreasing. *k* is a constant. The meaning of the variable distances, of which *Ps* is considered twice as important as the others, is indicated in the text (p. 332). The formula considers the possibility of non-reciprocity of the distances *E*, *St*, *P*, and *L* as viewed from each state by subtracting the distance as perceived by *b* to *a* from that as perceived by *a* to *b*. For a full explanation of the formula, see the unabridged edition of *A Study of War*, pp. 1484 ff.

ability of war between two democracies was about the same in the two estimates.

The greatest differences between the two estimates appeared in the cases of Italy-France and Great Britain–Japan and in the relations of the United States with Germany and Italy. In all these cases the probability of war had markedly risen. These differences may be accounted for by the influence upon relations of the polarizing tendency which resulted from the increasing tensions during the two years from 1937 to 1939. This augmented the probability of war between those states likely to be on different sides in a general war and decreased that probability for states likely to be on the same side.

The most notable error of this estimate, as judged by subsequent events, was its failure to foresee the Soviet-German non-aggression pact of August, 1939. This postponed war between those countries for two years and probably accounted for an overestimate of the chances of Japan's getting into war with the Soviet Union.

The formula here used considered only the bilateral relations of states and therefore neglected the potential influence of third states. A single unexpected change in relations, such as that of the Soviet-German pact, had an influence on many relations in a way which this method could not foresee. Probably such changes, altering at least temporarily the entire international configuration, are the least predictable elements in the probability of war between two states. By maneuvers of that type, leaders like Hitler can upset the calculations of both analysts and statesmen and create for themselves opportunities for temporarily successful aggression.

d) Probability of War for a Single State.—Similar methods might be used to study the probability that a single state will become involved in war within a given time. By adding together the distances of each great power from all the others, the degree of isolation of each may be indicated. No very clear correlations were indicated by this procedure, but there was a tendency for the psychic isolation of a power to be related to its political isolation, although the latter tended to be greater if legal status was relatively high. Psychic isolation also tended to be associated with war expectancy except when strategic isolation, as in the case of the United States, was relatively great. If it is assumed that a single state is likely to get into war in proportion as its average relations with all other states are unfriendly, then its prospects for peace are improving if its expectation of peace is increasing more rapidly than its vulnerability to attack and if its political relations are intensifying more rapidly than its legal status is rising.

e) *Probability of General War.*—Material of the kind discussed may also throw light on the prospect of general war by giving evidence of a rise or fall in the general tension level. Tabulation of relations between pairs of states showed a general flaring-out of the lines from 1937 to 1941, indicating more intense friendships with some and animosities toward others—a condition presaging general war. Analysis of the probable participants in general war would have to give consideration not only to the bilateral relation of all pairs of states but also to the tendency toward polarization of hostility about the two principal antagonists and toward a rapid change in bilateral relations during the course of such a war.

CHAPTER XXI

CAUSES OF WAR

Wars arise because of the changing relations of numerous variables —technological, psychic, social, and intellectual. There is no single cause of war. Peace is an equilibrium among many forces. Change in any particular force, trend, movement, or policy may at one time make for war, but under other conditions a similar change may make for peace. A state may at one time promote peace by armament, at another time by disarmament; at one time by insistence on its rights, at another time by a spirit of conciliation. To estimate the probability of war at any time involves, therefore, an appraisal of the effect of current changes upon the complex of intergroup relationships throughout the world. Certain relationships, however, have been of outstanding importance. Political lag deserves attention as an outstanding cause of war in modern civilization.

1. POLITICAL LAG

There appears to be a general tendency for change in procedures of political and legal adjustment to lag behind economic and cultural changes arising from technological progress. The violent consequences of this lag can be observed in primitive and historic societies, but its importance has increased in modern times. The expansion of contacts and the acceleration of change resulting from modern technology have disturbed existing power localizations and have accentuated the cultural oppositions inherent in social organization. International organization has not developed sufficiently to adjust by peaceful procedures the conflict situations which have arisen. This lag is related to the usual lag of value systems behind scientific and technological progress, accounting for the great transitions in civilizations referred to in chapter iv.

War tends to increase in severity and to decrease in frequency as the area of political and legal adjustment (the state) expands geographically unless that area becomes as broad as the area of continuous economic, social, and cultural contact (the civilization). In the modern period peoples in all sections of the world have come into continuous contact with one another. Although states have tended to grow during this period, thus extending the areas of adjustment, none of them has acquired world-wide jurisdiction. Their growth in size has increased the likelihood that conflicts will be adjusted, but it has also increased the severity of the consequences of unadjusted conflicts. Fallible human government is certain to make occasional mistakes in policy, especially when, because of lack of universality, it must deal with conflicts regulated not by law but by negotiation that must function within an unstable balance of power among a few large units. Such errors have led to war.

War tends to increase both in frequency and in severity in times of rapid technological and cultural change because adjustment, which always involves habituation, is a function of time. The shorter the time within which such adjustments have to be made, the greater the probability that they will prove inadequate and that violence will result. War can, therefore, be attributed either to the intelligence of man manifested in his inventions which increase the number of contacts and the speed of change or to the unintelligence of man which retards his perception of the instruments of regulation and adjustment necessary to prevent these contacts and changes from generating serious conflicts. Peace might be kept by retarding progress so that there will be time for gradual adjustment by natural processes of accommodation and assimilation, or peace might be kept by accelerating progress through planned adjustments and new controls. Actually both methods have been tried, the latter especially within the state and the former especially in international relations.

Sovereignty in the political sense is the effort of a society to free itself from external controls in order to facilitate changes in its law and government which it considers necessary to meet changing economic and social conditions. The very efficiency of sovereignty within the state, however, decreases the efficiency of regulation in international relations. By eliminating tensions within the state, external tensions are augmented. International relations become a "state of nature." War therefore among states claiming sovereignty tends to be related primarily to the balance of power among them.

Behind the power equilibrium are others, disturbances in any of which may cause war. These include such fundamental oppositions as the ambivalent tendency of human nature to love and to hate the same object and the ambivalent tendency of social organization to integrate and to differentiate at the same time. They also include less fundamental oppositions such as the tendency within international law to develop a world-order and to support national sovereignty and the tendency of international politics to generate foreign policies of both intervention and isolation. Elimination of such oppositions is not to be anticipated, and their continuance in some form is probably an essential condition of human progress. Peace, consequently, has to do not with the elimination of oppositions but with adequate methods of adjusting them.

The lag of adjusting procedures behind a change of conditions is a general cause of war. The persistence of this lag is due in part to the actual or presumed service of war to human groups. War has been thought (1) to serve sociological functions, (2) to satisfy psychological drives, (3) to be technologically useful, and (4) to be legally rational.

2. SOCIOLOGICAL FUNCTIONS OF WAR

Animal warfare is explained by the theory of natural selection. The behavior pattern of hostility has contributed to the survival of certain biological species, and consequently that behavior has survived. In the survival of other species other factors have played a more important role.

Among primitive peoples, before contact with civilization, warfare contributed to the solidarity of the group and to the survival of certain forms of culture. When population increased, migrations or new means of communication accelerated external contacts. The warlike tribes tended to survive and expand; furthermore, the personality traits of courage and obedience which developed among the members of these tribes equipped them for civilization.

Among peoples of the historic civilizations war contributed both to the survival and to the destruction of states and civilizations. Its influence depended upon the stage of the civilization and the type of military technique developed. Civilized states tended to fight for economic and political ends in the early stages of the civilization, with the effect of expanding and integrating the civilization. As the size and interdependence of political units increased, political and economic ends became less tangible, and cultural patterns and ideal objectives assumed greater importance. Aggressive war tended to become a less suitable instrument for conserving these elements of the civilization. Consequently, defensive strategies and peaceful sentiments developed, but in none of the historic civilizations were they universally accepted. War tended toward a destructive stalemate, disintegrating the civilization and rendering it vulnerable to the attack of external barbarians of younger civilizations which had acquired advanced military arts from the older civilization but not its cultural and intellectual inhibitions.

In the modern period the war pattern has been an important element in the creation, integration, expansion, and survival of states. World civilization has, however, distributed a singularly destructive war technique to all nations, with the consequence that the function of war as an instrument of integration and expansion has declined. Efforts to break the balance of power by violence have increasingly menaced the whole civilization, and yet this balance has been so incalculable that such efforts have continued to be made. Atomic weapons may have deprived war of any social function and made its consequences more calculable.

3. PSYCHOLOGICAL DRIVES TO WAR

Human warfare is a pattern giving social sanction to activities which involve the killing of other human beings and the extreme danger of being killed. At no period of human development has this pattern been essential to the survival of the individual. The pattern is a cultural acquisition, not an original trait of human nature, though many hereditary drives have contributed to the pattern. Of these, the dominance drive has been of especial importance. The survival of war has been due to its function in promoting the survival of the group with which the individual identifies himself and in remedying the individual problem arising from the necessary repression of many human impulses in group life. The pattern has involved individual attitudes and group opinion. As the self-consciousness of personality and the complexity of culture have increased with modern civilization, the drive to war has depended increasingly upon ambivalences in the personality and inconsistencies in the culture.

A modern community is at the same time a system of government, a self-contained body of law, an organization of cultural symbols, and the economy of a population. It is a government, a state, a nation, and a people.

Every individual is at the same time subject to the power and authority of a government and police, to the logic and conventions of a law and language, to the sentiments and customs of a nation and culture, and to the caprices and necessities of a people and its economy. If he fights in war, he does so because one of these aspects of the community is threatened or is believed by most of those who identify themselves with it to be threatened. It may be that the government, the state, the nation, and the people are sufficiently integrated so that there is no conflict in reconciling duty to all of these aspects of the community. But this is not likely because of the analytical character of modern civilization which separates military and civil government, the administration and the judiciary, church and state, government and business, politics and the schools, and religion and education. Furthermore, it may be that the threat is sufficiently obvious so that no one can doubt its reality, but this is seldom the case. The entities for whose defense the individual is asked to enlist are abstractions. Their relations to one another and the conditions of their survival are a matter of theory rather than of facts. People are influenced to support war by language and symbols rather than by events and conditions.

It may therefore be said that modern war tends to be about words more than about things, about potentialities, hopes, and aspirations more than about facts, grievances, and conditions. When the war seems to be about a particular territory, treaty, policy, or incident, it will usually be found that this issue is important only because, under the circumstances, each of the belligerents believed renunciation of its demand would eventually threaten the survival of its power, sovereignty, nationality, or livelihood. War broke out in 1939, not about Danzig or Poland, but about the belief of both the German people and their enemies that capacity to dictate a solution of these issues would constitute a serious threat to the survival of the power, ideals, culture, or welfare of the group which submitted to this dictation.

Even more remote from the needs of the individual and the state was the bearing of a campaign to expand the Roman frontier into Gaul, the Moslem frontiers into Africa, the Christian frontiers into Palestine, or the Communist frontier into central Europe. The meaning of Rome, of Islam, of Christendom, or of communism had to be understood by a considerable public. The importance of such increases in territory, population, and glory had to be inculcated by education of all those influencing policy, even though the prospect of immediate rewards to the active participants was obvious.

In the modern situation far more conceptual construction is necessary to make war appear essential to the survival of anything important. War, therefore, rests, in modern civilization, upon an elaborate ideological construction maintained through education in a system of language, law, symbols, and values. The explanation and interpretation of these systems are often as remote from the actual sequence of events as are the primitive explanations of war in terms of the requirements of magic, ritual, or revenge. War in the modern period does not grow out of a situation but out of a highly artificial interpretation of a situation. Since war is more about words than about things, other manipulations of words and symbols might be devised to meet the cultural and personality problems for which war offers an increasingly inadequate and expensive solution.

4 . TECHNOLOGICAL UTILITY OF WAR

The verbal constructions which have had most to do with war in the modern period have been those which center about the words "power," "sovereignty," "nationality," and "living." These words may be interpreted as attributes, respectively, of the government, the state, the nation, and the people. By taking any one as an absolute value, the personality may be delivered from the restlessness of ambivalence and from the doubts and perplexities which arise from the effort to reconcile duty to conflicting institutions and values, particularly in times of rapid change. Although the relation of war to the preservation of any of these entities requires considerable interpretation, the validity of the interpretation varies with respect to the four entities.

The power of the government refers to its capacity to make its decisions effective through the hierarchy of civil and military officials. In a balance-of-power structure of world politics even a minor change in the relative power position of governments is likely to precipitate an accelerating process, destroying some of the governing elites and augmenting the power of others. If a government yields strategic territory, military resources, or other constituents of power to another without compensating advantage, it is quite likely to be preparing its own destruction. The theory which considers war a necessary instrument in the preservation of political power is relatively close to the facts. The most important technological cause of war in the modern world is its utility in the struggle for power.

The sovereignty of the state refers to the effectiveness of its law. This rests immediately on customary practices and on the prestige and reputation for power of the state rather than upon power itself. Sensitiveness about departures from established rules about honor and insult to reputation has a real relation to the preservation of sovereignty. A failure to resent contempt for rights or aspersions on prerogatives may initiate a rapid decline of reputation and increase the occasions when power will actually have to be resorted to if the legal system is to survive. Thus in the undeveloped state of international law, self-help and the war to defend national honor had a real relation to the survival of states prior to the nuclear age.

Nationality refers to the expectation of identical reactions to the basic social symbols by the members of the national group. It has developed principally from common language, traditions, customs, and values and has often persisted through political dismemberment of the group. Although national minorities have usually resisted the efforts of the administration and the economic system of the state to assimilate them, these influences may in time be successful. Thus, the use of force to preserve the power of the government and the sovereignty of the state which supports a given nationality may be important to the preservation of the latter. War, however, has been less useful to preserve nationality than to preserve power or sovereignty.

Living refers to the welfare and economy of a people. The argument has often been made that war is necessary to assure a people an area sufficient for prosperous living. Under the conditions of the modern world, this argument has usually been fallacious. The problem of increasing the welfare of a people has not depended upon the extension of political power or legal sovereignty into new areas but rather upon the elimination of the costs of war and depression, improvements in technology and land utilization, and a widening of markets and sources of raw materials far beyond any territories or spheres of interest which might be acquired by war. Population pressure, unavailability of raw materials, and loss of markets are more often the effect of military preparation than the cause. Although it is true, in a balance-of-power world, that economic bargaining power may increase with political power, yet it has seldom increased enough to compensate for the cost of maintaining a military establishment, of fighting occasional wars, and of impairing confidence in international economic stability. Through most of modern history people, even if conquered, have not ceased to exist and to consume goods. Efforts toward economic self-sufficiency and toward the forced migration, extermination, or enslavement of conquered peoples have, however, added to the reasonableness of conventional war for the preservation of the life of peoples.

Modern civilization offers a group more alternatives to war in most contingencies than did earlier civilizations and cultures. Resort to war, except within the restricted conception of necessary self-defense, is rarely the only way to preserve power or sovereignty and even more rarely the only way to preserve nationality or economy. War is most useful as a means to power and progressively less useful as a means to preserve sovereignty, nationality, or economy. That economic factors are relatively unimportant in the causation of war was well understood by Adolf Hitler:

Whenever economy was made the sole content of our people's life, thus suffocating the ideal virtues, the State collapsed again. . . . If one asks oneself the question what the forces forming or otherwise preserving a State are in reality, it can be summed up with one single characterization: the individual's ability and willingness to sacrifice himself for the community. But that these virtues have really nothing whatsoever to do with economics is shown by the simple realization that man never sacrifices himself for them; that means: one does not die for business, but for ideals.

5. LEGAL RATIONALITY OF WAR

Which of the entities for which men fight is most important for men? Is there any criterion by which they may be rationally evaluated? Political power has been transferred from village to tribe, from feudal lord to king, from state to federation. Is it important today that it remain forever with the national governments that now possess it? The transfer of power to a larger group, the creation of a world police, under an international organization adequate to sanction a law against aggression, appears a condition for eliminating a major cause of war.

Legal sovereignty also has moved from city-state to empire, from baronial castle to kingdom, from state to federation. To the individual the transfer of authority over his language and law to a larger group, although it has brought regret or resentment, has assured order, justice, and peace in larger areas and has increased man's control of his environment, provided that authority has been exercised with such understanding and deliberation as to avoid resentments arising to the point of revolt.

Nationality, in the broadest sense of a feeling of cultural solidarity, has similarly traveled from village to tribe, city-state, kingdom, nation, empire, or even civilization; but when it has become too broad, it has become too thin to give full satisfaction to the human desires for social identification and distinctiveness. There is no distinctiveness in being a member of the human race. Few would contemplate a world of uniform culture with equanimity. Geographical barriers and historic traditions promise for a long time to preserve cultural variety even in a world-federation, though modern means of communication and economy have exterminated many quaint customs and costumes. The need of cultural variety and the love of distinctive nationality suggest that a world police power is more likely to be effective if controlled by a universal federation than by a universal empire.

The area from which individuals have obtained their living has expanded from the village to the tribal area to the kingdom and empire, until, in the modern world, most people draw something from the most remote sections of the world. This widening of the area of exchange has augmented population and standards of living. Diminution of this area, such as occurred when the Roman Empire disintegrated into feudal manors, has had a reverse effect. The economist can make no case for economic walls, if economy is to be an instrument of human welfare rather than of political power, except in so far as widespread practices on the latter assumption force the welfare-minded to defend their existing economy by utilizing it temporarily as an instrument of power.

It may be questioned whether a rational consideration of the symbols, for the preservation of which wars have been fought, demonstrates that they have always been worth fighting for or that fighting has always contributed to their preservation. The actual values of these entities as disclosed by philosophy and the actual means for preserving them as disclosed by science have been less important in causing war than popular beliefs engendered by the unreflecting acceptance of the implications of language, custom, symbols, rituals, and traditions.

CHAPTER XXII

CONDITIONS OF PEACE

War, springing from numerous factors bedded in human society and human nature, is "natural," but it is not beyond human control. Societies can do something about it, and although their efforts have never been successful in eliminating war generally and permanently, they have moderated the destructiveness and reduced the frequency of war in large areas for considerable periods of time. In the modern world-civilization, efforts to eliminate war have been more persistent and have a greater possibility of success than in earlier civilizations with uncontrollable civilizations on their periphery. This experience makes possible an analysis of general conditions of peace.

The conditions of peace are defined in the abstract by the law of the society within which war occurs. This law establishes the organization, rules of order, principles of justice, and procedures which the members of the society believe ought to be observed and enforced to maintain order and, as a consequence, peace. The conditions of order and peace are defined concretely by the functioning of a society as a whole and of its parts as determined by cultural patterns and forms of organization. These patterns and forms usually develop without conscious purpose before they are formulated in a legal system. The latter develops as a symbol and conceptualization of the existing practices, relating them to general purposes of the society. Once established, however, the law may influence the culture, organization, and practices to conform to its pattern, and in any case, it reflects the existing situation in considerable measure, but not entirely because law is never perfectly observed or enforced. There is always some gap between symbols and conditions.

At millennial intervals Western civilization has made an attempt to organize itself as a world-empire, as a world-church, or as a world-federation, always relapsing to a balance-of-power system in the intervals. Each form of organization has developed an appropriate legal system. Until World War I the modern family of nations professed a law which permitted sovereign states to initiate war, and so their basic security depended on the maintenance of a balance of power rather than on law. Attempts to organize the family of nations as an empire or as a federation have not been successful, though the League of Nations and the United Nations mark progress in the latter direction. It differs from earlier families of nations in that it has become world-wide. From the standpoint of effective political organization this novel situation has both disadvantages and advantages.

External opposition and internal uniformity have been among the most important inducements to intense political organization. Clearly, a universal community minimizes both of these conditions. There can be no external aggression against the world as a whole. The world's diversity of cultures militates against a general consciousness of kind.

A universal community of nations, however, has the advantage that, because of its freedom from external pressure, its members have less need for an intensive organization than do the members of the communities whose prime problem must be defense. The very diversity of cultural patterns assures a cross-fertilization of ideas and capacity for continuous adaptation to new conditions. Furthermore, in spite of the size and diversity of the world-community, inventions facilitating rapid communication and transport provide technical means for universal political organization, even more adequate than those upon which states and empires have rested in the past. Finally, the human will may be stimulated by appreciation of the need for permanent peace and the opportunity for advancing general welfare within a universal society. Such a society would be emancipated from the problem of external security which has absorbed much of the attention and energy of national societies claiming to be sovereign.

The aspirations and opportunities of a society are never wholly reflected in its practice. A body of law, if based on custom, as it is in primitive societies, is self-executing, but in the progressive societies characteristic of civilization, sanctions must be contrived. Law must be maintained by continuous effort. Communities of nations have made efforts to maintain their laws which limit or regulate war by controlling military power, by operating international organizations, and by educating people. Such efforts appropriate to the contemporary situation will be considered in the final part of this book, but a review of international law in modern history will indicate the basic conditions which statesmen, in their saner moments, have regarded as essential for maintaining that degree of peace which they consider possible and desirable and so accepted as law. The basic concept of that law has been the sovereignty of territorial states, subject, however, to rules, principles, and standards of international law springing from their express, tacit, or presumed consent.

1. GENERAL CHARACTERISTICS OF LAW CONCERNING WAR

In most civilizations theories have developed which are international but are not law, defining the circumstances in which war can properly be resorted to (*jus ad bellum*) and the methods which can properly be used in waging war (*jus in bello*). Each of the states in these civilizations has usually established rules which are law but not international, limiting private warmaking and regulating private profits from war. The first of these bodies of doctrine has served to reconcile war with the fundamental values of the civilization and the second to promote the sovereignty and efficiency of the states.

In the modern period the same two bodies of doctrine are observable. Modern states have made laws designed, with increasing comprehensiveness, to reduce or to eliminate private warmaking; booty and bounties to generals and soldiers; prizes and prize money to privateers, admirals, and sailors; war profits to traders, manufacturers, and financiers. These laws have had more or less success in making war a monopoly of the state to be used only for "reason of state" and not for private profit. They have also had an influence both on the development of international law and on the totalitarianization of war.

International standards have in the modern period achieved a more definitely juristic character than ever before. In spite of Cicero's aphorism, *inter arma leges silent,* this has been particularly true of the *jus in bello,* which has achieved a detailed exposition in adjudications of prize and other courts, army and navy regulations, bilateral treaties, and the general conventions and declarations of Paris (1856), Geneva (1864, 1906, 1929, 1941), St. Petersburg (1868), The Hague (1899, 1907, 1923), and London (1909, 1930). The rules thus prescribed and the degree of their enforcement have undoubtedly influenced the frequency and characteristics of war. Far-reaching regulation of military methods and instruments, if rigorously enforced, might do away with many of the evils of war, but such a result might reduce the reluctance to resort to war and so to make war more frequent. On the other hand, if rules of war are lax or unenforced, war is more severe if it comes but tends to come less frequently.

Modern civilization, like past civilizations, has tended during the past century toward an assertion of more and more rigorous rules of war but less and less observance of them in major wars. The latter result can be attributed to the decline of the conception of "military honor" with the reduction of the professional and mercenary elements in armies and the rise of universal military service; to the rise of the conception of "the nation in arms" with the growth in efficiency of propagandas of national fanaticism; to the breakdown of the distinction between combatants and noncombatants with the wide entry of the civilians into the supply services, with the increasing military regimentation of national economy and morale, and with the increasing technical possibility of attack behind the lines from the air and by blockade; and as a result of all these, to the development of the conception of "absolute war" and of broadened interpretations of "military necessity." These tendencies of modern civilization have been accompanied by a decreasing frequency and increasing seriousness of war.

The *jus ad bellum* retrogressed through most of the modern period. The medieval conception of just war was abandoned in the seventeenth century, and not until the establishment of the League of Nations was serious juristic attention again given to the problem. Prior to World War I international law provided no substantive and few procedural limitations upon resort to war and only certain vague qualifications upon lesser uses of force in reprisals, intervention, and defense. Although legal theory confined the latter to action necessary to prevent an immediately impending, irreparable injury to territory, government, or nationals, practice included in the concept of defense broad policies like the Monroe Doctrine and the balance of power. With respect to the initiation of war itself, the absence of any legal limitations was indicated by the doctrine of neutrality which asserted that third parties could not make a judgment of law on the legitimacy of such initiation and must act with formal impartiality.

Modern history, which coincides with the disintegration of Western Christendom and other historic civilizations and which constitutes the "heroic age" of the rising world-civilization, might be expected to be peculiarly unfavorable to an effective *jus ad bellum,* and this expectation has not been disappointed. Law effectively controlling or forbidding resort to war is, however, an essential condition, though by no means the only condition, of peace.

2 . DEVELOPMENT OF MODERN
    INTERNATIONAL LAW

a) *Period of Religious Wars* (*1492–1648*).—The medieval idea of chivalry and of a universal order both temporal and religious was in large measure scrapped by the *Realpolitik* of Machiavelli, the fanaticism of religious war, and the idea of territorial sovereignty. The humane spirit could not, however, be wholly suppressed, and sea-borne commerce could not continue without international law. Furthermore, law was necessary for the conduct and discipline of the diplomatic and military administrations in the new territorial states priding themselves on their efficiency. These two factors, the sentiment of humanity and reason of state, acting upon the institutions and practices developed by the maritime commerce and the inter-princely relations of the later Middle Ages, created modern international law.

The first formulator of this law was Francis of Victoria, a Dominican friar of Salamanca, whose humane spirit had caused him to become interested in the problem as a consequence of the conquest of Mexico by Cortez. Francis of Victoria delivered his lectures in 1532, and fifty years later the second important treatise on the subject was written by Balthazar Ayala, serving as judge advocate-general of the Spanish armies in the Netherlands "to keep that army in good discipline and justice."

The notion that sovereign princes, though supreme in their own domains, are bound by law in their external relations was emphasized by the habits of making treaties and exchanging diplomatic officers, by the mutually advantageous practices for the benefit of maritime commerce, by the tradition of common Christian civilization, by the conception of natural law and a state of nature, and perhaps also by the personal relationship and sense of common interest among the kings themselves. It was natural for them to act on the assumption that they should not become so hostile to one another that they could not assist one another in the common problem of preserving their positions against dissatisfied nobles and commoners.

The developing law of nations was handed on by the Spanish school and by the Italian Gentili to Hugo Grotius, Dutch lawyer, theologian, and diplomatist. Inspired by a humanitarian desire to ameliorate the practices he witnessed in the Thirty Years' War, Grotius gave international law a more systematic form. His whole treatment of the law of nations sprang from his original problem of determining the justifiability of military violence. He contemplated a family of Christian monarchs, each enforcing law in his own realm but ready to co-operate to punish the violator of the law of nations, especially the initiator of a war which was unjust according to the medieval conception. Neutrality was thus excluded unless it was impossible to determine which side in a war was just. The conduct of war itself, he realized, must be governed by military necessity, but he urged *temperamenta belli* when possible in the interest of humanity and of negotiating peace.

The purpose of the developing law was, therefore, justice and peace. It regarded war as a misfortune, generally unnecessary, and never justifiable except as a handmaid of law. The practice of statesmen, however, followed the precepts of Machiavelli rather than those of Grotius, although the concept of territorial sovereignty, permitting the prince to deal with religious problems as he saw fit in his own territory, was accepted by the Peace of Westphalia.

b) Period of Political Absolutism (1648–1789).—The period following the Thirty Years' War has been called by recent historians of international law "the Age of the Judge," referring to the legalistic character that international relations assumed under the influence of the increasing number of judicial opinions and text-writers on the subject, the multiplication of treaties, the enactment of national laws to maintain international obligations, the activities of diplomatic officers, and the resolutions of international conferences. The initiation and waging of war and the conduct of diplomacy became formalized, but the Grotian conception of a community of nations enforcing law was not accepted in practice.

The conception of war underwent changes. Instead of an instrument of justice, it came to be considered an instrument of policy. Vattel, who wrote in the middle of the eighteenth century, assumed that, although princes should satisfy themselves that they had a just cause before they initiated war, no one else could pass judgment on the matter. States with no direct interest in the controversy should be neutral, although qualification of that neutrality by treaties already in existence and by consideration of national interest was permitted. War was a trial by battle or duel whose results determined the merits of the controversy, not the execution of a judgment made after rational consideration of the merits, as it had been in the system of Grotius. The initiation of war became for third states, therefore, a question of fact, not of law. The legal interest of such states lay not in the circumstances of the war's origin but in the legal changes its initiation brought about.

The existence of war brought into operation new rules of law applicable to the relations of belligerents with one another and with neutrals. The latter found not only that their trading rights at sea were considerably limited but also that they were under obligations not to render any official assistance to either belligerent or to allow their territory to be used for belligerent purposes. The United States, geographically separated from the European wars which began after the French Revolution, contributed greatly to the concept of neutral status and of the rights and duties which flowed therefrom. The rules of war between belligerents tended toward a formalization of war, maintenance of the professional interests of officers, and exemption of civilians and their property from the hardships of war, both on land and on sea.

c) *Period of Industrial Nationalism* (*1789–1914*).—The "public law of Europe," as set forth, for example, in the treaties of Westphalia (1648), Utrecht (1713), and Paris (1783), and the customary rules of war and neutrality were given rude shocks by the enthusiasm of the French Revolution and by the absolute war of Napoleon.

The post-Napoleonic period was marked by an attempt at international organization inspired, on the one hand, by the Tsar Alexander's idealistic Holy Alliance and, on the other, by the diplomatic agreements for sustaining the system established by the Treaty of Vienna (1815).

The effort to preserve the status quo by identifying it with peace and international solidarity became progressively more difficult as the memory of the Napoleonic Wars receded. The system broke down in the liberal revolutions of 1848, followed by the nationalist revolt against the system in Italy and Germany. With the success of these revolts and the creation of two new "great powers" in the center of Europe, the system of the concert and of international law was revived. It even sought to deal with the general balance of power through disarmament at the Hague Conferences of 1899 and 1907. In the larger aim these conferences failed, although they contributed to the codification of the law of war and the development of international arbitration. Arbitration had been frequently resorted to for minor and some major problems since 1796.

As this period advanced, the diverse tendencies of nationalism and internationalism became more and more difficult to reconcile. Bismarck thought the political and economic interests of states could be dissociated; but as the economic foundations of effective war came to be recognized, the nationalistic spirit more and more sought to mobilize the internal and external economic activities of the state for the purpose of national power. On the other hand, the international spirit, favored alike by humanitarians and by bankers and businessmen who wished an opportunity for secure expansion of their operations, tended to qualify the freedom of national policy. This spirit sought to prevent war, which became more and more threatening to social and economic life as the latter became organized on a world basis.

The Grotian conception of a general law functioning for a genuine world-community appeared to be nearer to realization during the long periods of peace in the nineteenth century than it had ever been before. International law, however, in spite of its solidification and detailed development by international conferences and unions, general and bilateral treaties, international tribunals, diplomatic correspondence, and text-writers, had not grappled effectively with the problem of war. Although reprisals, intervention, and other forms of violence short of war were dealt with in the textbooks on international law, war itself was, throughout the nineteenth century, looked upon as a fact, and the propriety of recourse to it was considered not a legal question but an ethical question or a political question. Statesmen justified a war by its success in achieving its immediate objectives. The old idea of just war appeared less and less in the textbooks, which came to be characteristically divided into sections on peace, war, and neutrality. Thus international law condemned itself to deal only with minor controversies. The great controversies for which states were prepared to fight were in practice outside of its competence. Such a theory clearly could not assist in an institutional development for eliminating war. It could only define methods of pacific settlement in the hope that states would voluntarily use them rather than resort to the risk of using their unlimited power to convert a state of peace into a state of war. The relative peace of the nineteenth century was not in fact due to the functioning of an international law of peace but to the *Pax Britannica,* destined to survive only as long as British sea power was able to maintain the balance of power and British finance was able to maintain a high level of international trade and economic development.

d) Period of World Wars (1914–45).—The general wars which began in 1914 have been as disturbing to the continuity of legal development as were the wars which began in 1618 and 1789. Rules of war and neutrality were forgotten in mutual retaliations, ancient boundaries were discarded, and the doctrine of national self-determination was given legal effect by the creation of new states and of procedures for holding plebiscites, for protecting minorities, and for supervising mandatory administration.

An important change in the conception of war was developed as a consequence of the general acceptance of the League of Nations Covenant and the Pact of Paris in the 1920's. These instruments, springing from American opposition to war and confidence in international organization, British appreciation that its navy could no longer enforce peace alone, and French fear of a war of revenge, were based on the conception that the initiation of war is illegitimate, until such time as the specified peaceful procedures have been exhausted, according to the Covenant or, in any circumstances, according to the Pact. The latter permitted war only to a state which had already had war made against it and to others coming to its assistance. A state was never justified in initiating a state of war. The primary belligerent was always an aggressor.

This conception differed both from the Grotian conception, which considered war a suitable procedure for enforcing a just cause, and from the Vattelian conception, which considered war a fact, the initiation of which was outside law altogether. From being a right and then a fact, war had become a crime. On the basis of the latter conception definite progress was made toward the institutionalizing of procedures for defining and suppressing aggression.

In the great post–World War I documents—the Covenant of the League of Nations, the Statute of the Permanent Court of International Justice, the Constitution of the International Labour Organization, and the Pact of Paris for the Renunciation of War—the notions of the world-community, the system of international law, the liberties of nationalities and minorities, the protection of human rights, the perpetuation of peace, general disarmament, and progressive international legislation were all envisaged, and procedures of collective security and peaceful change were set up to realize them. Under this system the position of nonbelligerents became very different from that of traditional neutrals. Peace was thought of as indivisible; war was recognized as affecting the interests of all.

This system, however, was not immediately accepted. National politicians and public opinions, in greater or lesser degree, tended to resist encroachments upon national sovereignty and also upon the war system in so far as it might foster national solidarity. Many national economic interests, dependent on national preference, protection, or preparedness, were not ready to give up these advantages. National isolationists, imperialists, and reformers, accustomed to use the sovereignty of the state for preserving the peculiarities, spreading the blessings, or improving the character of the national culture, often hesitated to tamper with that symbol. Neither the

Soviet state seeking to protect and expand the communist revolution nor the conservative opponents which it engendered were ready to abandon the use of force. National lawyers and logicians, learned in a professional ritual, imbued with respect for traditions, and remote from the technological conditions of communication and war, which made the older conceptions of sovereignty, national interest, and neutrality inadequate, offered passive and sometimes active resistance to the new ideas.

With its institutions still young and not even formally ratified by all states, with its logic inadequately appreciated even by its protagonists, the organization of the world-community found both its power and its machinery insufficient when confronted by really dangerous crises. Political institutions, as Bagehot has pointed out, require both *dignified* parts to give them power and *efficient* parts to direct that power to appropriate ends, or in more recent terminology, they need a symbolic structure to attract opinion and an administrative machine to focus it on concrete problems. The League of Nations had not been able to develop the one, which is the by-product of venerable antiquity, or to perfect the other through long experience in adapting institutions to changing circumstances.

It is not surprising, therefore, that important groups revolted from the system after the serious depression of 1929. Japan, Italy, and Germany, desirous of territorial expansion, reverted to the Machiavellian conceptions of an anarchic world and the absolute sovereignty of the state.

These revolts gave an opportunity to test the new system. The definition of an aggressor as the state that refused to accept the invitation of consulting states to stop fighting was applied. The Stimson doctrine refusing to recognize the fruits of aggression was accepted as a necessary implication of the Covenant and the Pact. Moral opinion was mobilized against the aggressors. In the case of Italy, engaged in aggression against Ethiopia, economic sanctions were put into effect by most of the nations. The morale of the community of nations was not, however, sufficient to enforce the law.

Furthermore, the victors of World War I, overinterested in the perpetuation of a particular status quo, had given inadequate attention to the development of procedures for peaceful change. Grievances providing fuel for these revolts against the international system were not dealt with in time. It became clear that a working international polity must not only suppress aggression but must also prevent the development of political inferiority complexes.

General war was renewed following the German invasion of Poland in 1939 after a series of minor conquests by the "dissatisfied powers" and of "appeasements" in neglect of their obligations by the "satisfied powers." War on land and sea and in the air was conducted with little regard for the traditional immunities of noncombatants and neutrals. Nonbelligerent governments, seeking to avoid war, exhibited little confidence in the traditional law of neutrality and enacted regulations which renounced the exercise of some neutral rights, accepted new duties, or discriminated against the aggressors. The war, however, spread rapidly.

e) The Nuclear Age (1945——).—World War II ended with the first nuclear explosion in human history, the establishment of the United Nations, and the Nürnberg trials of war criminals. Although not fully realized at the time, the developments inherent in these events so accelerated existing tendencies as to revolutionize international law. The danger to human life on the planet from nuclear war came to be realized as the capacity to make nuclear weapons spread from the United States to the Soviet Union, Britain, and France, with the probability that others would soon follow; as the uranium bombs of Hiroshima and Nagasaki were superseded by the cheaper hydrogen bomb with a thousand fold greater destructive power; and as the means of delivery advanced from propeller planes to jet planes and intercontinental missiles.

In the United Nations Charter, although signed before the first nuclear explosion, the peoples who had fought the axis expressed their vigorous determination to end the scourge of war and agreed to settle their international disputes by peaceful means and to refrain from the threat or use of force in international relations except in defense against armed attack or under the authority of the United Nations. The obligations undertaken and the procedures for maintaining them were more far-reaching than in earlier instruments.

In the Nürnberg and other war crimes trials based upon the London agreement of 1945, the concept of aggressive war was defined and the responsibility of individuals for initiating such a war was enforced by criminal penalties.

The activities of the United Nations in interpreting and applying the Charter in the light of the changing military technology of the atomic age, of the changing composition of the international community with the emergence of over fifty new nations, of the changing values consequent upon the communist revolution and the revolution of rising expectations in the underdeveloped parts of the world, and of the changing concept of mankind consequent upon these developments and others—all communicated by instantaneous telecommunication, universal mass media, and satellites revolving in outer space—have developed a new international law in spite of the unfavorable atmosphere of "cold war." This law, although imperfectly developed, outlaws war, organizes collective security, protects human rights, punishes international crimes, facilitates the self-determination of qualified peoples, and provides procedures of international co-operation for economic and social welfare and the development of the codification of international law. This law is based on the sovereign equality of states and non-intervention in their domestic jurisdiction, but it has marked a revolutionary step in regulating state sovereignty so that it will serve the interests of man and of the world community, especially by establishing conditions of peace.

During its first two decades the United Nations did much to promote international co-operation in matters of social and economic welfare, developed the concept of human rights but not general procedures for maintaining them, and assisted many new states to independence. It stopped a number of wars, in most cases by "provisional measures" accepted by the belligerents soon after the hostilities had begun, but in the Korean and Congo hostilities only after serious fighting. In some cases (Germany, Vietnam, Formosa Strait), armistice lines, *de jure* or *de facto,* were established by other agencies. In the protracted hostilities of France in Vietnam and Algeria, the United Nations exercised little influence, and its demand that the Soviet Union withdraw from its aggression in Hungary was ignored. Its major weakness was shown by its inability to bring about a settlement of many of the disputes which had precipitated the hostilities leaving enemy forces facing each other across armistice lines. This situation has continued for more than a decade in Korea, Palestine, Kashmir, Germany, the Formosa Strait, and Vietnam. In the latter case, infiltration across the cease-fire line maintained active hostilities in which the United States has intervened.

3. INTERNATIONAL LAW AND MUNICIPAL LAW

While international law was developing in the world-community, a system of municipal law had been developing in each state. Originally such a system was embodied in the judgments and decrees handed down under authority of the sovereign prince but presumed to be applications of the traditional mores or customs of common law. With the rise of the concepts of democracy, constitutionalism, and nationality, municipal law came to be the fiat of the sovereign state. The latter was an abstract entity manifested in the union of a territory, a population, a government, and a recognized status. The monarch came to be but an agent of the sovereign state.

Wide acceptance of the absolute conception of sovereignty increased the difficulty of reconciling international law and municipal law. With Grotius the prince was the personal nexus between these two laws. He realized his responsibility under international law which flowed from the agreements which he himself or his dynastic predecessors had made or from the mutual interests of princes which his personal contacts continually impressed upon his attention. Because of this realization he was prepared to exert the powers which belonged to him in internal administration and adjudication to see that his subjects did not interfere with his meeting of these responsibilities. But when legislation came to be the expression of the sovereign will of an abstract state, enacted by legislators with little foreign contact or knowledge of international law, and when the sources of international law came to be the highly technical expositions of numerous text-writers in all languages, basing their conclusions upon a minute study of treaties, customs, general principles, commentary of judge and jurist, all of which was rather incomprehensible to the man in the street, the possibility of conflict between international law and municipal law became obvious.

The humanists were divided into two camps, one of which, with an eye to the dangers of war, sought to augment the authority of international law to the detriment of legislative omnipotence, and the other, with an eye to the needs of internal reform, sought to augment the absolutism of legislative sovereignty. The nationalists were also divided into two camps, one of which, fearful of war, sought to renounce the exercise of sovereignty and the pursuit of interests beyond the frontier, whereas the other, of more ambitious mold, sought to strengthen the state's capacity steadily to expand by military means. Jurists sought to solve the conflict, but they divided into three schools: the national monists who insisted upon the ultimate juristic dominance of municipal law, the international monists who insisted upon the ultimate dominance of international law, and the dualists who recognized the autonomy of each of the systems of law, the possibility of juristic conflict, and the necessity of adequate machinery of political adjustment to rectify such conflicts. On whether that machinery should be diplomatic or international in character, this school was again divided.

The problem was not solved, but there was a tendency to redefine sovereignty as superiority to municipal law and subordination to international law, thus making it possible for the abstract conception of sovereignty to serve the function which was formerly served by the personality of the prince. Sovereignty was to the state what liberty under law was to the individual, that is, full discretionary power within a sphere marked by the law of the wider community. That sphere, however, was increasingly conceived as defined by jural rather than by territorial boundaries. Furthermore, the law was not conceived as static, and the jural boundaries which it established for the sphere of sovereigns were not considered immutable but were subject to continuous adjustment through political procedures of diplomacy, recognition, treaty-making, conciliation, conference, and international legislation as well as through judicial procedures of arbitration and adjudication.

The development of the controversy with respect to the spheres of international law and municipal law has given to modern international law its outstanding characteristic. It has not been, like the Roman *jus gentium* and *jus naturale,* a body of principles of universal validity governing the relations of individuals of different states. Until the atomic age it considered the human individual to be subject only to the law of some state and international law to be confined to the relations of states as artificial personalities. States and, perhaps, unions of states, related to the state as, in systems of municipal law, artificial corporations are related to the individual, were considered the only subjects of international law. It had been hoped that international law might thus prove a system capable of reconciling nationalism with the world-community, sovereignty with law, progress with peace. It had been hoped that it might solve the dilemma of earlier civilizations which could find no road between universal empire and continuous war.

Observation of the excesses, both internally and externally, of national sovereignty grown into totalitarianism, however, stimulated a widespread opinion that international law could not command respect in the highly interdependent family of nations unless that community moved further toward true federalism. Such a development implies the establishment of a relationship between the individual and the world-community, making the individual a subject of international law with direct access to international procedures for protecting the rights guaranteed by that law.

Statesmen and analysts have concluded that effective federal organization must rest on the will of the people ultimately affected as well as on the will of the governments directly participating, and this conclusion was implied by the preamble of the United Nations Charter, which asserted that the UN resulted from the determination of "we the peoples of the United Nations." This recalled Webster's interpretation, that the Constitution of the United States was the will of "We the *People* of the United States" in opposition to Calhoun's thesis that it was merely a pact among "We the People of the united *States*." The full realization of Webster's interpretation appeared to be dependent not only upon effective protection against state encroachment upon the individual's rights defined by the federal law, but also upon the extension of effective guaranties of due process of law within the states. This was not effected in the United States until after the Civil War and has not been effected in the United Nations.

To summarize the juristic trend of the last four centuries, it appears that the anarchic theory of international relations, assumed by Machiavelli, tended to be modified as a world jural community became manifest through a network of treaties, a system of international law, permanent diplomatic missions, frequent international conferences, and numerous international organizations. The international lawyers generally assumed the existence of such a community, and a succession of international humanists had sought to promote it by proposals of more adequate institutions. International law was theoretically considered, not merely a convenience for solving unimportant problems or for justifying dubious policies, but a corpus of procedures and principles giving form and self-consciousness to the collectivity of varied but interdependent nations, so that the collectivity, in spite of its highly decentralized organization and its tendency to change with increasing velocity, would constitute a true society of nations.

There can be no doubt but that grave conflicts existed both in the fundamental assumptions of traditional international law and in the assumptions considered dominant in different parts of the world. Furthermore, these conflicts had increasingly serious practical results as greater interdependence of all sections of the world were accompanied by more rapid rates of social change and greater regional differentiation of political systems. Under these conditions the maintenance of international law presented grave difficulties. International law seemed doomed to become merely a description of the behavior of states or merely an ideal system without influence upon that behavior. In either case it would fail to establish conditions of peace.

Jurists and statesmen found it difficult to keep an eye on the values of the law in the abstract—continuity, good faith, order, and justice—and also on the realities of law in the concrete—objectivity of sources, consistency of rules, regularity of observance, and effectiveness of sanctions. They found it difficult to keep the law in advance, but not too far in advance, of state conduct and continually to encourage social and institutional construction to raise the community to its level.

The new international law began to develop before World War I and has been generally recognized in principle in the atomic age, but practice lags behind. The procedures and organization of the community are not yet adequate.

4. INTERNATIONAL PROCEDURES AND WAR

International law has attempted to rationalize the position of international violence by implicit if not explicit reference to various distinct bodies of standards—the code of the duel, medieval ethical doctrines, systems of private law, and the customs and practices of modern states. It has usually been possible to justify any war by application of one or the other of these bodies of material. It is not, therefore, surprising that the legal position of war has remained uncertain and that the contributions which international law has made to the elimination of war have been meager. During the nineteenth century, while British sea power and commercial policy maintained comparative tranquillity, doctrines of sovereignty, of nationality, of neutrality, of pseudo biology, and of pseudo sociology were developing which lowered even the feeble barriers which earlier concepts of international law had placed in the path of war.

In the period following World War I, however, conventions and practices did much to eliminate confusion by branding hostilities not in defense or under authority of international sanctions as illegal and requiring that international disputes, including those concerning pleas of defensive necessity, be settled by peaceful procedures. It has, however, been suggested that these principles can never be realized by international law because they conflict with the concept of sovereignty basic in that law. This argument appears to rest upon a misconception. International law has never conceived of sovereignty as a prerogative, freeing the state from the control of that law itself. It has regarded sovereignty as freedom to make and enforce municipal law, but only within a sphere which international law itself defines—a sphere which narrows with the growth of that law. Legal sovereignty is not, therefore, incompatible with the elimination of international violence. Nevertheless, the rule of recent international law proscribing war has not been observed by several important states. Substantive international law today opposes war, but procedural international law has not developed sufficiently to make the substantive law effective.

Procedural international law has not developed as rapidly as has substantive international law. It has consisted mainly in the description of practice with little influence from ethical and juridical theory. Substantive international law, on the other hand, although not unaffected by practice, has been greatly influenced by the theory of natural law and analogies drawn from developed systems of municipal law. The consequence has been that in international law rights have often been recognized and defined long before there have been adequate legal remedies to support them.

The practices of armies and navies have developed from considerations of internal discipline and military efficiency rather than from respect for international standards. Rules relating to discipline and efficiency can be effectively enforced by courts-martial, but for enforcing international standards, only such procedures are available as formal protest, neutral interposition, reprisals, or diplomatic claims after the war, the influence of which is doubtful or delayed. National courts-martial may punish soldiers who violate international standards, and national military commissions may punish persons in occupied areas or members of the enemy's armed forces whom they catch. But these procedures are primarily designed to promote discipline in the army and to govern occupied areas. They have not proved effective sanctions for the international law of war, nor have indictments before international criminal tribunes, which can function only after the war against persons of the defeated belligerent, although they may have some deterrent influence.

Consular courts in the late Middle Ages often had a genuinely international character, and the codes which emerged from and guided their practices, such as the Consolato del Mare, constituted rules of mercantile international law closely related to their remedies. The rise of sovereign states in the Renaissance, however, checked this development. Except in the Orient, consuls lost most of their judicial functions and became agents of national commercial policy. Extraterritorial consular courts in the Orient were agencies of imperialism rather than of international law and came to an end in World War II. Maritime law, although it retained much of its international character, came to be enforced by purely national courts of admiralty, influenced, it is true, by the possibility of diplomatic protest in the background. International maritime rights and international remedies ceased to be closely related.

The practices of foreign offices and diplomatic services were devoted primarily to the advancement of national policies, especially the maintenance of the balance of power. Even the legal advisers of foreign offices tended to advise their superiors on the best legal rationalization of a political decision instead of the state's international obligations as they would be interpreted by a disinterested court or jurist. Legal arguments were presented to defend national interests and acts, but the procedure was one of advocacy rather than of international adjudication. Procedures of mediation, conciliation, and arbitration, of conference and consultation, and of international administration, all of which grew out of diplomacy, were of a more international character. After the system of permanent missions had been established, assuring reciprocity and facilitating collective *démarche* by the diplomatic corps at a particular capital, diplomacy provided a quasi-international procedure for enforcing certain rules of substantive law, especially those defining the rights and privileges of diplomatic officers themselves. These rules were more closely related to their remedy than were most rules of international law.

Text-writers, although often in close contact with governments, could not directly enforce the precepts which they recommended. They appealed to the consciences of princes and peoples, but their rules had no other sanction in so far as they went beyond a mere classification of customs and treaty provisions. Whether a text-writer's background was juristic, philosophical, theological, or diplomatic (and often it was all four), he tended to emphasize the consistency and logical coherence of the rules of international law with only secondary regard to the procedures whereby these rules could be regularly applied and enforced. In recent times jurists have frequently criticized the action of governments, often their own, from the standpoint of international law and have published such criticism in journals soon after the event. The prospect of such public criticism may have some deterrent influence.

National tribunals, other than military tribunals, have had to deal with international problems in exercising admiralty jurisdiction, especially over prizes of war; in adjudicating controversies involving resident diplomatic, consular, and other foreign agents; in dealing with controversies involving the sovereignty of territory; in dealing with controversies involving aliens; in interpreting and applying treaties; in applying national legislation designed to enforce international obligations or to regulate foreign policy; and in dealing with controversies involving rights arising from or affected by foreign law.

A very large amount of case law has arisen under these heads, but although these precedents clearly indicate the procedures for enforcing the rules recognized, the rules have been in the main rules of municipal rather than of international law. Prize courts have, it is true, declared themselves courts of the law of nations, and common-law courts have from time to time espoused the doctrine of incorporation, which holds that international law is to be applied by national courts in appropriate cases, especially those concerning diplomatic officers. This doctrine, however, has almost invariably been subject to the exception that national legislation must be observed, even if contrary to international law or treaty, and that the courts will follow the political departments of the government on political questions, such as the recognition of states, governments, belligerency, and territorial changes, the limits of national domain, and the validity of treaties. Consequently, the theory that national judicial procedure should enforce international law within the national domain is subject in practice to important qualifications. National courts apply primarily national law, and their opinions on international questions, although less influenced by ephemeral policies than those of the executive, can at best be regarded only as national interpretations of international law. This is true even if national courts are obliged by the constitution to give priority to international law and treaties over national legislation, as has been provided by a few states since World War II. Upon the important questions of international law, involving issues of peace and war, national courts can seldom transcend the national policy as declared by the executive or the legislature.

International conferences have evolved rules for their own procedure which they themselves have power to enforce. The codification of substantive international law, which has occasionally been undertaken by such conferences, even if formally binding upon states because of subsequent ratification, has frequently lacked effective procedures of enforcement. Enforcement has usually been dependent upon action by national legislative and administrative authorities or upon presentation of diplomatic claims for reparation, perhaps supported by the threat of reprisal or denunciation of the treaty. General treaties have sometimes provided for their own interpretation and application by arbitration and for their own enforcement by guaranties whose execution has sometimes been intrusted to an international organization.

Efforts have been made to render the resolutions of international conferences or consultations immediately executable by constituting the national delegations of political or administrative officials, each with power to deal with the subject in his own territory. This practice proved effective among the Allies during the world wars but its development in normal times within the organs of the League of Nations and the United Nations proved more difficult. The practice whereby responsible ministers of state attended meetings of these bodies tended to render decisions immediately executable in the territories controlled by the ministers who had agreed.

The practice of mediation by third parties in a dispute has sometimes led to intervention by a powerful state, dictating the settlement in its own interests with little regard to law. Such intervention has, however, sometimes been collective, as by the Concert of Europe, and it has even been institutionalized in the procedures of the League of Nations and the United Nations when recommending on peaceful settlements of disputes. Mediation has also led to the practices of inquiry, conciliation, and arbitration whereby the mediator, with consent of the parties, defines facts, recommends a settlement, or makes an award. These procedures have tended to be institutionalized by converting the state or royal mediator into a technical or juridical body constituted by and acting according to accepted principles.

When the *ad hoc* arbitral tribunal selected by the parties to the dispute has been developed into a permanent court with established personnel as in the World Court and when that court has a compulsory jurisdiction, as it may under the optional clause, a procedure is at hand in which the development of the substantive law and the procedure for its application appear to be adequately linked. There is, however, a weakness. Many states have not accepted the optional clause, and those that have, have often made destructive reservations, as has the United States under the Connally amendment. Furthermore, the obligation of states to submit to the jurisdiction, even if they have accepted it by treaty, and to observe the award is sanctioned mainly by good faith. The procedure of the World Court rests primarily upon legal powers, not physical powers, and those legal powers rest upon the rule of substantive international law requiring the observance of treaties. Thus, although substantive law and international procedure are linked, the procedure is unable greatly to strengthen the substantive law which constitutes its chief sanction in spite of the competence of the veto-bound United Nations Security Council to take forcible measures to enforce an award.

International organization has grown through combining the practices of international conferences, treaty guaranties, intervention, inquiry, conciliation, arbitration, and judicial settlement with a permanent secretariat.

The post-Napoleonic European system depended on occasional conferences and guaranties. The nineteenth-century Concert of Europe proceeded by occasional conferences and collective interventions. The international administrative unions, although dealing in the main with non-political questions, utilized permanent conferences and permanent secretariats. The Hague system utilized the practices of periodic conferences, codification of international law, and a permanent court of arbitration, unified through an administrative commission consisting of the diplomatic representatives of the parties at The Hague, but it lacked the authoritative element which had been present in the interventions of the Concert of Europe.

The League of Nations combined all these aspects of international organization. Frequent periodic conferences were provided in the annual meetings of the Assembly and the more frequent meetings of the Council. These institutions had powers of inquiry, conciliation, and intervention in international controversies; of recommending changes in the status quo in the interests of peace and justice; and of initiating international legislation on numerous topics, such as armaments, international commerce and communications, native welfare, minority rights, health, and labor. Procedures of voluntary arbitration and judicial settlement were provided, as were guaranties and economic sanctions against war in violation of the Covenant.

The League suffered from the general requirement of unanimity and from the lack of political power. The ultimate sanction of the system was neither unified military power nor unified public opinion but the legal obligation of the member-states to observe their covenants. When the guarantors faltered in their legal duties, the whole structure fell.

The United Nations was similar, but the unanimity rule was qualified, antiwar obligations were made more comprehensive, sanctions to enforce them were strengthened, and co-operation was facilitated by the establishment of many specialized agencies, subject to supervision by the United Nations. In spite of these changes, state sovereignty was presumed, a state was not bound to submit to adjudication or to a change in its international obligations without its consent, sanctions were subject to great power veto, and, apart from the World Court in respect to disputing states that had accepted its jurisdiction, United Nations organs could only recommend the settlement of disputes or situations dangerous to peace or the modification of conditions deemed unjust or disturbing.

Without amending the Charter, the United Nations is gradually being strengthened by a process of universalizing its membership, interpreting the obligations of its members, improving its procedures for settlement of disputes, peaceful change, international legislation, and collective security, and establishing subsidiary agencies to contribute to these ends. The recruiting and functioning of peace-keeping forces, especially in Suez (1956) and the Congo (1960), have not increased the legal competence of United Nations organs but have increased their practical competence. The permanent maintenance of such forces would facilitate collective-security operations but would also greatly increase the United Nations budget. Such a force is dependent on general observance of the advisory opinions of the World Court, which held that the General Assembly is competent to make appropriations for the support of a force and to apportion them as obligations of the members.

Factors outside the law established by the Charter or the agencies of the United Nations itself have done much to strengthen the United Nations. Among these are diplomatic efforts to relax international tensions and educational efforts to increase attitudes favorable to internationalism, to broaden the concept of national interest, and to develop greater realization in world public opinion and the policy-making agencies of governments of the necessity for peace and international co-operation in the atomic age.

PART FOUR

CONTROL
OF WAR

CHAPTER XXIII

SYNTHESIS AND PRACTICE

In the analysis of war attempted in this study it has not been possible wholly to exclude consideration of the control of war and the objectives of that control, although the emphasis has been upon trends and prediction. This chapter deals with the practical problem of control, which involves synthesis rather than analysis.

Synthesis manipulates symbols and alters their relationship to the things symbolized and to the persons using the symbols so as to realize or to create phenomena. In dealing with physical and biological phenomena, applied science and art go hand in hand, but in such fields, including engineering, agriculture, and medicine, it is possible so to define objectives and conditions that a theoretical exposition can precede constructive activity. An engineer can produce a blueprint of a bridge with all details described before the work begins.

Planning a social construction in this sense is impossible for two reasons: the objectives may be expected to change with experience, and favorable opinion which is the major condition for success cannot be predicted far in advance. The social planner is faced by a problem like that of an architect asked to design houses in accord with specifications which will be changed every week, to be constructed of mud which will wash away with the rain, in a region where a heavy rain is expected every month. Under such conditions detailed engineering plans would not pay. The control of war involves, therefore, a synthesis of (1) planning and politics. In this synthesis, (2) principles of social action must be considered, and (3) ends and means must be intelligently discriminated.

1. PLANNING AND POLITICS

A recent proposal in large-scale international planning suggests an analogy between social and mechanical inventions. The user of an automobile, it is suggested, does not need to understand its mechanism. If he can see the completed machine in operation, he can appreciate its advantages and accept it. So, it is argued, the average man does not need to know about the process or principles of building a new international order. He can leave that to the social inventors and give his approval when he sees it working. The analogy fails because no large-scale social invention can work unless the people affected by it are convinced that it will work *before they see it working*. Otherwise their skepticism or hostility will kill it. No less important than the useful parts of social institutions, as Bagehot pointed out in reference to the British constitution, are the "dignified parts" which give "force" to the "efficient parts." Social inventions have little value unless in the process of developing them social interest is aroused and general confidence in their adequacy is established. Social innovation and planning are, in fact, arts—of which the arts of social education and propaganda are parts no less important than the arts of political organization and administrative management.

Jean Jacques Rousseau in 1763 extolled the Abbé Saint-Pierre's project for perpetual peace (1713), ostensibly based on the "grand design" of King Henry IV and Sully (1608). He added, however, that "there is only one thing the good Abbé has forgotten—to change the hearts of princes." Rousseau then compares the political method by which, he said, Henry IV and Sully had attempted to achieve their plan, cut short by Henry's assassination, with the literary method of Saint-Pierre, unfavorably to the latter. "There are the means which Henry IV collected together for forming the same establishment, that the Abbé Saint-Pierre intended to form with a book. Beyond doubt permanent peace is at present but an idle fancy, but given only a Henry IV and a Sully, and permanent peace will become once more a reasonable project."

Conditions have changed in a century and a half. The hearts of masses of men are now as important as those of princes. Archibald MacLeish in 1938 challenged the question, "Shall we permit poetry to continue to exist?" by discussing the question, "Will poetry permit us to continue to exist?" "The crisis of our time," he writes, "is one of which the entire cause lies in the hearts of men," and only poetry can cure this "failure of desire" because "only poetry, exploring the spirit of man, is capable of creating in a breathful of words the common good men have become incapable of imagining for themselves."

The social plan must always be desired by the influential affected by it. Before the prescription will help the patient, the social doctor must first convince the patient what it is to be well, that he wants to get well, and that the prescription will help him to that end. The plan must always be sufficiently flexible to permit of adaptation to changing social desires. A civilized society has many different potentialities of development. A social plan can, therefore, only include a broad statement of objectives, a brief exposition of conditions to be met and methods to be pursued, and a more detailed description of the personnel and powers of an organization to do the work. This organization must synthesize knowledge and persuade opinion as it progresses. Social synthesis is, therefore, history in the making. It is to be written in human behavior and social institutions, not in books.

2. PRINCIPLES OF SOCIAL ACTION

Certain postulates of social action so obvious as to be truisms are worth recording because, in constructing programs of international reform, they have often been forgotten.

a) We Must Start from Where We Are.—Neither nations nor international institutions which exist can be ignored, for the fact of their existence gives evidence of loyalties. Persons with loyalties will retaliate if their symbols are devalued. This retaliation may itself cause violence and failure of the program which is responsible for that devaluation. Action for peace should therefore proceed by the co-ordination rather than by the supersession of existing institutions. New institutions should only be established with the initial participation of all whose good will is essential for their functioning. Those left out at the beginning are likely to organize in opposition.

  b) We Must Choose the Direction in Which We Want To Go.— This cannot be discovered by science or analysis. It is an act of faith. Presumably, democratic societies wish the control of war to be in the direction of international peace—but of peace conceived as a state of order and justice. The positive aspect of peace—justice—cannot be separated from the negative aspect—elimination of violence. Peaceful change to develop law toward justice, and collective security to preserve the law against violence, must proceed hand in hand.

  The aim must be narrowed, however, if action is to be effective. No one organization or movement can embrace all reforms. International peace does not imply the elimination of all conflict or even of all violence. Forms of conflict, such as political and forensic debate, economic competition, rivalry in demonstrating the merits of different social, cultural, ideological, and political systems, may be essential to a progressive world. International justice requires that each nation be free to develop its own system in its own territory so long as it respects universally accepted human rights. Internal violence, such as crime, mob violence, and insurrection, is a local problem in the world as it is. The domestic jurisdiction of states must be respected or they will not co-operate in building peace. International peace might be achieved even though many economic and political ills remained. The elimination of war involves continual judgment as to the importance of abuses and of proposals for reform in relation to the objective of international peace.

  *c) Cost Must Be Counted.—*It is the vice of war that it seldom compares its costs with its achievements. Efforts to control war should not make the same mistake. Programs for dealing with war may be of varied degrees of radicalness. But every social change involves some cost. If a program for establishing international peace is to be effective, first things should be dealt with first. The degree in which the basic structure of international relations may be affected in the long run cannot be envisaged in the early stages, and attempts to envisage these changes would only arouse unnecessary opposition. Social costs are relative to social attitudes, and few reforms can progress if the changes which may be involved in the distant future are measured in terms of contemporary social values. Great changes may develop if those concerned calculate only the advantages and the costs of the step immediately at hand. When that is achieved, the advantages and costs of the next step can be appraised.

d) *The Time Element Must Be Appreciated.*—War might be defined as an attempt to effect political change too rapidly. Social resistance is in proportion to the speed of change. A moderate infiltration of immigrants or goods or capital will not cause alarm, but let a certain threshold be passed and violent resistance may be anticipated. Cherished institutions and loyalties can peacefully pass away through a gradual substitution of other interests, loyalties, and institutions, but gradualness is the essence of such a peaceful transition.

The establishment of international peace requires many important social changes, because war is an institution which penetrates comprehensively and deeply in the modern political world. Consequently, organizations working on the problem must not become impatient. This is not to say that on occasion it may not be expedient or necessary to seize a favorable tide for a long advance. Such an opportunity was presented by the plastic condition of many institutions after the world wars when the League of Nations and the United Nations were established. The appreciation of occasions and the adjustment of the speed of movement to the character of such occasions are the art of statesmanship.

3. ENDS AND MEANS

War may be explained from different points of view. What is treated as an unchangeable condition from one point of view may be a variable to be changed from another point of view. This results from the fact that few social conditions are really unchangeable; consequently, the distinction between constants and variables becomes a question of policy and strategy—a distinction between ends and means.

International peace has been sought by a more perfect balance of power, by a more perfect regime of international law, by a more perfect world-community, and by a more perfect adjustment of human attitudes and ideals. These different forms of stability cannot, however, be developed simultaneously or under all conditions. Policies promotive of one may be detrimental to another.

The military point of view assumes that international law, national policies, and human attitudes will remain about as they are. Attention should be concentrated on stabilizing the balance of power or mutual deterrence by maintaining the freedom of states to make temporary alliances, to perfect armaments, to deploy forces, and to threaten intervention as the changing equilibrium requires. Permanent alliances and unions, conceptions of aggression, disarmament obligations, systems of collective security, and economic interdependencies interfere with this liberty of state action and hamper the rapid political maneuvers necessary to maintain the balance.

The legal point of view, while assuming the permanent existence of states and the persistence of existing human attitudes, seeks to limit national policies, including balance-of-power policies, by rules of law. Such rules in the international field are certain to be influenced by the principles of justice and the procedures for administering justice accepted by the developed systems of private law. International law, therefore, tends to regard many actions essential to maintaining the balance of power as unjust and to develop international organization and adjudication in its place. This involves a reinterpretation of state sovereignty so as to permit rules of international law forbidding aggression, restricting intervention, and protecting human rights.

The sociological point of view tends to hold that law and armies are consequences of the more fundamental aspects of culture. Of the latter, nationalism is outstanding in present civilization. Efforts to increase the stability of the world-community should, therefore, be directed against the symbols of nationalism. Sociologists, however, are thoroughly aware of the obstacles which the processes of social integration and personality formation offer to plans and propaganda for substituting a world myth for national myths.

The psychological point of view considers armies, international law, and national policies as derivative phenomena and devotes primary attention to changing human attitudes by education. Educators, however, are aware that certain changes in international law are essential if education is to develop attitudes universally appropriate to peace, that the growth of economic and cultural internationalism tends to facilitate such a program, that wide diffusion of attitudes conducive to international peace involves important changes in the national cultures, and that educational efforts to promote peace can be regarded as successful only if they induce general reductions in national armaments and general abandonment of aggressive policies. The success of effective peace education tends to render the balance of power less stable and, therefore, requires the substitution of a very different world political structure.

Considering the general difficulties of large-scale social change and the particular conflicts of objectives and methods, of ends and means, in approaches to international justice and order, what should be the program of the statesman anxious to eliminate war? Reactions of statesmen to problems which confront them will be discussed in the next chapter.

CHAPTER XXIV

THE PREVENTION OF WAR

The analysis in this study suggests that the prevention of war involves simultaneous, general, and concerted attacks on educational, social, political, and legal fronts. Policies directed toward a military balance of power, toward political and economic separation of the great powers, or toward conquest of all by one give no promise of stability in the modern world. Policies directed toward these objectives are more likely to contribute to war than to prevent it.

The difficulty of finding points at which the results of theoretical studies might be injected into the onward rush of politics can be illustrated by considering the reaction of statesmen to certain practical problems which have confronted them in recent years—those of (1) the aggressive government, (2) the international feud, (3) the world crisis, and (4) the incipient war.

1. THE AGGRESSIVE GOVERNMENT

In a legal sense the word "aggressor" refers to a government which has resorted to force contrary to the international obligations of the state. Here the term is used in the sociological sense and refers to a government which, because of its internal structure or its environmental conditions, is likely to resort to force. Herbert Spencer distinguished the military state, which compels internal order and external defense by subordinating the economic, social, and political life to the needs of the army, from the industrial state, which uses persuasion to achieve internal order and external defense by subordinating the army to the needs of social service, economic prosperity, individual initiative, and international conciliation. The difference is only relative because all states have both productive and military organs, and in most the leadership is sometimes in one, sometimes in the other. Furthermore, aggressiveness is primarily a characteristic of a government rather than of a people. A people may rapidly substitute a peaceful for an aggressive government, but the type of government undoubtedly tends in time to infect the people.

How can aggressive governments be identified and eliminated? Statistical studies indicate that some governments have fought more frequently and have spent a larger proportion of their resources on war and armaments than have others. Political studies suggest that war and the army play a much larger role in the power-maintenance devices of certain governments than of others. Sociological studies suggest that military activities play a more important part in the culture of some governing elites than of others. Probably criteria could be set up to identify the aggressive governments at any time by utilizing figures of the kind mentioned, supplemented by analytic-descriptive materials relating to the degree of centralization and totalitarianism and to recent policies and acts.

A government may conclude that another is aggressive because of a traditional stereotype, interested propaganda, misinterpretation of acts or utterances, false analogies, popular prejudices, or cultural or ideological differences. The images of a state in the minds of its own people and in the minds of others usually differ greatly. Thus the first problem in dealing with an "aggressive government" is to be sure that objective evidence supports this characterization at the time it is made.

The problem of eliminating aggressive governments is less difficult than the sheep's problem of eliminating wolves, because no people is invincibly aggressive. The wolf cannot change its nature, but the people afflicted by an aggressive government suffer from a disease rather than from an inherent characteristic. This conclusion is suggested by the variability of the degree of aggressiveness in the history of all peoples. The disease is a result of the interaction of internal and external conditions. In time of general war, depression, and disorder all peoples tend to become aggressive; in long periods of peace most peoples tend to become peaceful and industrial; but stability may lead to despair of the underprivileged and boredom of the adventurous, inducing revolutionary and belligerent propagandas. Furthermore, the tradition of military prestige, aristocratic social organization, political autocracy, and a geographical situation inviting invasion render certain peoples more susceptible to the disease.

A people thus susceptible, after emerging from the despotism of a war, may for a time emphasize industry in order to recuperate, but with the inevitable postwar depression, its government will resort to saber-rattling as a method of diverting the attention of its people from "hard times." This will necessitate preparedness as a means of defense, of relieving unemployment, and of prestige, and parades to further divert attention from economic ills. Military preparedness, however, requires political preparedness by concentration of authority, economic preparedness by the diversion of trade to those areas capable of control in time of war, and psychological preparedness by censorship and propaganda of the military spirit among the population. All these factors augment the depression. The people must be told to draw their belts tighter, to give up butter for guns, and to prepare more intensively for war. All activities within the state tend to be evaluated in terms of their contribution to its military power. National power supersedes national prosperity as the goal of statesmanship. The vicious circle continues through the interaction of the forces making for internal revolution and those making for external war. If war can be staved off and the despotism has not become too inflexible, the vicious circle may be broken by the insistence of the population that conciliatory policies be pursued, that armaments be reduced, and that peaceful coexistence with states with diverse ideologies be established in order that production may increase and taxes decline.

Other governments should attempt to stave off war by skilful diplomacy which mollifies without yielding to threats and by a convincing expression of determination to apply sanctions against governments guilty of overt aggressions. Diplomacy should aim to isolate the aggressive government both from its own people and from other governments rather than to make a counteralliance to contain it. The latter policy tends to consolidate the aggressive government with its people and to group all the great powers into two hostile alliances. It may be more expedient to offer opportunities for external commerce to groups subject to the aggressive government than to isolate them economically if this can be done without greatly aiding the military preparation of that government. A program of political isolation of the aggressive government, economic collaboration with its people, and the threat of collective sanctions against overt acts of aggression is more likely to break the vicious circle than a program of counteralliances, economic isolation, and threats of preventive war. This statement suggests the importance of distinguishing among policies of sanctions, counteralliance, and containment; punitive, preventive, commercial, and discriminatory policies; conciliation, appeasement, and firmness. It also suggests that several appropriate policies should be co-ordinated in the light of the developing situation.

The distinction between international police or sanctions against aggression and counteralliances against aggressive states with threats of preventive war must be emphasized. This distinction is possible through the establishment, by general treaties, of clear juridical definitions and international procedures to identify and deal with acts of aggression. In the same way economic sanctions against *governments* found guilty of aggression must be distinguished from national policies of economic discrimination against *states*. In other words, aggressive states must be treated as sick or unsocial and brought back into normal life, unless the governments are proved to have committed acts of aggression, in which case international sanctions should apply, but so far as possible only against the government with the object of assisting the people to get rid of it.

The objection often made, that programs of continuing trade with a population whose government has an aggressive character will assist the aggressive government in its preparedness program and thus render it more powerful militarily, although important, is not always controlling. By becoming dependent upon distant sources of raw materials and markets, the aggressive government becomes more vulnerable to economic sanctions. Furthermore, internal interests against war will be established, not to mention the influence of foreign trade in raising the standard of living. The value of such a program in curing aggressiveness may therefore be greater than its disadvantages in contributing to the military power of the potential aggressor if that contribution is not large. The difficulty is often encountered that the aggressive government itself raises barriers to trade as a military preparation.

Appeasement should be distinguished from conciliation. The aggressor's success in utilizing threats of violence may stimulate him to utilize the same methods again. The argument is often made by non-resisters that generosity stimulates generosity and that the aggressor will reciprocate to such treatment by becoming docile and law-abiding. Doubtless, generosity may have that effect under certain circumstances, but it may be questioned whether either the aggressor or anyone else would characterize the sacrifice of someone else's rights under threats of violence as generosity. A spirit of conciliation leading to reasonable compromising of differences or voluntary rectification of inequities may prevent the development of potential aggression and stabilize the community of nations. But the same cannot be said of retreat before threats of violence at the expense of those who have right but not power on their side. Appeasement may not only stimulate the aggressor but may encourage his victim to prepare for aggression and may destroy belief in the possibility of a just and peaceful world. In his radio address of October 26, 1938, President Franklin D. Roosevelt commented on the consequences of the Munich agreement, which sacrificed Czechoslovakia by giving the Sudetenland to Hitler: "It is becoming increasingly clear that peace of fear has no higher or more enduring quality than peace of the sword. There can be no peace if the reign of law is to be replaced by a recurrent sanctification of sheer force. . . . You cannot organize civilization around the core of militarism and at the same time expect reason to control human destiny."

The use of the threat of nuclear weapons to deter a nuclear-equipped aggressor, however, is likely to be suicidal, as was concluded by the United Nations in the Hungarian uprising of 1956. Thus, no policy, whether of sanctions, containment, or counteralliance should contemplate such use, and it should not be assumed that any government plans aggression with nuclear weapons. To assume that a government is both intransigently aggressive and irrational may doom civilization.

2. THE INTERNATIONAL FEUD

It is obvious that certain pairs of states are more likely to get into war with each other than are other pairs. A war between Afghanistan and Bolivia would be more surprising than one between Albania and Bulgaria. That territorial propinquity is not the only factor influencing such expectation is suggested by the consideration that today no one anticipates a war between Canada and the United States or between Virginia and Pennsylvania, although both of these wars have occurred since 1812. Relations among geographic, commercial, cultural, administrative, ideological, and other aspects of distance, susceptible of statistical measurement, may throw light upon the probability of any given pair of states getting into war, as may the present policies of the states, the probable orientation of each on opposite sides or the same side in a general war, and factors of historic animosity.

The latter constitutes the problem of the international feud, a phenomenon exhibited in the state of intermittent war between Rome and Carthage for two centuries, between England and France for five centuries before 1815, between Great Britain and Ireland since the time of Henry II, between France and Germany from the Thirty Years' War to the period following World War II, and between China and Japan since 1894. The United States feuded intermittently with Great Britain from 1775 to 1900 and after that with Germany until 1945 and then with Russia.

These feuds grow in part from continuing ideological differences and political controversies, in part from the value to a government for internal political purposes of maintaining an external enemy against which the fears, ambitions, and military preparedness of its population can be mobilized, and in part from the sentiment of revenge natural in a population which has been the victim of war. This sentiment is often kept alive by dramatic accounts of the invasions and barbarities of past wars in popular histories, and by insistent demands for the recovery of unredeemed territories. Frequently, they become more intense with time because each successive war adds new fuel to the fire.

The feud between the United States and Russia has not been continuous and has never resulted in war. The Russian autocracy refused to recognize democratic America until 1813, but relations were on the whole friendly, especially during the Civil War period, until migrations from Russia in the 1890's provided the United States with more information about minority persecutions in Russia. In the twentieth century Tsarist pogroms and the Communist withdrawal from World War I after the Revolution of 1917 produced much friction. The Soviet government's confiscation of property, repudiation of international and financial obligations, and propaganda of communism abroad induced the United States to refuse recognition until 1933, and mutual suspicions continued even during the alliance against Hitler in World War II. The feud was continued in the "cold war" by acts of each which the other interpreted as evidence of aggressiveness, augmented by the process of retaliatory reactions by each and supported by ideological controversy, imperfect analogies to the Hitler regime which each made of the other, the formation of opposing alliances augmenting mutual fears, and an arms race.

International feuds have sometimes ended by conquest of one state, as in the case of Carthage; sometimes by a development of great disparity in the power of the two states, as in the case of England and Scotland; and sometimes by political union or federation, although the Anglo-Irish feud has withstood all these remedies. Sometimes they have ended by a shift in the balance-of-power situation so that both parties to the feud become more alarmed at a third state. The rise of Russia and Germany as military powers contributed greatly to the ending of the long Anglo-French feud. The rise of the German and Japanese navies contributed to the ending of the Anglo-American feud and fear of Soviet Russia to the ending of the Franco-German feud.

The Russo-American feud showed signs of abating after the death of Stalin in 1953 and Khrushchev's repudiation of "Stalinism." Its demise seemed assured by mutual recognition of the intolerable character of nuclear war after both sides were well equipped with missiles and satellites, by the weakening of the ideological alliances of both after China and France had manifested independent positions, by the disposition of both to spend less on armaments and more for economic development, and by their initiation of co-operation to control nuclear weapons through ratification of the "Atoms for Peace" and nuclear test-ban treaties. The end of a feud has often been marked by the conclusion of arbitration and disarmament agreements and the diplomatic settlement of claims following successful efforts to relax tensions. From the standpoint of peaceful international relations, it is clear that such methods should be utilized for terminating feuds in preference to the method of creating new feuds.

3. THE WORLD CRISIS

Statistical compilations of battles during the last four centuries disclose the gradual emergence of a fifty-year fluctuation in the intensity of war. This fluctuation has been attributed to fading social memory with the passage of a generation, to long economic fluctuations, to the lag of national policies and constitutions behind changing international conditions, and to the tendency of unsettled disputes to accumulate, aggravating the relations of states.

These fluctuations arise from many factors which vary from instance to instance, but they have a typical character because the critical points are determined by the political exigencies of governments. After a necessary period of postwar reconstruction, more protracted in modern industrial nations than formerly, there comes a secondary postwar depression producing internal unrest. All governments tend to seek a remedy in concentration of national authority for relief, programs of self-sufficiency for protection, and a preparedness program to relieve unemployment and to provide for defense. This characteristic is particularly evident in states traditionally susceptible to aggressiveness, but it is manifested to some extent in all states. This tendency toward military and isolationist programs is likely to produce a realignment of alliances and disturbances to the balance of power, marking the transition from a postwar period to an interwar period. The latter is likely to last for ten or fifteen years and to be characterized by fluctuations in the system of alliances, imperial wars, and minor civil wars. Gradually, however, the great powers tend to take positions on one side or the other of two hostile alliances, and with the solidification of such a bilateral balance of power, the interwar period changes into a prewar period. The political alignments being established, each group calculates the influence of time upon its prospects in a war which is now considered inevitable. The side against which time runs will sooner or later precipitate a war on the hypothesis that if it does not act now it will certainly be defeated. This course of development can be detected in the relations of European states from 1815 to 1854, in the relations of the states of the United States from 1815 to 1860, and in the relations of European states from 1870 to 1914.

There were similar developments from 1920 to 1939, but the course of events was greatly accelerated. In spite of numerous difficulties in adjusting to the Treaty of Versailles, a postwar era of peace and good feeling was ushered in by the Locarno agreements of 1926. But the failures of the economic conferences of 1927 and 1933, of collective action in the Manchurian case, and of the disarmament conference of 1932 aggravated the economic and political crises which had begun in 1930.

As a reaction to prolonged economic and political insecurity, economic and political nationalism and self-sufficiency developed in all countries with varying degrees of intensity. This reaction prevented recovery from the normal postwar depression and eliminated the usual interwar period. A prewar period at once began in which political alignments with a view to war rapidly shattered all effective action toward international political co-operation, augmented the expectation of war, and induced a panic flight of states into political and economic nationalism, manifested among the satisfied by policies of isolation and among the unsatisfied by policies of aggression.

The new world war really began with the Japanese invasion of Manchuria in 1931. It rapidly spread with the Italian invasion of Ethiopia (1935), the Spanish Civil War (1936), the renewed Japanese invasion of China (1937), the series of invasions in Europe by Hitler (1938, 1939), and the Japanese attack on Pearl Harbor (1941), until nearly all states were involved.

Wars involving great powers have always spread rapidly because they threaten the balance of power. It is very rare in the last three centuries that any great power has succeeded in keeping out of a war in which there was a great power on each side and which lasted for over two years. The position of lesser neutrals is different because, if in the vicinity of a great power, entry into the war might mean suicide; but even such states frequently have been drawn in. The United States was drawn into the Napoleonic Wars and into World Wars I and II. In the mid-century period of wars it fought its own Civil War.

In addition to cycles of some fifty years, history has exhibited a trend of war during the life of a civilization. Military strategy has tended to move from (1) the technique of agility and pounce to (2) the technique of momentum and mass charge, followed by (3) the technique of discipline and maneuver, which in time moves to (4) deadlock and the war of attrition ending the civilization.

The military history of modern civilization exhibits analogies to that of earlier civilizations. The highly trained but relatively small armies of the sixteenth, seventeenth, and eighteenth centuries capable of pouncing upon and paralyzing their enemies rapidly, especially when those enemies were Americans, Asiatics, or Africans without modern arms, grew gradually in size as population increased and methods of transportation and communication improved. When at war with one another they relied more and more upon defensive fortifications and siegecraft, but their basic strategy and tactics continued with little change until the French Revolutionary period.

Napoleonic doctrine, built on universal conscription and the revolutionary spirit, held that military power varies mechanically as the product of the mass and the mobility of the army, but it emphasized morale even more than matériel.

Acceptance of this doctrine of the nation in arms since the mid-nineteenth century may mark the transition to the second stage of modern warfare. National self-consciousness had been developed by Fichte, Mazzini, and Treitschke, and the doctrine of mass warfare had been developed by Clausewitz and his successors, especially in Germany. The practice of this type of warfare was facilitated by the use of the railroad for mass mobilization and of heavy mobile artillery for battering through. Its possibilities and tendencies were illustrated by the operations of Grant and Moltke, Kuropatkin and Oyama, and Hindenburg and Foch.

Throughout the entire modern period the doctrine of the strategic offensive had dominated, and it was hoped that the superior mobility of motor transport, the tank, and the airplane would present opportunities for maneuver. The forces, however, had become so great in World War I that they covered the entire front, presenting little opportunity for out-flanking surprises. The British effort to get around the front at the Dardanelles failed and the war became deadlocked in machine gun–lined trenches. It was eventually won by the vast industrial and population superiority of the Allies, especially after the entry of the United States, over the blockaded Central Powers.

Before World War II opportunities for maneuver were opened by the far-flung offensives of the axis powers in Manchuria, Ethiopia, Spain, the Rhineland, Austria, and Czechoslovakia. When full-scale war began in Europe, Hitler was at first successful with the blitzkrieg, combining airplane, tank, and infantry in surprise attacks, at separated points—Poland, Norway, the Netherlands, France, Russia, North Africa, and the Middle East—while strategic air raids were made on Britain and, by Japan, in Hawaii, the Philippines, Indonesia, southeast Asia, and the south Pacific. These initially successful efforts were eventually defeated by the superior capability of the United Nations to produce airplanes, high explosives, and the atom bomb.

The development of nuclear weapons and missiles, against which there is no direct defense, seems to assure that another major war will result in the destruction of both sides in a short time. Some experts in "deterrence" claim that victory is possible without intolerable losses if a "counterforce strategy" is pursued, supported by espionage to pinpoint enemy bases, by hardened mobile or submarine launching sites to protect missile capability, and by civil defense shelters to protect the population. Considering the blast, fire, and fallout from megaton nuclear bombs, this argument seems to convince nobody, apparently including the experts themselves. The heads of the leading states have united in regarding nuclear war as irrational.

In all civilizations the long-run trend of war has been toward increasing destructiveness of life and property in spite of its declining frequency. Each period of battle concentration in modern civilization has tended to be more serious than the previous one. Modern civilization differs from past ones in that it is world wide, and thus its destruction would be more catastrophic to the human race.

Efforts made by military men, statesmen, and international lawyers, even in the nuclear age, to limit methods of warfare or to localize war have had some success, but there is always danger of escalation. The "nuclear stalemate" is not a guaranty against accidental or even intentional use of nuclear weapons if defeat is believed to be the alternative. It does not seem likely that modern states will be able to revert to the old system of small professional armies whose activities might be kept within bounds. A nation in arms, goaded by suffering and propaganda, will tend toward absolute war when it fights. For similar reasons great states at war will pay little attention to neutrals. Large neutrals will be subjected to vigorous propaganda, and the war spirit will grow in response to inevitable indignities and apprehension of the possible effects of the war upon the balance of power until they enter on one side or the other. If small neutrals do not enter, they will be invaded or coerced into subordination to the needs of one or both belligerents.

Nations desiring peace must rely on prevention rather than on neutrality. As there seems little hope of smoothing out business cycles except through appropriate government control of currency, banking, taxation, and corporate organization to prevent privilege and monopoly and to preserve numerous competing units in industry, so there seems little hope of smoothing out the cycles and stopping the trend of war in the family of nations except through international organization to frustrate aggression, to provide peaceful machinery converting the balance of power from a military to a political equilibrium, and to prevent too great concentrations of political power.

There is another danger. Organized efforts to prevent economic crises may have sometimes staved off minor depressions only by so rigidifying economic processes that a more serious depression has eventually occurred. International organization, effective to prevent small wars and to stave off large wars, may so rigidify the status quo that eventually there will be a world war. In the past men may seem to have had a choice between frequent small wars or infrequent large wars. To avoid this apparent choice, international organization must be developed to facilitate peaceful change in political structure and the distribution of power when such changes are required because of differential rates of economic and social change in different parts of the world. An international organization devoted solely to the preservation of a given status quo cannot preserve permanent peace.

States which rely solely on their own resources for defense against potential enemies cannot be expected voluntarily to accept political readjustments which, however demanded by justice or economic conditions, will have the consequence of weakening their military position and strengthening that of potential enemies. Consequently, willingness to accept a system of peaceful change is dependent upon general confidence in a system of collective security. If the states are convinced that they cannot be deprived of their rights by violence, they may be willing to yield certain rights in the interests of justice, especially if the world-community is organized to exert political pressure to that end.

4. THE INCIPIENT WAR

At any moment observation of the policies of aggressive states which have morally revolted from the restraint of international law and treaty, of the course of international feuds perpetuating venom in the minds of populations, and of the gradual passage from an interwar to a prewar period may suggest points of tension which may easily become war. Diagrams indicating the changing attitudes of one people toward the symbols of other states have been made. A compilation of such diagrams for all the great powers might graphically exhibit the state of international weather at any moment.

Such indications of the rise and fall of hostile attitudes can be related to incidents and conditions in the cultural, economic, political, and juridical realm. As diplomatic controversies become more serious, incidents become more violent; political crimes are committed; merchant vessels are attacked; airplanes engage in espionage; civil strife is stimulated; frontier hostilities occur; and the graph of hostile attitudes of one population to the other, as indicated by the press, exhibits marked changes for the worse. A storm center is gathering. It is not possible to predict when war will occur precisely. Through the observation of such facts it is possible to see danger signs, but the diagnosis does not suggest a clear remedy.

The "natural" reaction of states not immediately involved in an emergency crisis is to scatter for shelter like a flock of chickens when two of their number get into a fight. This policy of pacifist isolationism was practiced by most of the states after the crisis of 1936, precipitated by Hitler's invasion of the Rhineland and the League's abandonment of sanctions in the Ethiopian case. The policy was especially defended by the northern neutrals of Europe and by the United States, which reverted to policies of neutrality. Former President Hoover, in an address of March 31, 1938, upon his return from Europe, explained the crisis situation there:

Every phase of this picture should harden our resolve that we keep out of other people's wars. Nations in Europe need to be convinced that this is our policy. . . . In the larger issues of world relations, our watchword should be absolute independence of political action and adequate preparedness.

This policy, by which each country seeks to preserve its own peace by isolating itself from the crisis, if pursued generally, tends both to intensify the crisis and to accentuate the characteristics of the world's political structure favorable to wars. The aggressors immediately responsible for the crisis will be stimulated to continue their aggressions because they will be convinced that no united opposition to them will be organized and that they can plunder their weaker neighbors without difficulty. In so far as aggressions have been the consequence of unredressed inequities in the past, the prospects of redress will be diminished, because the neutral powers, although ready to sacrifice weaker powers to the aggressor, will augment their armament and may even band together to defend their own possessions against the aggressor. If, instead of striving for isolation, those not immediately involved in a widespread crisis follow the lead of the dynamic aggressor like jackals, each hoping to share in the booty, the result will be war, because the wealthy intended victims will eventually resist. If, in a somewhat more sophisticated reaction, like a herd of quarreling apes, they momentarily forget their quarrels in accord with the precepts of balance-of-power politics and collaborate against an outside invader, little more contribution will be made toward a more peaceful world under present conditions of continuous material interdependence. The only policy which men have found capable of securing peace in times of crisis is that of rallying behind law and procedures of enforcement which have been prepared in advance. Conditions may have existed when states, because they lacked contacts, intelligence, and solidarity, could not do better than imitate the chicken, the jackal, or the ape; but conditions of communication now justify behavior more like that of men.

In crisis situations the policies of states not immediately threatened often fail to give adequate consideration to the long-run tendencies of action and, as the crisis deepens, to behave in ways which are considered necessary for the immediate security of each but which, like a panic in a theater fire or a stock-market collapse, actually involve all in common ruin. Crisis situations may, however, be used to promote united efforts to remedy genuine grievances and to establish universal principles and institutions, marking progress toward permanently stabilizing peace. Such efforts made before the two world wars were inadequate, but after World War I, the League of Nations was created, and although it did not prevent

World War II, its experience facilitated establishment of the stronger United Nations before the war was over. The service of these institutions in dealing with crises is indicated by the record.

Of the sixty-nine important political controversies from 1920 to 1939, fifty-five were dealt with successfully either by the League organs or by other international agencies. Seventeen of these instances involved the use of force in international relations: four were dealt with successfully by the League; three were settled by other agencies; and in ten the aggressor was not stopped until World War II.

Of seventy-eight political controversies from 1946 to 1964, United Nations recommendations had by 1964 been carried out in thirty but not in twenty-one cases; of the latter, twelve were under negotiation. Sixteen cases had been settled without UN action but in accord with recommendations of another international agency or by agreement or acquiescence of the parties. The remaining eleven were on the UN agenda or under negotiation in 1964. Thirty-four of these incidents involved the use of force and ten the threat of force in international relations. In seventeen of the former, the United Nations or some other international agency brought about a cease-fire, although in five (France-Indochina, India-Pakistan, Korea, Congo, France-Algeria), only after serious hostilities. In fourteen instances, hostilities ended without formal cease-fire, and in three they were continuing in 1964. In twenty-five instances the dispute which led to the hostilities had not been settled by 1964, leaving seven territories divided by armistice lines and fifteen quiescent; in two of the latter, territory occupied by the apparent aggressor remained in its possession. The United Nations has been more successful in stopping fighting than in settling disputes.

CHAPTER XXV

THE ORGANIZATION OF PEACE

Among the causes of war, attention was called in chapter xxi to the frequent lag of procedures of adjustment behind changes in conditions; to the role of war in serving certain social functions, psychological needs, political utilities, and legal claims; and to the objective and subjective conditions which made it difficult to find a substitute for these roles of war.

Among conditions of peace, attention was called in chapter xxii to the history of international law and its relation to municipal law, to procedures which have been used to develop and enforce it, and to the peace-keeping procedures and instrumentalities which have been utilized by the League of Nations and the United Nations.

Causes of war persist in 1964 and conditions of peace have not been established. How might international law, procedures, and organization be developed or new devices be established to meet the problem of war in the atomic age? Suggestions for preventing a war immediately pending have been made in chapter xxiv. This chapter will suggest steps for the more permanent organization of peace.

1. THE STRUCTURE OF PEACE

Richardson concluded from statistics that it is the "natural" tendency of governments to deal with immediate issues of war and peace by methods which make the general world structure less stable. The result has been the perpetual recurrence of war in the world. Statesmen have, when confronted by crises, usually turned the rudder the wrong way if their object was to bring the world to a harbor of political stability.

In this sense peace may be considered artificial and war natural. The ships of state have for so large a proportion of the time been tossed upon stormy seas that even the broadest characteristics of a peaceful port elude the imagination of statesmen. What are the characteristics of that port? Can it be sufficiently identified so that if the desire is present, progress can be made toward reaching it? What sort of a structure, to change the metaphor, should the engineers of peace try to build in order to increase stability?

Plans for improving European or world organization have been produced with increasing frequency for the last three centuries, all built upon appreciation of the need to decrease the lag of international solidarity behind technological interdependence. With this lag, Rousseau pointed out in the middle of the eighteenth century, the condition of the European people was worse than if they were completely isolated. The clock of science and technology cannot easily be turned back. The only way to close the gap is to develop international and supranational institutions able to adapt individual attitudes, social symbols, public opinions, and public policies in every part of the world to modern conditions. Political myths must be conformed to economic realities, political nationalism must be adjusted to technological internationalism; military policies must be adapted to the atomic age.

Most of the proposals for improving world organization have, according to the analysis here presented, suffered from both structural and functional defects. Structurally they have inadequately balanced educational and investigatory competencies, political and legal jurisdictions, legislative and executive powers, and regional and universal responsibilities. Functionally they have not provided adequate procedures for measuring and changing the representation of peoples and governments, for determining and dealing with basic offenses against a world-order, for assuring popular support to world institutions, and for relating the organization of peace to the basic values of modern civilization.

Every world crisis should be handled with an eye to progress toward a more adequate world organization. Times of peace and prosperity are adapted to solidifying the world institutions which have been established and the world symbols which have been accepted. Times of tension and depression produce crises and wars during which active efforts should be made not only to prevent a backsliding toward excessive localism, nationalism, and regionalism but to achieve new advances in the direction desired. When all symbols and institutions are being weighed in the balance and viewed with skepticism, an opportunity is offered to the forces of peace no less than to those of war. The League of Nations would not have been achieved at all had not Wilson seized the disillusionment of war to win acceptance for new symbols of world unity. In the period of peace which followed, statesmen did much, but not enough, to stabilize the meaning of these symbols and to augment their power. World War II presented another opportunity, and even before the war was over another long stride in advance was made by establishing the United Nations.

a) *Investigatory and Educational Competencies.*—It would generally be recognized that the Secretariat is the most indispensable agency of the United Nations. Its capacity to examine problems from a world point of view, to assemble information, and to recommend feasible plans of action has been demonstrated. No progress toward peace can dispense with such an agency, though it may be suggested that, useful as its economic and statistical investigations have been, its studies should be devoted in larger measure to objective examination of changes in the attitudes and opinions within the world's population. Solidarity of opinion is more important than progress in technology and should be developed first if peace is to be secured. Changing expectations of war, changing opinions in one nation about another, changing attitudes of unrest (local and general), and changing allegiances to the major social and political symbols—these things can be roughly measured. Up-to-date and accurate charting of these changes provides the indispensable data for peace action. To provide such materials, the Secretariat should have adequately equipped agents in all sections of the world analyzing the press, the activities of pressure groups, general opinion, and expert opinion and submitting their findings at frequent intervals by wire or radio to the central office. The opportunity to make such scientific investigations of attitude and opinion in all sections of the world should be a first requirement of an effective international organization.

Any official body tends to become juristic rather than scientific—to prepare briefs expository of existing rights and obligations rather than to prepare studies elucidating unsatisfactory or dangerous conditions and suggesting new methods and treatments. The analysis presented in this study suggests that common intellectual understanding of world problems will not contribute to peace unless accompanied by a broadening of attitudes from the national to the world horizon. International education must accompany or even precede international research. If it does not, the fruits of research may prove in part esoteric and in part grist for the nationalist lawyers. The world-secretariat must understand the problems of the world, but it must also educate the world in the attitudes necessary to solve these problems. It must also discover and inform the world of the consequences of alternative programs for handling problems as they arise.

To organize a world-secretariat that would be loyal to world interests, intellectually adequate, sufficiently representative to give all nations a sense of participation, and sufficiently alert to national attitudes to provide an inside liaison with the national governments is not easy. In general, it would appear that scientific and professional qualifications should take priority over geographical and representative qualifications, although efforts should be made to recruit personnel from as many divergent races and nationalities as possible.

The greatest weakness of the secretariats of the world organizations has been their want of access by right to the public in all sections of the world. Their publications have been distributed widely but in small quantities. Some important states were not members of the League. Even in 1964, nearly a third of the world's population was not represented in the United Nations. Some of the members have not permitted radio and television access to their publics. An effective world-order requires that scientific findings of the Secretariat as well as the political and legal findings of other principle organs of the United Nations be rapidly disseminated to the world public. UNESCO and other specialized agencies have contributed to the solution of this problem.

b) Legal and Political Jurisdictions.—Under the United Nations Charter and the World Court Statute, submission of controversies to legal adjudication is optional, but if controversies are not so submitted and are politically important, they must be submitted to political consideration by the Security Council or the General Assembly. These bodies may recommend adjudication, but determination of whether the dispute should be adjudicated is left to agreement of the parties in dispute unless they have accepted the compulsory jurisdiction of the Court.

The optional clause accepted by forty members of the United Nations, but generally with extensive reservations, reverses this procedure, requiring disputes concerning legal claims to be submitted to the Court, which can decide, upon unilateral application, whether a dispute is within that category. The Court, however, is open only to states, not to individuals, corporations, or international organizations. The latter are occasionally able to bring their problems before the Court through the device of advisory opinions.

Neither of these systems for distinguishing between political and juridical questions is without difficulty. The system of the Charter makes it possible for an intransigent state to avoid adjudication altogether, thus preventing law from acquiring authority. The system of the optional clause, on the other hand, makes it possible for a state with a good legal case to oppose modification of its rights, although in equity its case may not be good. It is true, systems of private law are thus weighted in favor of the status quo, since every plaintiff is entitled to bring his case to court. But in such systems the community has a more intense social solidarity than does the community of nations. Under such conditions it is possible to develop legislative procedures assuring that law will approximate justice as the latter is interpreted by the public opinion of the community.

The difficulty might be moderated by utilization of the Court's competence to decide cases *ex aequo et bono,* to modify rights under strict law when justice demands, and gradually to liberalize the law. This competence, however, can be exercised by the Court only with the consent of the parties. It may be doubted whether many states would accept the compulsory jurisdiction of a world-court with such an enlargement of the sources upon which it could base its decisions. It cannot be supposed, however, that, even with such a change, the court could greatly modify the application of the law in a particular case. Equitable jurisdiction could scarcely be a substitute for political legislation.

A further development would be to open the court to individuals who claim that national legislation or administrative action has deprived them of rights to which they are entitled under treaties or international law. Such a procedure, analogous to that found in many federal constitutions, would tend toward acceptance of juristic monism. This position holds that national laws contrary to international law are null and void and recognizes the international status of individuals. Although it may be expected that such a principle will develop slowly, a start has been made by the creation of the Court of Human Rights among many west European states and by the United Nations administrative tribunal which is competent to deal with complaints of members of the Secretariat. Further progress might be made by permitting resident aliens to appeal directly to an international court on alleging a "denial of justice." Maritime cases arising under the general law might also be subject to appeal to the international tribunal, thus extending to civil admiralty cases the international jurisdiction proposed for prize cases in the Hague Convention of 1907. The suggestion has been made that the Court's jurisdiction be extended to permit the United Nations, as a juridical person, to defend by judicial procedure the human rights of individuals that have been denied by a state.

Unquestionably, a judicial development of international law would proceed much more rapidly if the principles of that law could be authoritatively established in connection with claims of individuals which usually have less political importance than cases between states. Furthermore, such a procedure would make states more continuously aware of international law and less likely to encroach upon it by legislative or administrative acts whose purpose is primarily domestic.

With certain changes strengthening the position of law in the community of nations and thereby stabilizing the status quo, the states with grievances for which the law clearly offers no relief should have adequate opportunity to bring their cases before the political organs of the United Nations. It is unfortunate that the political competence of the United Nations organs is based on a phraseology which implies that a state must endanger international peace and security before this political procedure can be invoked. Furthermore, unless the fundamental values of modern civilization are widely understood and accepted, such a political organ has no standards of policy or ethics to justify it in transcending the existing law. In such circumstances it would tend in serious cases to yield to the demands of the more powerful state. The Charter, by setting forth broad standards of human rights, self-determination of peoples, co-operation for economic and social welfare, and maintenance of justice, contributes to this end and improves upon the Covenant. The United Nations, however, cannot decide, but can only recommend, on political settlement. Thus, many serious disputes have remained unsettled and threats to the peace. Since it is improbable that states will formally confer full legislative power upon the United Nations, this situation can be remedied only by the development of a world public opinion insistent that United Nations recommendations be carried out. With such a development, the authority of the United Nations would gradually increase.

c) Executive and Legislative Powers.—The Charter provides less explicit guaranties against territorial changes by violence than did the Covenant, but it imposes broader obligations against the use or threat of use of force in international relations. It favors political adjustments rather than preservation of the status quo. The Charter even provides for political consideration of any situation which may "impair the general welfare or friendly relations among nations." An authoritative decision, however, cannot be made.

The development of controversy before World War II between "revisionist" and "status quo" powers resulted in a wide discussion of this problem without definite conclusions. The controversy centered around territorial changes, though its scope was actually much broader. It dealt, in fact, with the relative roles of legislative and executive authority in international government.

So long as there is no adequate collective-security system and the existence of states depends solely on their own armaments and the balance of power, any state subject to demands for territorial cession will pay more attention to the influence of such a change upon its power position than upon the equity of the demand per se. Poland was obliged to subordinate consideration of the justice of Germany's demands for Danzig, based upon the principle of self-determination, to consideration of the influence this cession would have in augmenting the prestige and aggressiveness of Germany, weakening the morale of Poland, sapping confidence in French and British guaranties, and thus leading to further demands and a gradual dismemberment of Poland. In short, no system for peaceful territorial change appears to be possible until states are assured that collective security is so reliable that only claims which are based on justice as interpreted by international bodies can ever be successfully promoted.

General pledges of collective action, economic or military, against states guilty of aggression are not likely to be sufficiently reliable to give a sense of security in times of tension. This is because the hazards of states, especially those neighboring a powerful aggressor, are likely to be so great that they will neglect their obligations.

On the other hand, no system of world order is possible without some protection of the members against violent breaches of that order. The problem of determining the aggressor has not proved difficult when international procedures have been available under which provisional measures, such as an armistice, can be promptly proposed and when states have pledged themselves to recognize that the state refusing to accept such measures is the aggressor. Resorts to violence contrary to specific international obligations are thus considered aggression, irrespective of the merits of the claims which the aggressor has sought to promote. Aggression, therefore, does not refer to the objectives of a state's policy but to the methods used in promoting those objectives. It is not determined by the offensive or defensive character of a state's tactics or strategy at a particular moment. Aggression in the legal sense differs, therefore, from the meaning of the word in either the political or the military sense. Aggression is "a resort to armed force by a state when such resort has been duly determined, by a means which that state is bound to accept, to constitute a violation of an obligation," according to a report of the Harvard Research in International Law in 1939.

Qualitative and quantitative disarmament, inspection to prevent sudden attack; development of procedures authorizing provisional measures with respect to military movements; development of the theory that aggressions are acts of governments, not of states, and that sanctions should be directed only against governments and those elements of the population which support them; more effective use of propaganda to unify the forces of the world-order and to disunite the population subject to the aggressor government; immediate and general embargoes of all war materials destined for the use of the aggressor government; establishment of a world peace force; and exclusion of nuclear weapons and missiles from as many regions as possible are steps which together might render aggression impracticable. Aggression would be further deterred if the moral urge for it were reduced by the development of equitable and political procedures capable of modifying rights in cases of substantial merit. These measures have been discussed in the United Nations, but progress is dependent on a relaxation of tensions and genuine as well as formal acceptance of peaceful coexistence.

There will always be some dissatisfaction in the distribution of the world's territory because of historical grievances, changing economic needs, and the sentiment of minorities in areas of mixed population. Any system of collective security which so stabilizes the territorial status quo that peaceful rectification is deemed impossible will be subjected eventually to attack by coalitions of the dissatisfied. Practical security, therefore, requires effective procedures for changing the status quo when justice or wise policy demands. Such changes are essentially political, and the justice of demands cannot be based on any precise rule or principle but on vague standards accepted by world opinion and on a practical appreciation of the changing technological, economic, social, and political conditions of the world and of the area in question. They must, therefore, be determined by a body representative of contemporary world opinion rather than by any sort of equity tribunal or expert commission, however helpful the advice of such bodies may be. Some such procedure as that envisaged in Article 14 of the Charter is therefore suggested. There might, however, be a possibility of authoritative decision in case the vote is adequate (probably more than a mere majority) and in case certain legal safeguards have been observed, such as compensation to the ceding state and perhaps acceptance of the change by the population of the area in question.

Changes in general law are, however, to be preferred to changes in specific rights. The latter type of change, particularly when the rights in question are territorial, is at best disturbing, and consequently the demands for such changes should be reduced to a minimum. Effort to reduce the psychic, political, economic, and technological importance of territorial boundaries, if in the order named, might increase international solidarity and reduce tensions. This might be done by international guaranties of basic human rights, including rights of minorities, by facilitating travel, by lowering barriers to trade and capital movements, and by assuring defense through international sanctions and police. International legislation should attempt to attain such objectives by general rules rather than by transferring particular rights or regulating particular boundaries. The functioning of the International Law Commission and the General Assembly of the United Nations, and of conferences under their aegis, to codify aspects of international law have contributed to bringing international law up to date. More adequate procedures for such general legislation might be developed by modifying the *liberum veto* in international conferences, as is done in the General Assembly, and by increasing the legislative authority of such bodies, but progress in this direction requires considerable modification of prevailing conceptions of state sovereignty.

By gradual modification of procedures for securing rights against violence, for preventing aggression, for transferring rights, and for modifying international law, a better balance between law and change might be established in the world-society.

d) Regional and Universal Responsibilities.—The importance of geography has been reduced by modern inventions, decreasing the time of travel, transport, and communication, but the significance of geography with respect to cultural distinctiveness, military strategy, political interests, and public adiminstration is likely to continue indefinitely. There will continue to be nationalities giving distinctiveness to areas whose population has cultural characteristics and historic memories in common. The military action of states will continue to be most effective in regions within or just outside their frontiers. Land armies will continue to be difficult to transport to distant areas. This geographical limitation has always applied less to navies, though they have become increasingly dependent upon bases, and it is even less important with the development of aviation, missiles, and satellites. It is likely, however, that effective military action by most states will for a long time be confined to limited geographical regions.

Tradition, trade, and strategy will continue to induce each state to be more interested in the events in some external areas than in others, and in those areas where its interest is greatest it will be prepared to assume larger responsibility than in others. Consequently, regional organizations are desirable, but as provided by the Charter, they should be supervised by the United Nations. It is, however, true that many services, especially postal and electrical communication, maritime and aviation regulation, epidemiological and narcotics control, scientific standards and statistics, and peace and the prevention of war should be world wide. With the present range of armed forces the absolute sovereignty of continental regions or of a union of democracies, a union of Soviet republics, or a union of developing states would prove as dangerous to peace as is the absolute sovereignty of nations today. Any absolute sovereignty which is less than universal must have frontiers on land or sea, on either side of which will lie potentialities of war.

In principle, therefore, an organization of peace must be world wide. Realistic consideration must, however, be given to the geographic variations referred to. Responsibilities in respect to sanctions and power in respect to legislation must be varied according to such regional interests.

2 . THE FUNCTIONING OF PEACE

World organizations cannot acquire vitality unless their functioning is important to people. People want recognition, security, response, and new experience. World institutions may be related to these wants by according appropriate representation to groups, by preventing fundamental transgressions against world order, by becoming identified with the larger self of the individual, and by facilitating the search for, and diffusion of, new values.

a) World Representation.—Although there has been a trend toward a development of minimum world standards in science, in law, and even in cultural and political institutions, there are still vast variations in the economic standards, intelligence, culture, and awareness of world problems of the masses in different sections of the world. A system of world legislation, giving equal votes to units of population, would not be more satisfactory than one giving equal votes to states. It is unlikely that a universal pattern of representation can for a long time be recognized.

Until the virulence of nationalism has been reduced, nation-states will continue to be the only units of representation when major political problems are dealt with, though even on such problems some of the representatives might be elected by the peoples or the parliaments instead of being appointed by the governments. On many matters, organizations other than states with a technical, political, or economic interest in the subject matter might have independent representation to debate if not to vote. Important groups would thus be given a satisfying sense of world recognition.

Special conferences appropriately organized functionally and in some cases regionally might prove more satisfactory for dealing with some subject matters than a universal parliament of man. For the central problem of peace, however, which concerns political controversies, political change, and sanctions against aggression, a universal organization is necessary, though its function should extend to supervising regional and functional organizations and to maintaining peace among them, as does the United Nations. Whether the transition from security by balance of power to security by collective police should be made all at once or gradually is controversial. Gradual development of collective security, applied among countries especially vulnerable to invasion, may prove a delusion and a snare. Those who rely on collective security prematurely may cease to exist. On the other hand, states are not likely to submit their security to international agencies until they have had experience of their effectiveness. The United Nations operations in the Suez area and in the Congo provide such experience. Gradual development of policing agencies by such operations may in the long run affect the transition.

b) World Crimes.—A world-community cannot function without widespread awareness of its existence, but that awareness cannot be maintained without an objective definition of the acts, whether by individuals or by governments, deemed to threaten the existence of the members and of the community. A state in large measure defines its character by the way in which it convinces its members that their security depends upon it. The most important manifestation of this method is the criminal code, in which the state announces the transgressions against which it protects its activities and its subjects and thereby asserts what acts threaten its own existence.

Although beginnings have been made in the legislation of most states toward defining "offenses against the law of nations" by individuals, such as piracy, attacks upon public ministers, insults to foreign sovereigns, offenses against foreign currency, and offenses against neutrality and the peace of foreign states, these have in the main developed for purposes of national security rather than for defending the individual and the community of nations. Further progress was made in the Nürnberg and other war crimes trials in holding individuals criminally responsible for crimes against peace, against humanity, and against the laws of war.

There has also been progress, in the League of Nations Covenant, the Pact of Paris, the United Nations Charter, and other general treaties, toward defining acts of governments and states deemed to be breaches of the peace of the world or threats thereto, subjecting the violator to international sanctions. In general, only military aggression has been denounced. Although it constitutes a distinct offense, alone justifying military defensive action, other acts may be equally destructive of world order. Arbitrary raising of commercial barriers, sudden mobilization or augmentation of military forces, warmongering propaganda inciting to disturbances of international peace, and subversive intervention inciting or assisting one faction in civil strife have been discussed in the United Nations and the Nürnberg trials. They should be clearly defined and met by appropriate sanctions.

Such a code should relate to the behavior of individuals and governments as well as of states, and it should be confined to acts whose noxious influence is immediate. An attempt to denounce all acts which might eventually endanger the world-community would unduly encroach on the domestic jurisdiction of states. Future historians may record that negligence by the United States government during the 1920's and by the British government during the 1930's crippled the world order and encouraged aggression by the Japanese, Italian, and German governments in the latter decade, and that negligence by both the United States and the Soviet Union maintained a dangerous situation during the "cold war." A world criminal code should condemn acts of criminal negligence as well as of criminal aggression.

c) *World Citizenship.*—The basic defect in the structure of the world before World War II was the lack of consciousness in the minds of individuals that they were related to the world-community. They lived in a world in which the way of life of most people was affected by economic, political, and cultural conditions in the most distant countries. It was true that any war would modify the cost and availability of goods, would modify national laws and liberties, and would spread eventually, leading to regimentation of one's activities, to conscription of one's self or neighbors to overseas combat, or even to subjection to destruction by bombing enemy aircraft overhead.

The world was a unit in that events in every part of it affected each individual in it; but social, economic, and political thinking and institutions regarded the individual not as a member of the world-community but only of his own country. He was bound only by its laws and conceived of himself as responsible only for its behavior. His country was a member of the community of nations, governed by international law, but he himself was a member only of his national community.

The fact that the political attention of each individual was concentrated on his own country alone meant that politically he ignored the profound effect of the behavior of his own country and other countries upon the life of the world-community as a whole. He looked upon the world outside of his own nation as an environment which, like the weather, could only be submitted to and could not be controlled or which, like a wild beast, could only be hunted but not tamed. He did not conceive of it as part of the great community to which he belonged—as, indeed, part of his larger self.

National governments, though responsible for the foreign policies of their countries to the community of nations and international law, were responsible for their offices only to their own people. They were obliged to be more concerned with the source of their power than with the source of their responsibilities, and in any crisis they naturally preferred the wishes of the national constituency to the welfare of the human race. With the conditions of thought, symbolic structures, and institutions which limited the political horizon of the average individual to the home territory, these wishes of the national constituency were usually narrow, self-centered, and unaware of the tendency of world events. Under such conditions adherence of governments to international law and treaties was at best precarious. Governments had to respond to the immediate fears, greeds, habits, and fantasies of a parochial-minded population and could not be relied upon to observe international law, to respect international agreements, or to pursue foreign policies for the long-run welfare of the world-community, especially when hard-pressed by economic crises and threats of war.

It was primarily this situation which caused the failure of the institutions of world order created after World War I. The League of Nations, the International Labour Organization, the Pact of Paris, and the Permanent Court of International Justice respected the legal sovereignty of states, but they assumed that the community of nations was superior to the nation. They instituted advanced procedures of international action but were unable to function adequately because the governments did not consider the authority of their principles and procedures superior to the authority of national tradition and national opinion.

This attitude can be observed historically, deplored morally, and condemned legally, but politically, statesmen in many instances could not do otherwise. Those from democratic countries were bound to put the conception of the national welfare held by the masses of their populations ahead of any conception of world welfare which they or the general opinion of the literate from all countries might have held. Both democracies and autocracies sometimes imprisoned themselves by their own propaganda. They had to pursue the goals they had taught their people to demand.

The United Nations Charter professes to have emerged from the determination of the peoples of the United Nations to achieve certain purposes, and the United Nations and specialized agencies have attempted to develop a sense of world citizenship and responsibility for mankind among people and governments everywhere. Leading statesmen have professed a sense of such responsibility. Peace requires that the identification of peoples and governments with mankind as well as with the nation be more than verbal. The United States had to go through nearly a century of political controversy and civil war before it was certain that the union was of the people as well as of the states.

d) *World Welfare.*—This discussion rests upon the assumption that peace cannot be approached directly but is a by-product of a satisfactory organization of the world. The direct approach to peace is likely to result in retreats before threats of violence, grave injustices, and the perpetuation or aggravation of conditions in which permanent peace is impossible. Peace movements often go into reverse in times of crisis. They strive for isolationism or tolerate injustice and thus, instead of strengthening, weaken the world-community. On the other hand, support for particular treaties or institutions may result in creating overconfidence in the efficacy of those institutions, when, in fact, they lack authority because the opinion of the world is not behind them. China, Ethiopia, and Czechoslovakia, by overreliance upon the League of Nations and collective security, suffered injustices which a more correct appreciation of the actual reign of nationalism in world affairs might have prevented. Institutions, however desirable in themselves, cannot undertake responsibilities beyond their power to achieve. Responsibility without power is as dangerous as power without responsibility. The League suffered from one and the nations from the other. The United Nations has attempted to develop into a world structure in which power and responsibility go hand in hand.

In proportion as individuals in all the countries of the world rise to an appreciation of their own interest in and relationship to the world-community, institutions suitable for performing the functions of the world-community will be created and will develop a power which will enable them to meet their grave responsibilities. The idea that individual welfare and human progress are ideals to be striven for was challenged by the axis, but since the era of world contact began at the time of the Renaissance with the discoveries and the spread of knowledge by the printed word, governments have usually attempted to justify themselves on the ground that they were increasing the freedom and welfare of those whom they governed and that, by their co-operation with other governments, they were advancing the freedom and welfare of the human race. Nationalism itself was supported on the theory that it enabled governments better to accomplish these results. On the other hand, there have from time to time been governments which have denied these premises and have asserted that they exist not to advance the welfare of the governed or of the human race but only to advance the power of a particular nation, race, or class and to maintain the position of those who at the moment are controlling that group.

This latter philosophy, asserted by powerful governments in World War II, cannot be reconciled with a peaceful world. The wide acceptance of this philosophy was a consequence of the grave crises of revolution, economic disorganization, and fear of invasion which developed after World War I. This acceptance, however, tended to perpetuate these conditions. The United Nations, by endorsing President Roosevelt's Four Freedoms in World War II, asserted that they fought the war to restore general allegiance to the philosophy of human progress and human welfare which the great thinkers—religious, philosophical, and political—of all regions and all ages of civilization have accepted. An organization to prevent war must accept the philosophy that institutions are to be judged by the degree in which they advance human freedom and welfare and that the special aims of nation, state, government, or race are subordinate.

At the same time it need not deny that progress requires that the nations have sufficient independence to experiment with different systems of economy, polity, and ideology. Such an organization need express no preference for uniformity over variety but must assert that whatever group distinctiveness is to be prized and augmented must be justified because of its contribution to the progress of humanity as a whole. Such a balance is inherent in the United Nations Charter. In the continuous struggle to realize the philosophy of unity in diversity, under changing conditions, individuals and groups may satisfy the wish for ever newer experience.

3. EDUCATION FOR PEACE

Peace cannot be maintained unless the world is adequately organized to prevent governments from initiating war, and such an organization cannot function unless it serves the wants and needs of the world's population. Beyond this, however, people must believe that this organization does actually serve their requirements. The survival and functioning of the United Nations depends not merely on the commitments of governments or on publicly professed opinions of peoples but also on the attitudes of people on which the continued maintenance of commitments and expressed opinion rests.

Attitudes, though originating in the drives of the individual organism, are given form by education, the process by which the culture of a group is developed and passed on to the rising generation. Propagandas are addressed to the group, educational procedures to the individual. Propagandas may influence public opinion and stimulate immediate social action through superimposing group objectives upon the individual conscience. Education seeks to influence private attitudes, thus building the individual personality and the group culture into an organic unity.

If peace is synonymous with the general use of reason rather than impulse in organizing society and in dealing with conflict situations, peace education would be education supporting and transmitting the ideal of the rational man. The technique for accomplishing this end will not be dealt with, but attention may be given to the difficulties faced by any society in applying such an educational program.

The rational ideal usually seeks a *via media* between contending goods. The desire for individual freedom, secured by the possession of wealth and power, is often in conflict with the desire for freedom of the group with which the individual has identified himself. The security, wealth, and power of the group may require subordination of the individual in the internal system of values. The claims of the personality and of the culture, though their sources overlap, may in a given situation be in conflict.

In times of peace and prosperity, when the position of the group is not in question, philosophies of liberalism have flourished and activities have been predominantly economic as the individual has striven to improve his position in the established system. During such periods groups have tended, on the one hand, to disintegrate internally and, on the other, to unite with one another to form larger groups. In times of revolution and depression, on the other hand, the position of the group and the entire scheme of values are threatened. Philosophies of authoritarianism, whether revolutionary or reactionary, are in the saddle, activities are predominantly political, and the individual blindly follows the leader of the group with which he has identified himself. This tends to integrate established groups, to differentiate each group from the others, and to develop intergroup hostility. Such fluctuations, characterized by opposite movements in the extensity and the intensity of group life, have been observed not only in civilized societies but also among such primitive people as the Murngin of north Australia. The life of the latter fluctuates between periods of warfare which unify the clan and periods of intergroup ceremonials tending to merge the clans in larger associations. An organization of the world-community which will avoid violent fluctuations between general social disintegration and general war and a composition of cultures which will give adequate satisfaction to both the aggressive and the affectionate impulses of all the people all of the time is not easy to attain.

War may, for a time, offer this dual satisfaction to many. The soldier senses to the full and in all satisfaction his participation in the group's great task, but at the same time he is free, without inhibitions of conscience, to satisfy his individual aggressiveness against the persons and property of the enemy. The elation which usually marks the early stage of war results from the complete reconciliation offered by war to the individual's psychic conflicts, but the unreality of this adjustment gives it the character of a collective psychosis and renders its participants impervious to rational appeal until the illusion is dissipated.

The occurrence of such group psychoses suggests that the ideal of the rational man cannot survive in a society where the countermores and the mores are continually in conflict. The strain of continuous compromises between the passions and the conscience is too much for human nature. The education of children must be so conducted as to reduce the need for scapegoats to provide relief for suppressed aggression. The activities of adults must include opportunities for sport, adventure, relaxation, economic competition, political controversy, and self-expression, satisfying all the organic drives and at the same time approved by the conscience. Such activities might provide substitutes for the charm of war. War for a short time may permit a balanced expression of ambivalent impulses such as hate and love. It encourages a free expression of suppressed hates in the service of public loyalties. Peace requires an equally balanced expression of both, but the hates which to a certain extent are inevitable in any culture might be displaced upon the impersonal evils and might be expressed in forms less destructive of civilization than modern war.

To exorcise the charm of war is not enough, if war is occasionally useful for important groups. Special-interest groups benefiting directly from war, such as armament-makers hungry for markets, professional military men looking for promotions, and sensational newspaper managers pressing for increased circulation, might be regulated in the public interest. It cannot be expected that such special interests will voluntarily sacrifice a certain opportunity for individual gain because of a probable group loss. War itself cannot satisfy the more speculative interest of business in wider markets and sources of raw materials under the flag, of younger sons and intelligentsia in colonial jobs, or reformers in the expansion of cultural or religious ideals. These anticipations depend upon victory, and the prospects of that for any state in a balance-of-power war is uncertain and in a nuclear war negligible. The prospect of economic gain for the general public even from victory becomes increasingly remote as the costs of war increase. Education might influence attitudes by elucidating the relation of war to economy under present conditions.

Fear of war has functioned in preserving internal peace and in keeping rulers in power. It is difficult to find a substitute method for performing this service. Can the state attract the loyalties of its subjects sufficiently to maintain internal order if those subjects no longer feel that the state is necessary to protect them from invasion? Can custom, reason, and sentiment, unsupported by necessity, preserve the individual's love for his political group in just relation to his love for himself and for the other groups with which he is associated? To provide an answer is the task of civic education in the modern state. The individual's loyalties must be sufficiently centered to give strength to the social order and sufficiently divided to provide a basis for criticizing that order and better adapting it to changing conditions. The individual must be so educated that he himself may assume responsibility and exercise critical judgment in the solution of conflicting demands.

Even if the charm, utility, and social function of war could be eliminated, war would still occur as long as people devoted major attention to preparing for it. The expectation of war has been an important cause of war. If statesmen *generally* should abandon hope that war or the threat of war might solve their problems, belief in the inevitability of war might be undermined.

The attitude conducive to peace is neither that popularly attributed to the ostrich, which denies the possibility of war, nor that of the cynic, who considers war inevitable, but that of the rational man, who appraises the opinions and conditions tending to war and the direction of human effort which at a given point in history might prevent it. In the present age planetary comprehensiveness of vision and the utmost foresight are essentials of such rationality. These imply guidance by a central investigatory organization with capacity, free from possible impairment by national states, to communicate with individuals throughout the world. So long as control of education and communication is a monopoly of national states, it is not to be expected that attitudes conducive to war can be prevented from developing in certain areas, and the virus once developed in one section of the human population, like a cancer in the human body, will under present conditions spread to other sections and involve the whole in war.

4. PEACE IN THE ATOMIC AGE

Even if the consequences of initiating war in the atomic age are suicidal, perception of those consequences may not be sufficient actually to deter the initiation of such war in all circumstances. If a government believes the growing power or the public threats of a rival present an immediate danger, especially when the state of military technology gives great advantages to the first strike, it may make a pre-emptive strike. Disarmament and abandonment of threats as an instrument of policy are the answer, an answer impossible of realization, however, unless governments and people regard war as intolerable and threats as incredible. The development of nuclear weapons facilitates such beliefs and has induced an increasing number of serious writers and prominent statesmen to foresee the obsolescence of war. Is such prediction rational? Should such a belief be encouraged by education?

In spite of the instability of the balance of power, the arms race, the unparalleled expenditures for defense by the great powers, the frequent threats of war, the subversive interventions, and the minor wars which have afflicted the world since World War II, nuclear war has not occurred, tensions have diminished, and a nuclear test-ban treaty has been accepted by all important states except France and China. Statesmen seem to have realized, especially after the crises in Korea, Hungary, Berlin, and Cuba, that employment of the new weapons between powers which both have them would make war too dangerous.

This belief, however, though widespread, has been attacked by a school of thought which foresees not a declining but an increasing role for arms and threats in international politics through policies that utilize nuclear threats not only to deter nuclear aggression but to deter other objectionable actions. This school of thought concentrates on how nuclear threats made to deter such action can be made credible, even though carrying out the threat would certainly result in retaliation injuring the threatener as much or more than the threatened. Some conclude that the threatened must be induced to believe that the threatener is insane and that to encourage this belief either he must make himself insane by arousing a fanatical opinion that honor requires "standing firm" or he must create a situation by "burning his bridges" in which there is no alternative to abject surrender or fighting. The conclusion that war may become obsolete unless the nuclear powers become insane perhaps supports the belief that it will become obsolete, but it is well to recall Rousseau's comment that "it is a kind of folly to be wise in the midst of fools." Some governments may be insane.

Some in this school of thought, however, believe that nuclear war may be rational and that victory may be possible if suitable preparations are made for overwhelming counterforce attacks supported by a civilian-defense program to reduce losses from retaliation and by protected bases to assure the possibility of making subsequent attacks. Some suggest that with such preparation, victory might be won even after a nuclear first strike by an enemy. This school of thought, however, while continuing to assert that states must retain nuclear arms as an instrument of diplomacy, contends that the balance of terror may be stabilized by the assurance that all nuclear powers have a second-strike capability of sufficient power that no state will risk a first strike because no civilian-defense program can adequately protect its population from the second strike. They thus assert that in the interest of power politics threats of nuclear attack must be made *credible* but in the interest of international stability they must be made *incredible!*

Though statesmen have continued to operate on such contradictory opinions, they have increasingly declared their allegiance to peaceful coexistence, general and complete disarmament, and the United Nations Charter prohibiting use or threat of force in international relations. Such confusions in logic, action, and declaration may manifest a transition in which the balancing of power and war itself are on the way out but the minds of neither statesmen nor people have yet been adjusted to that situation. It appears that education should deal with the conditions which make general belief in the obsolescence of war rational and even necessary, but at the same time it should stress that conditions change, that the improbable may happen, and that continuous awareness of these contingencies make peace a condition which must be continually maintained by human effort.

INDEX

Absolute war, 364, 402

Absolutism, meaning of, 158–59

Act of war, 6, 10

Acton, Lord, on the nation, 216

Adjudication, conditions favoring, 334

Administration, and group integration, 244–45

Africa, 37–38; new nations of, 79; partition of, 130

African states: non-alignment of, 140; participation of, in campaigns, 53, 58

Aggrandizement, policies of, 128

Aggression: and alliance, 132; and appeasement (see Appeasement); and cease-fire, 184; concept of, 184, 193; conditions unfavorable to, 143; consequences of, 16; crime of, 183, 198; cumulative process of, 345; definition of, 185–86, 371, 397, 416; and disarmament, 141; fruits of, 200, 209–10; instances of, 16; liability for, 197; and military invention, 126; motives for, 69; and nationalism, 220, 222; and neutrality, 136–37; and non-alignment, 140;

and pacifism, 261; and peace propaganda, 267; and population pressure, 292; prevention of, 334, 416–17; response to, 130; suppression of, 325, 370; target of, 113

Aggressive governments: problem of, 394 ff.; sanctions against, 397

Aggressive intentions, evidences of, 129

Aggressive policies: causes of, 158; and war, 106

Aggressive weapons, 148

Aggressiveness, 6, 15–16; among animals, 27, 32; philosophies of, 326; and war, 107; see also Warlikeness

Aggressor: collaboration against, 135; determination of, 416; meanings of, 394–96

Agrarianism, and war, 300–301

Agriculture, and warlikeness, 165

Air war: bombardment in, 152; object of, 141

Aircraft carriers, 149

Airplane: and balance of power, 125; and

PHOENIX BOOKS
in History

PHOENIX BOOKS
in Sociology

PHOENIX BOOKS
in Anthropology

PHOENIX BOOKS
in Archeology

PHOENIX BOOKS
in Political Science

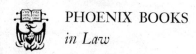

PHOENIX BOOKS
in Law